Advanced Oxidation Processes for Contaminants of Emerging Concern

In Focus – Special Book Series

Advanced Oxidation Processes for Contaminants of Emerging Concern

Editors

P. V. Nidheesh, Milad Mousazadeh, Mika Sillanpää and Zacharias Frontistis

IWA PUBLISHING

Published by **IWA Publishing**
Unit 104-105, Export Building
1 Clove Crescent
London E14 2BA, UK
Telephone: +44 (0)20 7654 5500
Fax: +44 (0)20 7654 5555
Email: publications@iwap.co.uk
Web: www.iwapublishing.com

First published 2023

British Library Cataloguing in Publication Data
A CIP catalogue record for this book is available from the British Library

ISBN: 9781789064339

Contents

doi: 10.2166/wst.2023.163

Editorial: Advanced oxidation processes for contaminants of emerging concern

Water contamination is a global problem that affects millions of people worldwide. Contaminants of emerging concern (CECs) are a particular challenge, as they are not yet regulated and can pose significant risks to human health and the environment. Conventional water treatment approaches are not very effective at removing contaminants of emerging concern; however, these can sometimes be chemically, energetically, and operationally intensive. In this context, advanced oxidation processes (AOPs) have emerged as sustainable technologies for stopping pollutants from reaching the environment, offering different solutions to many environmental problems. In this book, we present recent developments in AOP technology for treating CECs.

The use of AOPs to treat water contaminants has gained significant attention in recent years. AOPs involve the generation of highly reactive species, such as hydroxyl and sulfate radicals, which are capable of degrading organic compounds into simpler and less toxic products. This book explores the use of AOPs for the treatment of antibiotics, pharmaceuticals and personal care products, and other organic pollutants. The chapters cover a range of AOP technologies, including photocatalysis, electrochemical oxidation, catalyst, persulfate oxidation, and air plasma.

This book highlights the potential of AOPs for treating CECs. However, it is essential to note that the effective treatment of CECs requires a holistic approach considering the entire water treatment process. AOPs are just one component of a comprehensive water treatment strategy, and their effectiveness depends on a range of factors, including the characteristics of the contaminants, the water matrix, and the treatment conditions.

One of the key themes that emerges is the use of novel nanomaterials and processes for the treatment of CECs. Due to their high surface area and unique electronic properties, nanomaterials, such as Cu-N/TiO_2 and zinc oxide, are effective catalysts for AOPs.

The book examines the importance of considering real water matrices, including palm oil mill effluent waste activated sludge, wastewater treatment plants, and pharmaceutical wastewater, in developing and optimizing AOPs for the treatment of CECs. The characteristics of the water matrix can significantly affect the effectiveness of AOPs, and it is crucial to consider these factors when designing and optimizing AOP treatment systems.

In conclusion, the studies included here demonstrate the potential of AOPs for removing a wide range of organic pollutants from water and emphasize the importance of process optimization and integration of other processes for the effective treatment of CECs. While AOPs are a promising technology for treating CECs, it is important to recognize that effectively treating these contaminants requires a holistic approach considering the entire water treatment process.

Guest Editors

P. V. Nidheesh

Environmental Impact and Sustainability Division, CSIR – National Environmental Engineering Research Institute, Nagpur, Maharashtra, India

Milad Mousazadeh

Department of Environmental Engineering, Qazvin University of Medical Sciences, Qazvin, Iran

Mika Sillanpää

Department of Biological and Chemical Engineering, Aarhus University, Nørrebrogade 44, Aarhus C 8000, Denmark

Zacharias Frontistis

Department of Chemical Engineering, University of Western Macedonia, Kozani GR-50132, Greece

doi: 10.2166/wst.2022.243

Photocatalytic and adsorptive performance of polyvinyl alcohol/chitosan/TiO_2 composite for antibiotics removal: single- and multi-pollutant conditions

Neghi N[a] and Mathava Kumar[b,*]

[a] Department of Civil Engineering, National Institute of Technology, Tadepalligudem, Andhra Pradesh 534101, India
[b] Environmental and Water Resources Engineering Division, Department of Civil Engineering, IIT Madras, Chennai 600036, India
*Corresponding author. E-mail: mathavakumar@gmail.com

ABSTRACT

A polymer-TiO_2 macro composite (i.e., PVA-CS-TiO_2) was synthesized via chemical precipitation of PVA-CS-TiO_2 blend in alkali/solvent medium and applied for the removal of three model antibiotics (i.e., metronidazole (MNZ), ceftiofur (CEF) and tetracycline (TET)), as single compound and multi-compound conditions. The photocatalytic and adsorptive removals of antibiotics (concentrations of 0.1, 1 and 10 mg L^{-1}) by the composite in an UV reactor system (32 W UV-C power, 0.3 g L^{-1} of composite) was assessed through kinetic models. Antibiotics adsorption followed pseudo-second-order kinetics, and the order of adsorption was MNZ > TET > CEF. On the other hand, the hydrophilic MNZ was degraded faster compared to hydrophobic CEF and TET drugs. Moreover, UV reactor system exhibited antagonistic behavior under multi-compound condition. Micro-toxicity of antibiotics was performed using bioluminescent bacterium *Vibrio fischeri* and EC_{50} of CEF, TET and MNZ were found to be 18.25 mg L^{-1}, 173.8 mg L^{-1}, and 668.6 mg L^{-1}, respectively. However, the relative toxicity levels of PVA-CS-TiO_2 and treated effluent were well with the limits as inferred from the microtoxicity analysis. Thus, synthesized biocompatible composite exhibited structural stability, consistent performance for three photocatalytic cycles for all antibiotics at a minimal catalyst loading, easily retained using metallic tea strainer and does not exhibit microtoxicity has a scope for real-time applications.

Key words: antibiotics, chitosan, micro-toxicity analysis, photocatalysis, polyvinyl alcohol, TiO_2

HIGHLIGHTS

- PVA-CS-TiO_2 was stable and produced good antibiotics removal under single and multi-compound conditions.
- Removal rates of antibiotics by adsorption and photocatalysis were modelled.
- Hydrophilic MNZ exhibited faster degradation than hydrophobic CEF and TET.
- EC50 of antibiotics was determined using *Vibrio fischeri* bacterium.
- Hydrophobic antibiotic exhibited greater toxicity to *V. fischeri*.

GRAPHICAL ABSTRACT

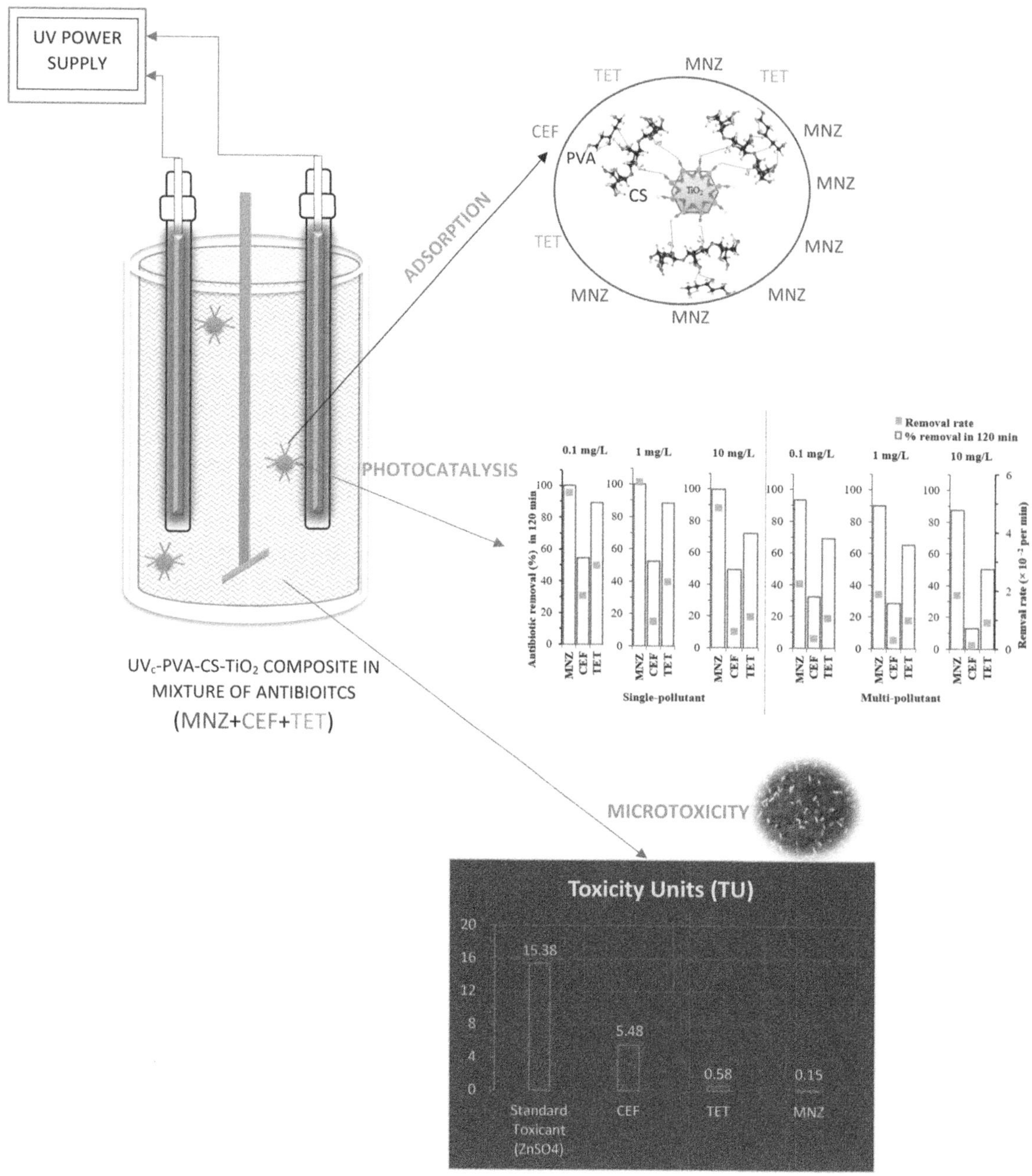

1. INTRODUCTION

Photolytic and photocatalytic processes are gaining attention in water purification and wastewater polishing steps. TiO_2 has been extensively inspected as a photocatalyst owing to its chemical stability, superhydrophilicity, mineralize organic contaminants by the production of oxidative hydroxyl radicals, low cost and non-toxicity (Liu *et al.* 2017). The photocatalytic activity of nano-TiO_2 has been utilized in various areas such as environment and energy fields that include air and water purification systems; antimicrobial TiO_2 based nanocomposites; self-cleaning coatings; hydrogen production and photo-electrochemical conversion. UV-TiO_2 photocatalytic system is one of the extensively researched advanced oxidation systems for the removal of emerging contaminants from aqueous systems. Upon exposure to UV irradiation, TiO_2 (band gap of 3.2 eV) promotes an electron to the conduction band (e^-_{CB}) and formation of a positive hole in the valence band (h^+_{VB}). The photo-generated h^+_{VB}

oxidizes either water to produce $HO^{\cdot}$ or directly the organic contaminants that are exposed to TiO_2 surface. The generated $HO^{\cdot}$ non-selectively oxidize almost all electron-rich organic contaminants, eventually converting them to CO_2 and water. The charge carries, if not utilized recombine in a short reaction time limiting the photocatalytic activity of TiO_2. The main hindrance in the application of this system is the post-recovery of used slurry-TiO_2 from the treated effluent and inefficient utilization of visible light (Dong *et al.* 2015). Therefore, recent investigations have a lot of interest in developing sustainable TiO_2 photo-catalytic composite materials in the light of following aspects: (a) to increase the light harvesting ability of the photocatalysts, (b) to reduce the recombination of the photo-generated radicals, and (c) to enhance the ease of re-use when compared to application of nano-TiO_2 in slurry form.

In the context of developing recoverable/reusable TiO_2 photocatalyst, two approaches are practiced. One is coating TiO_2 on substrates such as glass/silica fibers (Barton *et al.* 2016), glass (Ganesh *et al.* 2012), quartz tubes (Koo *et al.* 2012), stainless steel (Zhu *et al.* 2017), carbon materials (Singh *et al.* 2016), plastic materials (Rokhmat *et al.* 2017), etc. The other approach is embedding TiO_2 in matrices such as polymers (Wang *et al.* 2019), cement (Mendoza *et al.* 2015), carbon fiber film (Chang *et al.* 2018), membranes (Behboudi *et al.* 2016), etc. Each approach has its own advantages and disadvantages. The main drawback of coating process is it can reduce the available surface area of TiO_2 and leaching of TiO_2 happens from the coated substrates. On the other hand, the main limitation of the embedding process is that based on the method of synthesis the crystallinity of TiO_2 might be affected. Some of the recent investigations include embedding TiO_2 in polymer matrices such as cellulose (Ali *et al.* 2018), polyethylene (Jafarzadeh *et al.* 2015), chitosan (Zhao *et al.* 2017), polyaniline (Zhu *et al.* 2015), polyvinyl alcohol (Pham *et al.* 2018), etc., in order to retain TiO_2 NPs in the matrix and reuse it for multiple photocatalytic treatment cycles. Based on the matrix chosen, the the synergy of adsorption and photocatalysis greatly improves the performance of photocatalytic composites attained through embedding process. The particle size, chemical nature, specific surface area of the nano-filler (for example TiO_2), modifies the properties of the polymer such as electrical and thermal conductivity, phase behaviour, mechanical properties, etc. The choice of the polymer considered for the study was based on the availability, stability, cost-effectiveness, and biocompatibility.

In the present study, TiO_2 is embedded in the polyvinyl alcohol-chitosan (PVA-CS) polymer blends. CS has been extensively reported as a bio sorbent material for metals, dyes, organic compounds, etc. It possesses amino (-NH_2) and hydroxyl (-OH) groups that contribute to its adsorptive properties as well as act as potential sites for modification (Kyzas & Bikiaris 2015). However, when it comes to application in the form of hydrogels/beads, it exhibits poor stability with respect to pH and poor mechanical resistance. Therefore, several polymeric blends (two or more polymers) are investigated to overcome the limitations of individual counter parts and complement each other. One such combination is the 'PVA-CS' blends that possess better hydrophilicity, mechanical properties and biocompatibility than the individual components. PVA-CS blends have been applied for food packaging (Liu *et al.* 2018), ion exchange membrane (Gopi *et al.* 2019; Gu *et al.* 2021), wound healing (Zhao *et al.* 2019; Abbas *et al.* 2022) and adsorption (Cesar Filho *et al.* 2018; Vatanpour *et al.* 2020). To the best our knowledge, the application of TiO_2 embedded in PVA-CS blend (PVA-CS-TiO_2) for the photocatalytic removal of antibiotics from aqueous systems is very limited. The antibiotics considered for the study are metronidazole (MNZ; nitroimidazole class), ceftiofur (CEF; cephalosporin class) and tetracycline (TET; tetracycline class). The synthesis, characterization, stability and reusability of PVA-CS-TiO_2 was reported in our earlier investigation (Neghi *et al.* 2019).

The objective of the present study is the application of PVA-CS-TiO_2 for antibiotics removal based on the following parameters: (a) analyze the adsorptive ability for the considered antibiotics, (b) analyze the photocatalytic efficiency for both single and multi-pollutant condition, (c) calculate antibiotic removal using modified Langmuir-Hinshelwood kinetic model developed and (d) micro-toxicity associated with the considered antibiotics, composite, and the treated effluent.

2. MATERIALS AND METHODS

Chitosan (CS, deacetylation: ≥95%, viscosity 200–600 mPa.s), poly (vinyl alcohol) (PVA, alcoholysis degree: 86–89%, M_w ~124,000 g mol^{-1}), methanol and TiO_2 (mean particle size ~25 nm) were procured from Tokyo Chemical Industries Ltd, Japan, SD Fine Chemicals Ltd, India and Avra Synthesis, India, respectively. Metronidazole ($C_6H_9N_3O_3$; 171.15 g mol^{-1}), ceftiofur ($C_{19}H_{17}N_5O_7S_3$; 523.55 g mol^{-1}) and tetracycline ($C_{22}H_{24}N_2O_8$; 444.44 g mol^{-1}) were received from Sigma Aldrich, India. The properties of antibiotics considered for the study has been provided in Table S1 (Supplementary Material). All other chemicals used were of analytical grade and used as received.

2.1. PVA-CS-TiO_2 composite synthesis

The optimization and synthesis of PVA-CS-TiO_2 composite beads (1.5–3 mm) was reported in our earlier work (Neghi *et al.* 2019). Accordingly, PVA-CS-TiO_2 ultra-sonicated gel (8:2:2% m/v; 8 g of PVA beads in 100 mL of water, 2 g of CS in 100 mL of acetic acid (2.68% v/v) and 2 g of TiO_2 in 100 mL water) was dropped in 0.25 M NaOH-methanol solution. The synthesized beads were rinsed and oven-dried for two repetition cycles to remove water-soluble products from the surface of the flakes. The oven-dried composite flakes (size range of 1.5–3 mm) were taken for adsorption and photocatalytic experiments carried out for antibiotic removal in both single and multi-pollutant condition. The composite flakes were easily retained by a metallic tea strainer. The composites are made spherical shaped rather than sheets. This ensures exposure of high reaction surface for TiO_2.

2.2. Methods

2.2.1. Analytical techniques

The selected antibiotic compounds (i.e., metronidazole (MNZ), ceftiofur (CEF), and tetracycline (TET)) were analyzed using HPLC (Dionex, USA) fitted with UV detector. A silica-based reverse phase Acclaim C-18 column (220 mm × 4.6 mm (I.D); 5 μm) was used for the separation of compounds. For MNZ and CEF, 10 mM of phosphate buffer (adjusted to pH 5.5 by 10% ortho-phosphoric acid) and acetonitrile were mixed at 70:30 (v/v) and used as the mobile phase. The analysis was conducted at 310 nm. The retention times of MNZ and CEF in the column was identified as 3.8 and 4.2 min, respectively. For TET, 10 mM phosphate buffer (adjusted to pH-2.5 by 10% ortho-phosphoric acid) and acetonitrile was used in the ratio of 70:30 and analyzed at 260 nm. Samples and buffers were filtered through 0.45 μm prior to the analysis. The mobile phase was pumped at a flow rate of 1 mL min^{-1} for MNZ and CEF, whereas for TET it was operated at 1.5 mL min^{-1}. pH and temperature of the reaction mixture were measured using a multi-parameter analyzer (OAKTON, India) connected to the probes of the batch reactor. Organic carbon contents (TOC) contributed by the composites, MNZ and its degradation products were analyzed in the sample using total organic carbon analyzer (Shimadzu, Japan) equipped with NDIR analyzer. The cations and anions present in the sample were identified using Ion chromatography (Thermo Scientific, Dionex, USA).

2.2.2. Adsorption

Adsorption experiments were conducted in a round bottom flask with 100 mL of antibiotic (MNZ/CEF/TET or mixture) solution of varying concentrations, i.e., 10, 1 and 0.1 mg L^{-1} and 0.03 g of photocatalyst as adsorbent. The sorbed amount of antibiotic per unit weight of PVA-CS-TiO_2 (q_e, mg g^{-1}) at equilibrium time (t) was calculated using Equation (1).

$$q_e = (C_0 - C_e)\,\frac{V}{W} \tag{1}$$

where, C_0, C_e, V and W represent initial antibiotic concentration (mg L^{-1}), antibiotic concentration at equilibrium (mg L^{-1}), reaction volume (L) and weight of PVA-CS-TiO_2 (g), respectively. Moreover, the data was used to estimate maximum antibiotic adsorption capacity of PVA-CS-TiO_2 using linearized expressions of Langmuir (Equation (2)), Freundlich (Equation (3)), Temkin (Equation (4)) and Dubinin-Radushkevich (Equation (5)) isotherms.

$$\frac{1}{q_e} = \frac{1}{q_{mL}} + \frac{1}{q_{mL}\,K_L C_e} \tag{2}$$

$$\log q_e = \log K_F + \frac{1}{n}\log C_e \tag{3}$$

$$q_e = B\,\ln(A_T) + B\ln(C_e);\ B = \frac{RT}{b_T} \tag{4}$$

$$\ln q_e = \ln q_m - (K_{DR}\varepsilon^2) \tag{5}$$

In Equation (2), q_{mL} is the amount of adsorption corresponding to the complete monolayer coverage and K_L is the Langmuir isotherm constant. In Equation (3), 1/n and K_F are the adsorption intensity and Freundlich isotherm constant, respectively. In Temkin adsorption isotherm, A_T is the isotherm equilibrium binding constant (L mg^{-1}), b_T is the Temkin isotherm constant, R is the universal gas constant (8.314 J mol^{-1} K^{-1}), T is the temperature at 298 K and B is the constant related to heat of sorption (J mol^{-1}). K_{DR} refers to mean free energy of adsorption ($mol^2\,kJ^{-2}$) and $\varepsilon = RT\ln[1 + 1/C_e]$ is Polyani potential (Equation (5)). Antibiotic adsorption kinetics on PVA-CS-TiO_2 were analysed by plotting Lagergren's pseudo-first-order (Equation (6)), pseudo-second-order (Equation (7)) and intraparticle diffusion (Equation (8)) models.

$$\log(q_e - q_t) = \log q_e - \frac{k_1}{2.303}t \tag{6}$$

$$\frac{t}{q_t} = \frac{1}{K_2 q_e^2} + \frac{1}{q_e}t \tag{7}$$

$$q_t = K_{id}t^{1/2} + c \tag{8}$$

where, q_e and q_t are the adsorption capacities at equilibrium and at time (t) (mg g^{-1}). k_1 is the pseudo-first-order rate constant (min^{-1}), k_2 is the pseudo-second-order rate constant (g mg^{-1} min^{-1}), k_{id} is the intra-particle diffusion constant (g mg^{-1} $min^{-0.5}$) and C is the intercept that were obtained by plotting Equations (6)–(8), respectively.

2.2.3. Photocatalysis

Photocatalytic experiments were conducted using PVA-CS-TiO_2 composite flakes (size range of 1.5–3 mm) in a batch reactor as designed in our previous work (Neghi & Kumar 2017). Antibiotic solution (2 L) having the required concentration (10, 1 and 0.1 mg L^{-1}) was added with 0.3 g L^{-1} of PVA-CS-TiO_2. The reactor was provided with 32 W UV power ($\lambda_{max} = 254$ nm), continuous stirring (200 rpm) and operated for 120 min without pH and temperature controls. The pH of the model wastewater (drugs and composite in distilled water) was around 6 to 7 and temperature before and after the UV reactor system (32 W UV power, 0.3 g L^{-1} of the composite) was 31.5 and 33 °C, respectively. At regular intervals of 30 min, samples were collected from the reactor and analyzed for antibiotic concentration. Using the data, antibiotic removal rates were calculated based on Langmuir-Hinshelwood (L-H) kinetic model as shown in Equation (9).

$$\ln\frac{C_t}{C_0} = -k_1\,t \tag{9}$$

where, C_0, C_t and k_1 are initial antibiotic concentration (mg L^{-1}), antibiotic concentration at 120 min (mg L^{-1}) and pseudo-first-order-rate constant (min^{-1}), respectively, at time (t). The slope of the plot between ln (C_t/C_0) and t gives the value of k_1. The antibiotic removal rate was also calculated using modified Langmuir-Hinshelwood (L-H) kinetic model as shown in Equation (10) as developed in our earlier work (Neghi *et al.* 2019).

$$\ln\frac{C_t}{C_0} = -k_c.C_T.t \tag{10}$$

where, k_c is the modified-pseudo-first-order rate constant and C_T is the concentration of the composite (i.e. PVA-CS-TiO_2) for a fixed UV energy power.

2.2.4. Micro-toxicity analysis

The micro-toxicity analysis was performed using a bioluminescent bacteria *Vibrio fischeri* (NRRL number B-11177) obtained from Modern Water Inc., UK. The analysis was conducted based on the procedures mentioned in 11348-3:2007 standards (ISO 1998). This assay is widely applied in many investigations owing to its sensitivity towards wide range of chemicals, short reaction time and reproducibility. Therefore, this method was used to evaluate the toxicity associated with the antibiotics, PVA-CS-TiO_2- composite and the treated effluent from the batch reactor.

The 81.9% basic (consisting of six test dilutions) was carried out to evaluate the toxicity unit of antibiotics considered for the study. Accordingly, 900 μL of the samples were exposed to 100 μL of bacterial suspension that contains 1 million cells of *Vibrio fischeri*. The bacterial luminescence readings before and after toxicant exposure was measured at different exposure times, i.e. 5, 15, 30 and 60 min using Microtox analyzer (Model 500). The percentage inhibition effect of the samples to the bacterium is calculated as toxicity units (TU) using Equation (11). In Equation (11), EC_{50} represents 50% inhibition concentration as percentage.

In order to understand the toxicity level of antibiotics, the measured toxicity values of antibiotics was compared with the toxicity of a known standard toxicant $ZnSO_4$ (ISO 1998), i.e. equivalent-toxicity approach.

$$\text{Toxicity unit (TU)} = 100/EC_{50} \tag{11}$$

Similarly, in order to assess the relative toxicity of the composite and the treated samples 81.9% screening protocol was carried out. The toxicity contributed by the composites in the water sample was assessed by exposing the composite flakes to the number of days that they were stable in water, i.e. PVA-CS-TiO_2 was stable in water (pH ~ 7) for 35 days. Therefore, the exposed water taken after 35 days was considered as the sample for microtoxicity analysis of PVA-CS-TiO_2 system.

3. RESULTS AND DISCUSSION

3.1. Adsorption of antibiotics on PVA-CS-TiO_2

The composite obtained in the form of beads (Figure 1) was oven-dried to flaky form and used for adsorption studies. The effect of type and initial antibiotic concentrations on the adsorption kinetics of PVA-CS-TiO_2 system was evaluated. The adsorptive removal of the contaminants followed pseudo-second-order kinetics on PVA-CS-TiO_2. The contaminant removal (%) for MNZ, CEF and TET are shown in Figure 2. A maximum removal of 75.3% was observed for MNZ followed by TET (45.2%) and CEF (33.2%) at the lowest contaminant concentrations considered (0.1 mg L^{-1}). Both Langmuir and Freundlich isotherm models have well represented individual antibiotic adsorption on PVA-CS-TiO_2 ($r^2 > 0.99$) (Table 1). The value of Freundlich constant (1/n) was found to be between 0 and 1, which indicates the favorable antibiotic adsorption on PVA-CS-TiO_2. Based on the residual MNZ concentration profiles, the maximum antibiotic adsorption capacity for PVA-CS-TiO_2 were found out using Equations (2)–(5) and the values are given in Table 1. The maximum adsorption capacity for MNZ, CEF and TET on PVA-CS-TiO_2 was found to be 1.673, 0.306 and 0.885 mg g^{-1}, respectively. The BET surface area of PVA-CS-TiO_2 was found to be 0.1813 m^2 g^{-1}(Neghi *et al.* 2019). From the data, it is inferred that adsorption efficiency of PVA-CS-TiO_2 is not governed by porosity or the surface area. This is further validated by the E-value calculated using Dubinin-Radushkevich isotherm, i.e., 3,535.54 to 2,886.8 KJ mol^{-1}. This indicates that adsorption of the antibiotics onto the composite was mainly driven by chemisorption.

From the intraparticle diffusion plots it can be understood that rate of mass transfer was uniform throughout the process time as the intercept is close to zero except MNZ at 10 mg L^{-1} concentration, i.e., there is no influence of the boundary layer of the composite. The K_{id} values are lower for CEF and TET, compared to MNZ for 1 mg L^{-1} concentration. This indicates that the molar mass of the drug plays a major role in the diffusion process. Overall, the K_{id} values are higher at higher

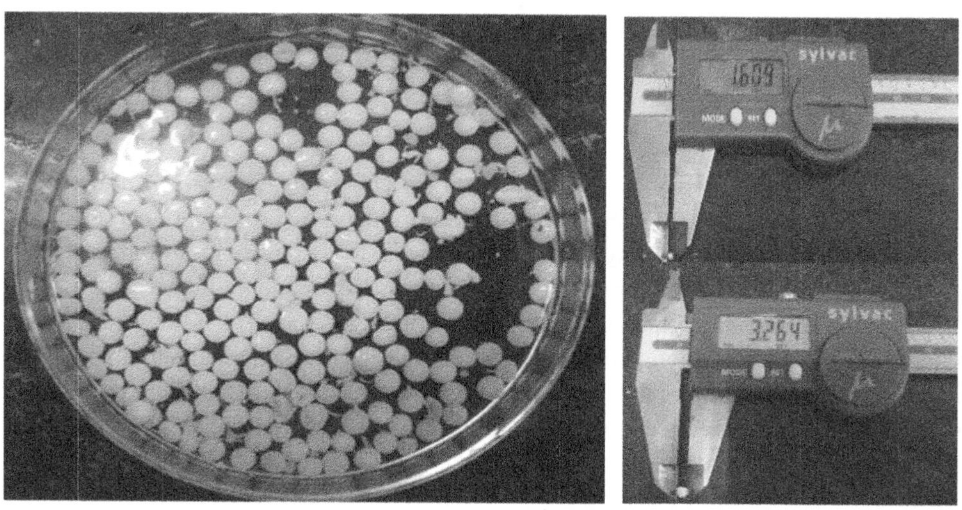

Figure 1 | Photo of PVA-CS-TiO_2 beads (average composite size (dried flakes) is 1.5–3 mm).

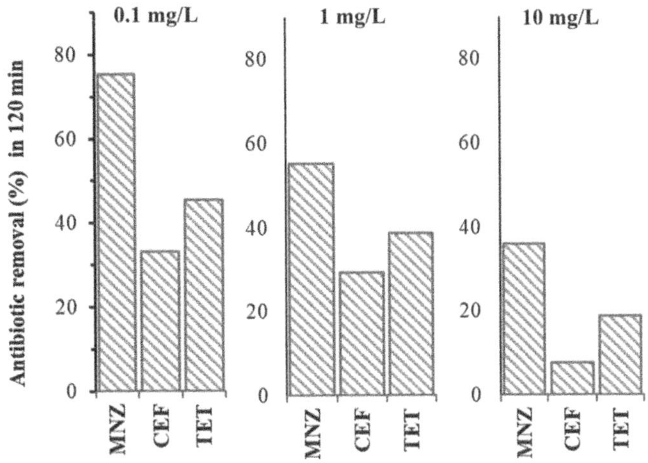

Figure 2 | Adsorptive pattern of MNZ, CEF and TET on PVA-CS-TiO_2.

Table 1 | Adsorption isotherm parameters calculated for MNZ, CEF and TET on PVA-CS-TiO_2

Isotherm models	Parameters	MNZ	CEF	TET
Langmuir	q_m (mg g^{-1})	1.673	0.306	0.885
	K_L (L mg^{-1})	0.382	0.615	0.296
	r^2	0.9269	0.9995	0.9969
Freundlich	K_F (mg g^{-1})	0.198	0.02	0.076
	n	0.689	0.648	0.74
	r^2	0.9998	0.9439	0.9898
Temkin	A_T (L mg^{-1})	0.506	0.874	0.732
	B (J mol^{-1})	3.897	18.682	7.1085
	r^2	0.8418	0.987	0.9046
Dubinin-Radushkevich	q_m (mg g^{-1})	0.578	0.19	0.352
	K_{DR} (mol^2 KJ^{-2})	4.00E-08	6.00E-08	6.00E-08
	E(KJ mol^{-1})	3,535.54	2,886.8	2,886.8
	r^2	0.8233	0.9479	0.8793

antibiotic concentrations indicating that there is an ease of contact of the antibiotic onto the surface of the composite facilitating the intra-particle diffusion process.

The pH_{zpc} of PVA-CS-TiO_2 composite was found to be 5.82 (Fig. S6). Moreover, considering the neutrality of the antibiotics considered for the study (pK_a range of 2.4 to 3.7 (Table S1)) in the operating pH range of the aqueous solutions, i.e., pH = 6 to 7, the adsorption would be the maximum. For example, PVA-CS-TiO_2 possess positive charge below pH_{zpc} which makes the surface charge of the composite positive. At neutral pH, there is no repulsive force between the antibitoic and the composite, therefore, better adsorption for MNZ followed by TET and CEF.

The extensive hydroxyl groups from PVA and the amino/hydroxyl groups from CS are the major functional groups that could interact with the antibiotic as evidenced from the DFT analysis previously done for different combinations of parts of PVA-CS-TiO_2 composite and MNZ adsorption (Neghi *et al.* 2019). It was anticipated that other antibiotics would have the same behaviour as PVA-CS-TiO_2 and the functional groups of the antibiotics considered for the study are similar. Overall, the antibiotic adsorption on PVA-CS-TiO_2 is probably governed by H-bonding and electrostatic forces interactions. The antibiotic adsorption kinetics follows pseudo-second order for PVA-CS-TiO_2 for all the antibiotics ($r^2 > 0.85$). From the values of k_2 (Table 2), it can be inferred that the rate of adsorption was faster in hydrophilic antibiotics (MNZ) compared to hydrophobic antibiotics (CEF and TET).

Table 2 | Adsorption kinetic parameters calculated for MNZ, CEF and TET on PVA-CS-TiO_2

		Kinetic models								
		Pseudo-first order			Pseudo-second order			Intraparticle diffusion		
Type of contaminant	Conc (mg L^{-1})	q_e (mg g^{-1})	k_1 (min^{-1})	r^2	q_e (mg g^{-1})	k_2 (g mg^{-1} min^{-1})	r^2	k_{id} (g mg^{-1} $min^{-0.5}$)	C	r^2
MNZ	10	0.969	0.239	0.919	1.518	0.061	0.968	0.134	0.122	0.919
	1	0.975	2.02	0.856	0.185	0.577	0.969	0.015	0.02	0.871
	0.1	0.948	3.78	0.945	0.03	8.52	0.996	0.003	0.006	0.8
CEF	10	0.975	0.9	0.851	0.31	0.11	0.856	0.25	0.007	0.988
	1	0.938	4.23	0.701	0.108	0.348	0.878	0.009	0.001	0.995
	0.1	0.976	4.5	0.964	0.014	6.53	0.976	0.001	0.001	0.954
TET	10	0.91	1.35	0.617	0.637	0.061	0.873	0.052	0.005	0.989
	1	0.919	0.363	0.619	0.019	17.79	0.855	0.01	0.004	0.996
	0.1	0.922	2.53	0.645	0.14	0.239	0.929	0.006	0.0008	0.997

3.2. Photocatalytic removal of antibiotics using PVA-CS-TiO_2 for single pollutant condition

The characterization results of the synthesized composite have been provided in the supplementary material. Accordingly, scanning electron microscopy (SEM) and transmission electron microscopy (TEM) images, X-ray diffraction (XRD) analysis, ultraviolet diffuse reflectance spectroscopy (UV-DRS) analysis and Fourier transform infrared (FTIR) spectrum of the composite have been provided in Figs. S1 to S5. From the SEM analysis, it can be summarized that TiO_2 nanoparticles (NPs) are distributed on the surface of the composite and energy dispersive X-ray (EDX) spectrum confirms the presence of C, Ti, O and N in PVA-CS-TiO_2 (Fig S1) TEM analysis indicates the presence of both spherical/hexagonal shaped TiO_2 NPs with well-defined fringes of lattice spacing (d = 0.36 nm) and limited agglomeration was seen on the polymer matrix (average particle size between 50–100 nm) (Fig S2). XRD analysis indicates that crystallinity of TiO_2 is undisturbed in the composite with the Miller indices of reflections indicated for anatase phase TiO_2 for the planes at 2θ values of 25.24, 37.84, 48.03, 53.88, 55.08, 62.57, 68.72 and 74.9 degrees (Fig. S3). UV-DRS spectra indicate the band gap of TiO_2 and its composite remains the same at 3.16 eV (Fig. S4). It indicates the polymeric moiety has no influence in the band gap energy of TiO_2 (Mohamad Idris *et al.* 2021).

The photocatalytic experiments with PVA-CS-TiO_2 were conducted under a catalyst dosage of 0.3 g L^{-1} and 32 W UV power. The antibiotic removal followed pseudo-first-order kinetics and the removal (%) are provided in Table 3. It can be noticed in Table 3 that a near complete MNZ removal was achieved for concentrations up to 10 mg L^{-1}. The maximum % removal obtained for MNZ, CEF and TET were 99.9, 54.5 and 89.3% at the lowest concentrations operated. However, the lowest removal (49.2%) was observed for CEF at 10 mg L^{-1}. It is known that $OH\cdot$ react more rapidly to electron-rich carbon-carbon double and triple bonds compared to aliphatic compounds (Ulliman *et al.* 2018). This might be the reason for the enhanced removal of the considered antibiotics in the UV reactor. Moreover, the synergistic action of both adsorption and photocatalysis by $OH\cdot$ led to the enhanced removal of antibiotics using PVA-CS-TiO_2.

Table 3 | Photocatalytic removal kinetics of PVA-CS-TiO_2 composite for MNZ, CEF and TET in UV reactor

Single pollutant condition

	Antibiotic removal (%) in 120 min			Antibiotic removal rate ($\times 10^{-2}$ min^{-1}) calculated using Equation (9)			Antibiotic removal rate ($\times 10^{-3}$ min^{-1}) calculated using Equation (10)		
Type of antibiotic	0.1	1	10	0.1	1	10	0.1	1	10
MNZ	99.9	99.9	99.7	4.2[a]	4.6	4.8	1.4[a]	1.5	1.6
CEF	54.5	52.5	49.2	0.66	0.68	0.55	0.2	0.2	0.18
TET	89.3	88.7	72.1	1.7	1.8	1.05	0.6	0.6	0.4

[a]MNZ removal rates calculated for 90 min.

Antibiotic removal rates were calculated using Equations (9) and (10) and the values are provided in Table 3. MNZ removal rates are higher followed by TET and CEF. The comparison of MNZ removal rates (1 mg L^{-1}) using TiO_2 slurry (0.0032 min^{-1}) (Neghi & Kumar 2017) and PVA-CS-TiO_2 (0.0046 min^{-1}) indicate that MNZ removal rate was increased by 1.44 times. This was further accomplished with lesser catalyst loading in PVA-CS-TiO_2 (0.3 g L^{-1}) whereas it was 2.5 g L^{-1} in case of TiO_2 slurry system. This indicates that when an adsorbent matrix is chosen to embed the photocatalyst, i.e. TiO_2, the photocatalytic rate of the composite is considerably increased when compared to slurry system. Moreover, the presence of PVA increase the conductance of PVA-CS film due to intra- and intermolecular hydrogen bonding in the composite (Islam *et al.* 2016) as PVA is a conducting polymer. So, PVA-CS films conducts electrons which makes the recombination of electrons and holes slower and thereby improved charge dynamics of PVA-CS-TiO_2. Thus, PVA-CS-TiO_2- composite is highly photoactive in the UV reactor system.

The reusability of the composite for three consecutive photocatalytic treatment cycles have been studied for the considered antibiotics in the study (Figure 3). It is inferred that there is no change in the photocatalytic activity of PVA-CS-TiO_2 for the first three consecutive cycles. The prepared PVA-CS-TiO_2 composite have a consistent performance under UV-C radiation. It was also reported that PVA-TiO_2 blends showed greater stability in all three bands of UV light exposure that further validates the performance of PVA-CS-TiO_2 composite (Moustafa *et al.* 2021). Moreover, the FTIR spectra indicates the structural integrity of PVA-CS-TiO_2 before and after three consecutive photocatalytic treatment cycles for MNZ removal (Fig. S5). It has also been reported that TiO_2 enhances the thermal and UV-resistant properties of polysaccharide-based films including chitosan, which emphasizes the structural stability of PVA-CS-TiO_2 (Anaya-Esparza *et al.* 2020).

The mineralization potential of the reactor system was assessed through TOC analysis (Figure 4). It was revealed that with the progress in the treatment cycle there was improvement in % removal for all antibiotics at a concentration of 10 mg L^{-1} after 120 min. This trend might be due to the contribution from loosely attached parts of the polymer at the first run. Moreover, for CEF antibiotic, TOC removal (%) was remaining constant (~ 31%) and a progressing TOC removal (66%) for TET with the three consecutive treatment cycles. MNZ antibiotic showed a maximum of 76% TOC removal at the third run. In addition, the ion chromatography analysis revealed the presence of NO_3^-(0.34 mg L^{-1}) and Cl^- (3.82 mg L^{-1}) for MNZ removal, NO_3^- (0.95 mg L^{-1}), Cl^- (4.12 mg L^{-1}) and SO_4^{2-}(4.4 mg L^{-1}) for CEF removal and NO_3^- (1.2 mg L^{-1}) and Cl^- (3.95 mg L^{-1}) for TET removal.

3.3. Photocatalytic removal of antibiotics using PVA-CS-TiO_2 for multi pollutant condition

To evaluate whether the reactor exhibited a synergistic or antagonistic behavior for the multi-pollutant degradation process, experiments were conducted at varying initial antibiotic concentrations. The effect of initial concentration of the antibiotics was also examined for concentration ranging from 0.1 to 10 mg L^{-1}. The pattern of photocatalytic removal profile (as shown in Figure 5) was directly related to the adsorption profile of the pollutants onto the composite (Figure 2). The photocatalytic removal kinetics of the mixed pollutants followed pseudo-first-order. It was observed that irrespective of the initial pollutant concentration MNZ showed steady removal rates of around 0.045 min^{-1}. However, the removal rates drastically decreased

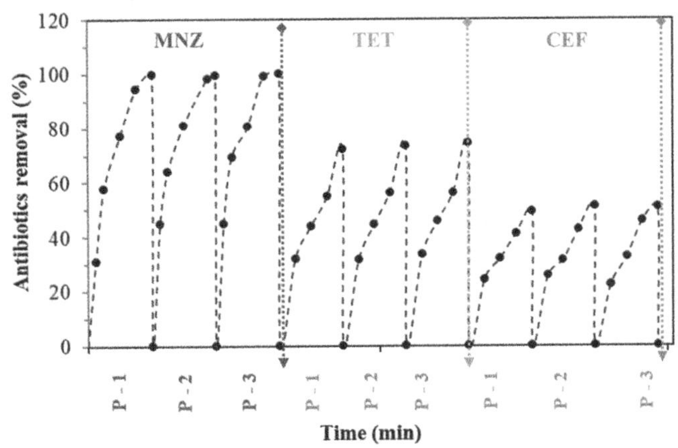

Figure 3 | Reusability studies of PVA-CS-TiO_2 for photocatalytic removal of 10 mg L^{-1} of MNZ, TET and CEF for 3 consecutive cycles.

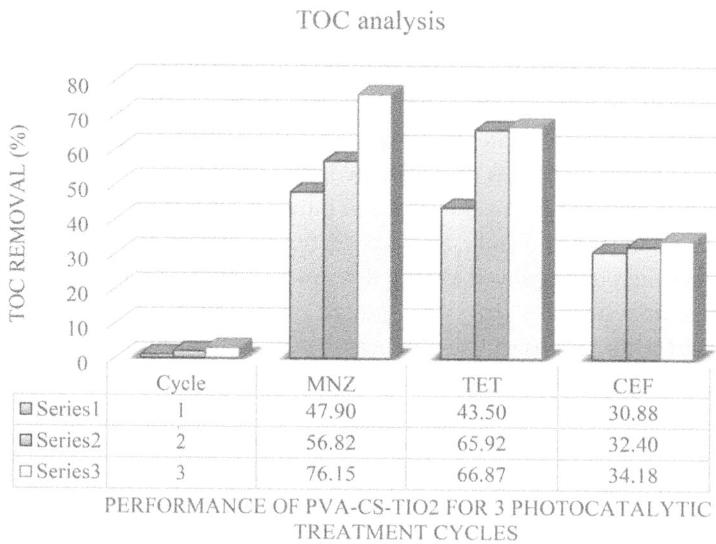

Figure 4 | TOC analysis for the performance of PVA-CS-TiO_2 composite in UV reactor system for photocatalytic removal of 10 mg L^{-1} of antibiotics.

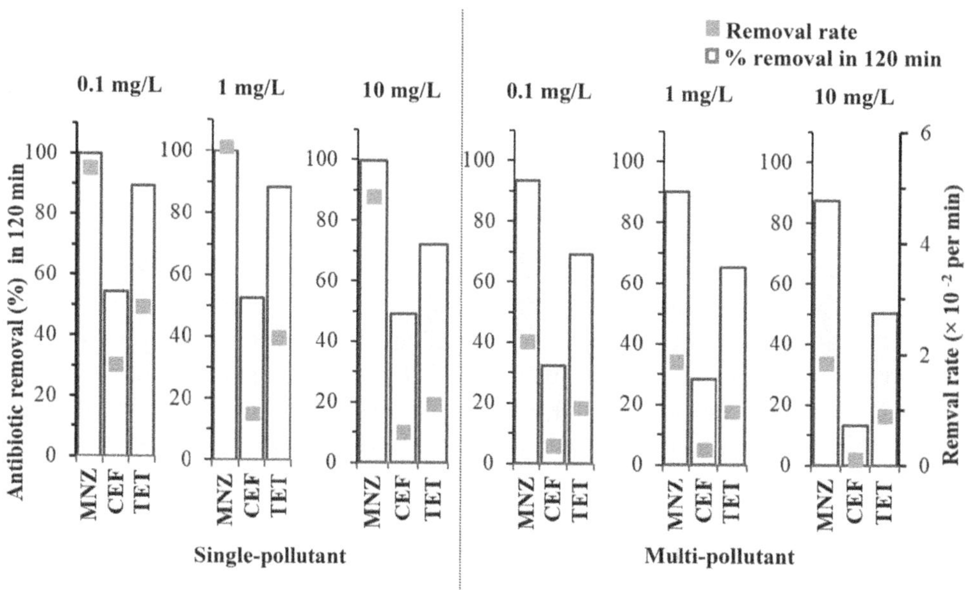

Figure 5 | Effect of initial concentration of antibiotics on removal efficiency and removal rates using PVA-CS-TiO_2 for single and multi-pollutant condition (UV = 32 W, pH ~7).

with increase in initial antibiotic concentration compared to single pollutant condition. For example, k_1 decreased from 0.0032 to 0.0011 min^{-1} when the initial concentration was increased from 0.1 to 10 mg L^{-1} for CEF whereas in the single pollutant condition it was from 0.0066 to 0.0055 min^{-1} (Figure 5). The log k_{ow} of antibiotics considered for the study, i.e., MNZ, CEF and TET are −0.1, 1.6 and −1.37, respectively. The hydrophilic compounds are easily hydrolysed in the UV reactor compared to hydrophobic compounds. This had a greater impact compared to the adsorptive tendency of the compounds to the composites. CEF was the least removed and is attributed to its hydrophobicity and also the highest molecular weight compound amongst others considered for the study. It was envisaged that during UV photocatalysis using PVA-CS-TiO_2 composite, some parts of the antibiotic molecules gets mineralized completely and some parts may generates numerous degradation intermediates. The adsorption pattern of these intermediates on the surface of the composites also needs to be investigated. In mixed pollutant conditions, these new intermediates might have quenched *HO·* radicals to a larger extent compared to *HO·* quenched by the intermediates in single antibiotic condition. This could be the reason for antagonistic degradation behavior of multi-pollutant system in the UV photocatalytic reactor which needs further investigation.

3.4. Micro-toxicity analysis

The micro-toxicity assay was carried out using bioluminescence freeze dried bacteria *Vibrio fischeri* to analyze the toxicity induced by the antibiotics used for the study. The 81.9% basic (consisting of six test dilutions) protocol was carried out to assess the toxicity of MNZ, CEF and TET. In order to understand the toxicity level of MNZ, the measured toxicity values of MNZ was compared with the toxicity of a known standard toxicant $ZnSO_4$, i.e. equivalent-toxicity approach. The trend of inhibition effect (%) with respect to different concentration of MNZ, CEF and TET is provided in Figure 6.

For MNZ, 50% inhibition effect was found at an MNZ concentration of 668.6 mg L^{-1}. The toxicity units (TU) of $ZnSO_4$ and MNZ were found out as 15.4 and 1.5, respectively. The corresponding EC_{50} concentrations were found out as 6.5 mg L^{-1} and 668.6 mg L^{-1}, respectively (Figure 6(a)). However, the low TU value observed in this study could be attributed to the sensitivity of *V. fischeri* to MNZ and the mode of action of MNZ on the bacterium. The factsheet of MNZ indicates that

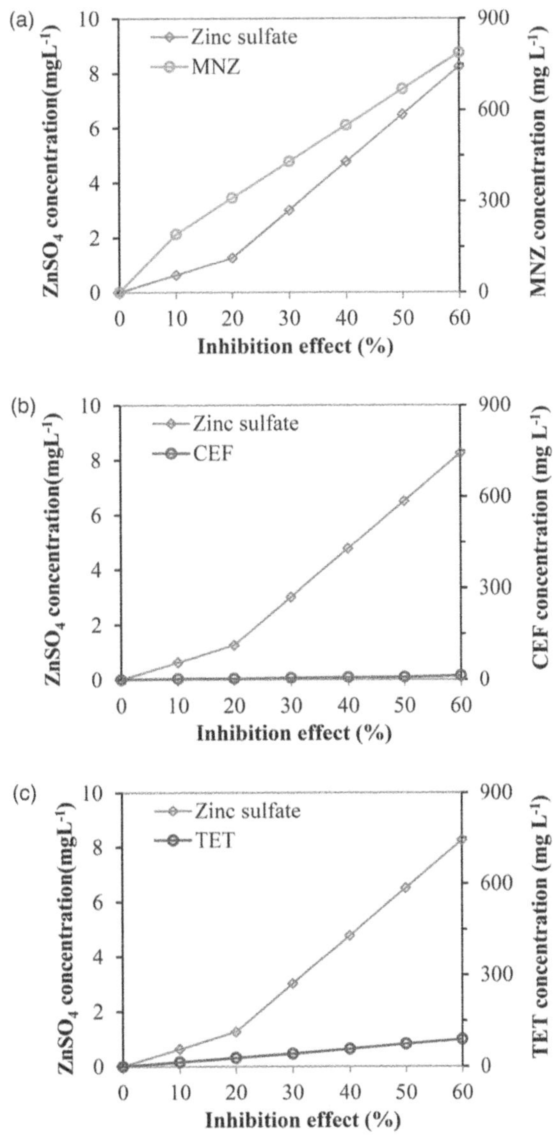

Figure 6 | EC_{50} of (a) metronidazole, (b) ceftiofur and (c) tetracycline used for the study against *Vibrio fischeri*.

MNZ after consumption may transfer to acid and hydroxyl metabolites. The hydroxyl metabolite, formed by reduction of nitro group from MNZ to a hydroxyl/amine group in the cytoplasm of anaerobes/protozoa, accounts for 65% activity of the parent drug. This has the potential to react with DNA and stop the synthesis of nucleic acid in the species. Since, MNZ targets only anaerobic bacterium it did not show deleterious effect to *V. fischeri*. For CEF, TU and EC_{50} were found out to be 54.79 and 18.25 mg L^{-1}, respectively (Figure 6(b)). The factsheet of CEF indicates that it is a broad-spectrum antibiotic target both gram-positive and gram-negative bacteria by inhibiting the cell wall synthesis and repair. Owing to the hydrophobicity of CEF, it can easily diffuse through the selective permeable membrane of the cell wall and impair the microbial cell function. Thus, it has shown profound lethal effect to the bacterium. For TET, TU and EC_{50} were found out 5.75 and 173.8 mg L^{-1}, respectively (Figure 6(c)). The factsheet of TET indicates that these are protein synthesis inhibitors by binding to the 30S ribosomal subunit and thus hinder the cell viability.

In order to understand the toxicity level of different antibiotics, the trend is compared with the standard toxicant ($ZnSO_4$). Accordingly, in the order of highly toxic antibiotics considered for the study (which can also induce toxic effects to exposed microbial biota in the environment) is as follows: CEF > TET > MNZ.

The toxicity of the composites developed in the water was assessed by exposing the composites to the number of days that they were stable in water. Correspondingly, PVA-CS-TiO_2 was stable in water (pH ~ 7) for 35 days. Therefore, the exposed water taken after 35 days was considered as the sample for micro-toxicity analysis of PVA-CS-TiO_2 system. The relative toxicity of the composite and the treated sample from the reactor used for removal of 10 mg L^{-1} of MNZ is found to be 22.8 and 33.1%, respectively. This indicates that there is a slight increase in toxicity after treatment due to the presence of intermediates of MNZ. However, the sample equivalent toxicity values were considerably within the limit. Additionally, ecological risk of the considered antibiotics and similar other compounds can be evaluated at different taxon levels such as fish embryo toxicity test, algal test, yeast estrogen screen test, etc., to get a complete understanding of the toxicity induced by the antibiotics to the environment. Moreover, it is well reported that long-term exposure of the antibiotics considered for the study will accelerate the evolution of antibiotic resistance to the exposed microorganisms.

4. CONCLUSIONS

A recoverable PVA-CS-TiO_2 composite has been successfully developed and applied for the photocatalytic removal of antibiotics. The photocatalytic ability of the composites was assessed through kinetic models. The antibiotic removal by adsorption followed pseudo-second-order. The hydrophilic compounds are degraded at a much faster rate compared to hydrophobic antibiotics considered for the study. The antagonistic behavior of the UV photocatalytic reactor towards mixture of pollutant condition needs further investigation. The toxicity units observed for the antibiotics considered for the study indicate that hydrophobic antibiotics have a profound deleterious effect on the marine bacterium *V. fischeri*. The micro-toxicity analysis revealed that ceftiofur was highly toxic followed by tetracycline and metronidazole. The relative toxicity of the effluent samples as indicated by % inhibition values provides preliminary information on the toxicity of the effluent that contains intermediates of the parent compound and the ions present in it.

ACKNOWLEDGEMENT

The present study is funded by the ICSR, IIT Madras (Grant number: CIE/14-15/832/NFIG/SMAT) and WTI, DST, India (Grant number: DST/TM/WTI/2K15/192).

DATA AVAILABILITY STATEMENT

All relevant data are included in the paper or its Supplementary Information.

CONFLICT OF INTEREST

The authors declare there is no conflict.

REFERENCES

Abbas, M., Arshad, M., Rafique, M., Altalhi, A., Saleh, D., Ayub, M., Sharif, S., Riaz, M., Alshawwa, S. & Masood, N. 2022 Chitosan-polyvinyl alcohol membranes with improved antibacterial properties contained Calotropis procera extract as a robust wound healing agent. *Arabian Journal of Chemistry* **15** (5), 103766.

Ali, F., Khan, S. B., Kamal, T., Alamry, K. A. & Asiri, A. M. 2018 Chitosan-titanium oxide fibers supported zero-valent nanoparticles: highly efficient and easily retrievable catalyst for the removal of organic pollutants. *Scientific Reports* **8** (1), 6260.

Anaya-Esparza, L. M., Villagrán-de la Mora, Z., Ruvalcaba-Gómez, J. M., Romero-Toledo, R., Sandoval-Contreras, T., Aguilera-Aguirre, S., Montalvo-González, E. & Pérez-Larios, A. 2020 Use of titanium dioxide (TiO2) nanoparticles as reinforcement agent of polysaccharide-based materials. *Processes* **8** (11), 1395.

Barton, I., Matejec, V. & Matousek, J. 2016 Photocatalytic activity of nanostructured TiO2 coating on glass slides and optical fibers for methylene blue or methyl orange decomposition under different light excitation. *Journal of Photochemistry and Photobiology A: Chemistry* **317**, 72–80.

Behboudi, A., Jafarzadeh, Y. & Yegani, R. 2016 Preparation and characterization of TiO2 embedded PVC ultrafiltration membranes. *Chemical Engineering Research and Design* **114**, 96–107.

Cesar Filho, M., Bueno, P. V., Matsushita, A. F., Rubira, A. F., Muniz, E. C., Durães, L., Murtinho, D. M. & Valente, A. J. 2018 Synthesis, characterization and sorption studies of aromatic compounds by hydrogels of chitosan blended with β-cyclodextrin-and PVA-functionalized pectin. *Rsc Advances* **8** (26), 14609–14622.

Chang, G., Ullah, W., Hu, Y., Lin, L., Wang, X. & Li, C. Z. 2018 Functional carbon nanofibers with semi-embedded titanium oxide particles via electrospinning. *Macromolecular Rapid Communications* **39** (14), 1800102.

Dong, H., Zeng, G., Tang, L., Fan, C., Zhang, C., He, X. & He, Y. 2015 An overview on limitations of TiO2-based particles for photocatalytic degradation of organic pollutants and the corresponding countermeasures. *Water Research* **79**, 128–146.

Ganesh, V. A., Nair, A. S., Raut, H. K., Walsh, T. M. & Ramakrishna, S. 2012 Photocatalytic superhydrophilic TiO 2 coating on glass by electrospinning. *RSC Advances* **2** (5), 2067–2072.

Gopi, K. H., Dhavale, V. M. & Bhat, S. D. 2019 Development of polyvinyl alcohol/chitosan blend anion exchange membrane with mono and di quaternizing agents for application in alkaline polymer electrolyte fuel cells. *Materials Science for Energy Technologies* **2** (2), 194–202.

Gu, L., Zhang, Z., Yang, S., Liu, X., Zhang, M., Gao, L. & Xiao, G. 2021 Chitosan-modified polyvinyl alcohol membrane high performance in biodiesel/methanol pervaporation separation. *ChemistrySelect* **6** (34), 9052–9059.

Islam, A., Yasin, T., Akhtar, M. J., Imran, Z., Sabir, A., Sultan, M., Khan, S. M. & Jamil, T. 2016 Impedance spectroscopy of chitosan/poly (vinyl alcohol) films. *Journal of Solid State Electrochemistry* **20** (2), 571–578.

ISO W 1998 Determination of the Inhibitory Effect of Water Samples on the Light Emission of Vibrio Fischeri. (Luminescent Bacteria Test). *Iso 11348-1, 2 and* **3**, 1–21.

Jafarzadeh, Y., Yegani, R. & Tantekin-Ersolmaz, S. 2015 Effect of TiO2 nanoparticles on structure and properties of high density polyethylene membranes prepared by thermally induced phase separation method. *Polymers for Advanced Technologies* **26** (4), 392–398.

Koo, H.-J., Jur, J. S., Woodroof, M., Kalanyan, B., Lee, K., Devine, C. K. & Parsons, G. N. 2012 Stable anatase TiO 2 coating on quartz fibers by atomic layer deposition for photoactive light-scattering in dye-sensitized solar cells. *Nanoscale* **4** (15), 4731–4738.

Kyzas, G. & Bikiaris, D. 2015 Recent modifications of chitosan for adsorption applications: a critical and systematic review. *Marine Drugs* **13** (1), 312–337.

Liu, Y., Li, Z., Green, M., Just, M., Li, Y. Y. & Chen, X. 2017 Titanium dioxide nanomaterials for photocatalysis. *Journal of Physics D: Applied Physics* **50** (19), 193003.

Liu, Y., Wang, S. & Lan, W. 2018 Fabrication of antibacterial chitosan-PVA blended film using electrospray technique for food packaging applications. *International Journal of Biological Macromolecules* **107**, 848–854.

Mendoza, C., Valle, A., Castellote, M., Bahamonde, A. & Faraldos, M. 2015 Tio2 and TiO2–SiO2 coated cement: comparison of mechanic and photocatalytic properties. *Applied Catalysis B: Environmental* **178**, 155–164.

Mohamad Idris, N. H., Rajakumar, J., Cheong, K. Y., Kennedy, B. J., Ohno, T., Yamakata, A. & Lee, H. L. 2021 Titanium dioxide/polyvinyl alcohol/cork nanocomposite: a floating photocatalyst for the degradation of methylene blue under irradiation of a visible light source. *ACS Omega* **6** (22), 14493–14503.

Moustafa, H., Karmalawi, A. M. & Youssef, A. M. 2021 Development of dapsone-capped TiO2 hybrid nanocomposites and their effects on the UV radiation, mechanical, thermal properties and antibacterial activity of PVA bionanocomposites. *Environmental Nanotechnology, Monitoring & Management* **16**, 100482.

Neghi, N. & Kumar, M. 2017 Performance analysis of photolytic, photocatalytic, and adsorption systems in the degradation of metronidazole on the perspective of removal rate and energy consumption. *Water, Air, & Soil Pollution* **228** (9), 339.

Neghi, N., Kumar, M. & Burkhalov, D. 2019 Synthesis and application of stable, reusable TiO2 polymeric composites for photocatalytic removal of metronidazole: removal kinetics and density functional analysis. *Chemical Engineering Journal* **359**, 963–975.

Pham, N. K., Vu, N. H., Van Pham, V., Ta, H. K. T., Cao, T. M., Thoai, N. & Tran, V. C. 2018 Comprehensive resistive switching behavior of hybrid polyvinyl alcohol and TiO 2 nanotube nanocomposites identified by combining experimental and density functional theory studies. *Journal of Materials Chemistry C* **6** (8), 1971–1979.

Rokhmat, M., Wibowo, E. & Abdullah, M. 2017 Coating TiO2 nanoparticles on the surface of transparent plastic granules using combined electrostatic and heating methods for the photocatalytic degradation of organic pollutants in water. *Environmental Nanotechnology, Monitoring & Management* **8**, 1–10.

Singh, P., Vishnu, M., Sharma, K. K., Borthakur, A., Srivastava, P., Pal, D., Tiwary, D. & Mishra, P. K. 2016 Photocatalytic degradation of acid Red dye stuff in the presence of activated carbon-TiO2 composite and its kinetic enumeration. *Journal of Water Process Engineering* **12**, 20–31.

Ulliman, S. L., Miklos, D. B., Hübner, U., Drewes, J. E. & Linden, K. G. 2018 Improving UV/H 2 O 2 performance following tertiary treatment of municipal wastewater. *Environmental Science: Water Research & Technology* **4** (9), 1321–1330.

Vatanpour, V., Salehi, E., Sahebjamee, N. & Ashrafi, M. 2020 Novel chitosan/polyvinyl alcohol thin membrane adsorbents modified with detonation nanodiamonds: preparation, characterization, and adsorption performance. *Arabian Journal of Chemistry* **13** (1), 1731–1740.

Wang, S., Xu, M., Peng, T., Zhang, C., Li, T., Hussain, I., Wang, J. & Tan, B. 2019 Porous hypercrosslinked polymer-TiO 2-graphene composite photocatalysts for visible-light-driven CO 2 conversion. *Nature Communications* **10** (1), 676.

Zhao, Y., Tao, C., Xiao, G. & Su, H. 2017 Controlled synthesis and wastewater treatment of Ag_2O/TiO_2 modified chitosan-based photocatalytic film. *RSC Advances* **7** (18), 11211–11221.

Zhao, Y.-F., Zhao, J.-Y., Hu, W.-Z., Ma, K., Chao, Y., Sun, P.-J., Fu, X.-B. & Zhang, H. 2019 Synthetic poly (vinyl alcohol)–chitosan as a new type of highly efficient hemostatic sponge with blood-triggered swelling and high biocompatibility. *Journal of Materials Chemistry B* **7** (11), 1855–1866.

Zhu, J., Liu, X., Wang, X., Huo, X. & Yan, R. 2015 Preparation of polyaniline–TiO2 nanotube composite for the development of electrochemical biosensors. *Sensors and Actuators B: Chemical* **221**, 450–457.

Zhu, X., Gu, P., Wu, H., Yang, D., Sun, H., Wangyang, P., Li, J. & Tian, H. 2017 Influence of substrate on structural, morphological and optical properties of TiO2 thin films deposited by reaction magnetron sputtering. *Aip Advances* **7** (12), 125326.

First received 25 May 2022; accepted in revised form 1 August 2022. Available online 6 August 2022

doi: 10.2166/wst.2022.304

Combined ultrasound cavitation and persulfate for the treatment of pharmaceutical wastewater

Karan Pandya[a], Anantha Singh T. S.[a,b,*], Pravin Kodgire[c] and Saji Simon[b]

[a] Department of Civil Engineering, School of Technology, Pandit Deendayal Energy University, Gandhinagar, Gujarat, India
[b] Department of Civil Engineering, National Institute of Technology Calicut, Calicut, India
[c] Department of Chemical Engineering, School of Technology, Pandit Deendayal Energy University, Gandhinagar, Gujarat, India
*Corresponding author. E-mail: singh87@nitc.ac.in

AST, 0000-0002-7735-497X

ABSTRACT

In recent years industrialization has caused magnificent leaps in the high profitable growth of pharmaceutical industries, and simultaneously given rise to environmental pollution. Pharmaceutical processes like extraction, purification, formulation, etc., generate a large volume of wastewater that contains high chemical oxygen demand (COD), biological oxygen demand, auxiliary chemicals, and different pharmaceutical substances or their metabolites in their active or inactive form. Its metabolites impart non-biodegradable toxic pollutants as a byproduct and intense color, which increases ecotoxicity into the water, thus this requires proper treatment before being discharged. This study focuses on the feasibility analysis of the utilization of ultrasound cavitation (20 kHz frequency) together with a persulfate oxidation approach for the treatment of complex pharmaceutical effluent. Process parameters like pH, amplitude intensity, oxidant dosage were optimized for COD removal applying response surface methodology-based Box–Behnken design. The optimum value observed for pH, amplitude intensity and oxidant dosage are 5, 20% and 100 mg/L respectively with 39.5% removal of COD in 60 min of fixed processing time. This study confirms that a combination of ultrasound cavitation and persulfate is a viable option for the treatment of pharmaceutical wastewater and can be used as an intensification technology in existing effluent treatment plants to achieve the highest amount of COD removal.

Key words: Box–Behnken design, cavitation, optimization, persulfate, pharmaceutical wastewater, ultrasound

HIGHLIGHTS

- Synergistic effect of ultrasonication and persulfate oxidation studied on pharma wastewater.
- Box–Behnken method applied for process parameter optimization.
- Significant reduction in the COD from the pharma wastewater within one hour of the reaction.
- Potential choice for the implementation of the process as a pre-treatment option for wastewater treatment.

1. INTRODUCTION

Environmental pollution is one of the global challenges of today's world (Spina *et al.* 2012; Singh & Prashant 2017). Industrial effluent and other hazardous discharges from industries are one of the prime concerns for developing countries like India. In India, one-third of water pollution in natural water bodies and marine pollution are induced by industrial wastewater (Kansal *et al.* 2013; Singh & Prashant 2017). At this time, high amounts of pharmaceutical substances are used for the protection and cure of diseases for humans and animals, which lead to the generation of huge quantities of wastewater from pharmaceutical industries (Mohapatra *et al.* 2014; Ghafoori *et al.* 2015). The USEPA reported the daily effluent produced from the pharmaceutical unit as 1.0068×10^9 L (Adishkumar *et al.* 2012). These effluents contain highly biologically active compounds which can cause damage to health in humans and animals, as well as promoting the development of antibiotic resistance genes, and interrupts the ecological balance of the aquatic environment by initiating irreversible transformation to aquatic fauna (Ng *et al.* 2014; Tiwari *et al.* 2020). Industrial wastewater quality is estimated, based on the amount of organic matter present, as COD, total carbon, BOD, and wastewater quality parameters. The wastewaters produced from pharmaceutical sectors are complex and hazardous with high COD, BOD, solids containing supplementary chemicals, and pharmaceutical secondary metabolites leading the wastewater to be classified under the 'red category' (Gadipelly *et al.* 2014; Changotra *et al.* 2017, 2019; Martínez *et al.* 2018).

Pharmaceutical wastewater contains a large amount of recalcitrant or bio-refractory molecules which are very difficult to degrade through conventional treatments. Hence, most is converted into other complex by-products as fluoxetine and endocrine disruptors, which has resulted in a multitude of undesirable problems like harm to reproduction and metabolism of aquatic organisms and feminization of fish populations (Ford & Fong 2016; Huang *et al.* 2016; Thanekar & Gogate 2019). Thus it is necessary to treat pharmaceutical wastewater efficiently before discharging it into any water bodies to avoid further damage to the environment and ecosystem, it also pollutes water bodies (Gadipelly *et al.* 2014; Changotra *et al.* 2017, 2019; Singh & Prashant 2017; Martínez *et al.* 2018).

Treatment processes like biological oxidation require longer time because they have a slower reaction rate (Crini & Lichtfouse 2019). Chemical coagulation is useful for removal of pollutant in the form of colloidal or suspended particles that do not settle rapidly, it requires a longer retention time and more sludge production (Verma *et al.* 2012). Adsorption reactors rapidly becomes less effective due to clogging and their regeneration are not cost-effective, hence, this decreases the acceptance of a certain type of metal molecule (Crini & Lichtfouse 2019). These demerits of treatment, make it not efficient to remove all hazardous compounds effectively from pharmaceutical wastewater (Jeworski & Heinzle 2000; Ayare & Gogate 2019). In recent times advanced oxidation processes (AOP) such as Fenton oxidation (Singh *et al.* 2013), ozonation (Cortez *et al.* 2010), ultraviolet/hydrogen peroxide (Hu *et al.* 2011), and ultrasound cavitation (Gągol *et al.* 2018) have been considered as a useful process for treatment of bio-refractory pollutants (Yang *et al.* 2015). Among the different AOP the ultrasound cavitation-based process has been seen as a promising process for the oxidation of various pollutants in industrial wastewater. The principle behind the ultrasonication process is the pressure variation in liquid caused by inducing ultrasound which produces cavitation for the treatment of wastewater (Thanekar & Gogate 2019). Cavitation occurs due to pressure difference in the outer flow and the pressure induced inside the system by sound waves as system pressure decreases or increases in flow velocity, which produces small cavities; with time these start to grow longer compared with higher system pressure (Dular *et al.* 2016). Furthermore, variation in pressure leads to cavity formation, growth, and collapse at the micro-scale duration. These three stages occur during the ultrasound cavitation process which leads to the generation of free radicals and releases a large amount of energy. This energy can be up to 5,000 K temperature and 1,000 Atm pressure in wastewater known as a 'local hotspot' which can be extremely suitable for the oxidation of pollutants (Leighton 1995; Thanekar & Gogate 2019). The main advantage behind this treatment is it does not require any chemicals to promote oxidation, no sludge formation, no visible light requirement (Vega & Peñuela 2018) and it breaks up the agglomerates (Jordens *et al.* 2016; Dong *et al.* 2020). In the same way, the presence of reactive nitrogen such as NO_3, NO_2, NO or degradation of other organic pollutants that contain nitrogen species forms nitrate radicals ($NO_3\bullet$), nitrite radicals ($NO_2\bullet$), and nitrogen oxide radicals ($NO\bullet$), etc. These reactive nitrogen radicals participate in redox reactions as well and help in the nitration reaction of the pollutants (Rayaroth *et al.* 2022). Ultrasound cavitation is useful for the treatment of active pharmaceutical compounds (Tran *et al.* 2015; Serna-Galvis *et al.* 2019; Estrada-Flórez *et al.* 2020), such as carbamazepine, diclofenac, ciprofloxacin (Beckett & Hua 2001; Rayaroth *et al.* 2016), and the tetracycline antibiotic (Soltani *et al.* 2019). Ultrasound cavitation accelerates the degradation of pollutants in combination with other oxidation processes like Fenton's reagent (Zhang *et al.* 2016), zinc oxide, the zinc oxide nanotube catalytic process (Soltani *et al.* 2019), hydrogen peroxide (Chandak *et al.* 2020), carbon tetrachloride, and persulfate ions (Anipsitakis & Dionysiou 2003) to generate radicals (Rayaroth *et al.* 2016). The persulfate dissociation can also occur due to high localized temperature developed by the ultrasound system (Wei *et al.* 2017). The combination of persulfate oxidation with the ultrasound cavitation system has been less studied by other research groups (Monteagudo *et al.* 2015, 2018; Wang & Zhou 2016). An alternative to hydrogen peroxide, persulfate ion has gained more attention due to the increased interest in the higher redox potential of sulfate radicals (2.5–3.1 V) compared with hydroxyl radicals. The hydroxyl radicals prefer to react by addition, hydrogen atom abstraction, as well as through electron transfer mechanisms, while $SO_4^{-\cdot}$ prefer to react through electron transfer mechanisms, which makes them selective towards electron-donating groups such as amino ($-NH_2$), hydroxyl (-OH), alkoxy (-OR), amino ($-NH_2$), alkoxy (-OR), π electrons of aromatic compounds, and unsaturated bonds (Fedorov *et al.* 2020):

$$S_2O_8^{2-} + activator \rightarrow SO_4^{-\cdot} + (SO_4^{-\cdot} \text{ or } SO_4^{2-}) \tag{1}$$

The formation of sulfate radical occurs by breakage of the O-O bond of persulfate ions using different activation methods like heat (Tan *et al.* 2012), transitional metals (Liu *et al.* 2016), ultraviolet light (Xie *et al.* 2015; Arvaniti *et al.* 2020), ultrasound (Zou *et al.* 2014; Yang *et al.* 2019), and other means as shown in Equation (1) (Arvaniti *et al.* 2020). The homogeneous and heterogeneous catalytic activation of persulfate by using cobalt or manganese oxides, etc., has been studied by many researchers (Liang *et al.* 2012; Yuan *et al.* 2020). Among them, activation by ultrasound acts as an emerging method and is typically used for pharmaceutical wastewater treatment (Zou *et al.* 2014; Yang *et al.* 2019). The synergistic effect of combined cavitation and the persulfate process has been studied by Fedorov *et al.* (2022) and has computed the synergistic index for different combined process. The synergistic index of combined cavitation and persulfate varied from 1.52 to 4.43 for different pollutants (Fedorov *et al.* 2022). Therefore, it is significant to use a combined process for the treatment of pharmaceutical wastewater to achieve increased removal.

In the combined process multiple operating parameters influence or affect the removal efficiency. For complex systems governed by several parameters, generally, use of one parameter optimization at a time could provide misinterpretation due to lack of interaction effects, thus the Design of Experiment (DoE) is an effective tool for optimization. There are various DoEs available for optimization of parameters by reducing the number of experiments and cost (Dopar *et al.* 2011) like the Box–Behnken method for response surface methodology (RSM) which is a powerful design tool for modeling complex conditions (Tak *et al.* 2015). This tool estimates the relation between manageable input parameters and response variables (Khorram & Fallah 2018).

This study focuses on the combined effect of ultrasound cavitation and persulfate oxidation on the treatment of pharmaceutical wastewater. The synergistic effect of the combined process was investigated along with the effect of biodegradability of the combined process. The major process influencing factors of the combined ultrasound–persulfate process like pH, amplitude intensity, persulfate dosage, reaction time were investigated for their individual and the combined effects by using Box–Behnken design as an RSM for the treatment of pharmaceutical wastewater. The effect of reactive radicals on the removal of pollutants was investigated through scavenging tests.

2. MATERIAL AND METHODOLOGY

2.1. Industrial wastewater

A wastewater sample of 40 L was collected from the pharmaceutical industry from Bharuch, Gujarat, India. Collected samples were stored in a refrigerator at 4 °C to avoid further biodegradation.

2.2. Chemicals

In this study, chemicals of analytical grade were used for the entire experimental process: potassium persulfate purity 98% purity, potassium dichromate 99% purity, mercury sulfate 98% purity, silver sulfate 98% purity, ammonium ferrous sulfate 99% purity, manganous sulfate 98% purity, ferric chloride 98% purity, di-potassium hydrogen phosphate 99% purity, potassium iodide 99% purity, potassium hydroxide 85% purity, sodium thiosulfate 98% purity were procured from M/s Merck India. Sulfuric acid of 98% purity, ferroin indicator, calcium chloride 96% purity, magnesium sulfate 99% purity, starch as an indicator were procured from M/s Finar India. All the required reagents were prepared using double-distilled water. For ultrasonic cavitation, a 20 kHz frequency probe-type sonicator made of Ti6Al4V titanium alloy with maximum power output of 750 W was used with a microprocessor-based programmable probe (Model No. VCX 500 with 17 mm solid probe diameter) and procured from Sonics Vibracell, USA.

2.3. Design of experiment (DoE)

Design-expert software was used to prepare DoE. The removal of organic matter from pharmaceutical wastewater was optimized using Box–Behnken design as RSM. Table 1 represents its coded and real value for lower level (−1), mid-level (0), high level (+1) for removal of organic matter as COD. Table 2 provides the details of a different combinations of experimental runs for the considered parameters pH (A), amplitude intensity% (B), persulfate dosage (C). The experimental range for the parameters was decided based on published literature.

Mathematical representation of selected process parameters to the estimation of the dependent response y (COD removal) by coded terms is as mentioned below, by general model regression Equation (2):

$$y = \beta_0 + \sum_{j=1}^{k} \beta j X j + \sum_{j=1}^{k} \beta jj X j^2 + \sum \sum_{<j=2}^{k} \beta ij X_i X_j \tag{2}$$

Table 1 | Coded and real values of the factors selected for Box–Behnken design

Factors	Low level −1	Mid-level 0	High level 1
pH (A)	2	5	8
Amplitude intensity % (B)	20	50	80
Dosage (mg/L) (C)	100	250	400

Table 2 | Box–Behnken design

Experimental runs	pH	Amplitude (%)	Persulfate dosage (mg/L)	COD remaining (mg/L)
1	2	50	100	11,200
2	2	50	400	12,480
3	5	80	100	10,944
4	2	80	250	10,880
5	5	50	250	9,440
6	5	50	250	9,440
7	5	20	100	8,320
8	8	50	100	12,640
9	8	80	250	12,640
10	5	20	400	12,800
11	8	20	250	12,800
12	5	80	400	11,680
13	8	50	400	12,960
14	5	50	250	12,320
15	2	20	250	11,200

where, y is a response, β_0 is the linear, quadratic and regression constant, βj, βjj and βij represent the coefficients for linear, quadratic, and second-order interaction effects, respectively. x_j and x_i represent the coded independent process variables.

2.4. Experimental procedure

All experiments were conducted in a cylindrical batch glass reactor with 250 mL of working volume (capacity of 500 mL). The schematic diagram of the experimental process reactor is shown in Figure 1. The temperature of the sample was maintained at 50 ± 10 °C for the heat-based activation of persulfate (Ahmadi *et al.* 2019), with ultrasound and the pH was adjusted using nitric acid (1 N) and sodium hydroxide (1 N). Ioannidi *et al.* (2022) reported an acceleration in degradation of pollutant when the temperature of the wastewater was increased from 40 °C to 60 °C, whereas in the current study the external heating of wastewater was not done, instead the temperature of the water was maintained within the 50 ± 10 °C range through a water circulation method. The addition of sulfate ions in the advanced oxidation process will convert the ion to sulfate radicals. The sulfate radicals generated through the addition of sulfate ion may oxidize the organic matter over the persulfate-generated sulfate radical. Thus, to study the efficiency of persulfate radicals on wastewater treatment, the addition of sulfate ions (sulfuric acid) was restricted and instead nitric acid was used. The persulfate dosage was added according to the real value of factors suggested by DoE. All 15 experimental runs were conducted according to values suggested by DoE. All the experiments were duplicated twice to check their reproducibility, a ±10% change in results was accepted and the experiments were repeated when the deviation in results was beyond ±10%. Treated samples were withdrawn during the experiment at a time interval of 15, 30, and 60 min and analyzed for COD. The COD of the sample was analyzed as per the standard method using the closed reflux method. The synergistic effect of the combined process (ultrasound cavitation and persulfate oxidation) was analyzed using Equation (3) (Rokhina *et al.* 2008; Li & Zhu 2016). The synergistic index of

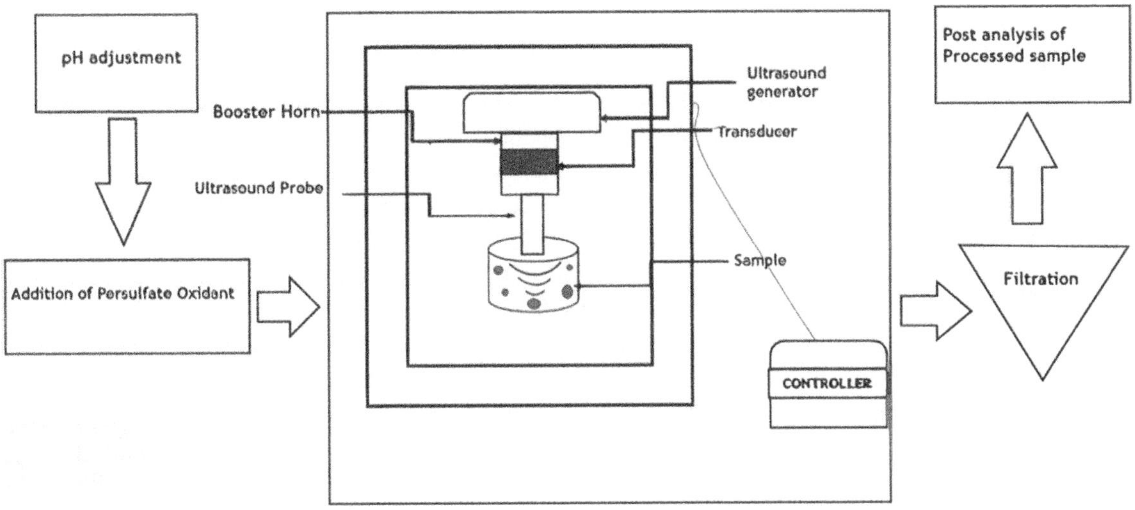

Figure 1 | Schematic diagram of the experimental process.

the combined process can be determined by comparing the removal efficiency of the stand-alone process ($RR_{ultrasound\ cavitation}$, $RR_{persulfate\ oxidation}$) with that of the removal efficiency of the combined process (RR):

$$\text{Synergistic index, SI} = \frac{RR}{(RR_{ultrasound\ cavitation} + RR_{persulfate\ oxidation})} \tag{3}$$

where,
RR – Removal rate of COD for the combined process
$RR_{ultrasound\ cavitation}$ – Removal rate of COD for stand-alone ultrasound cavitation treatment
$RR_{persulfate\ oxidation}$ – Removal rate of COD for stand-alone persulfate oxidation treatment.

If the synergistic index of the combined process is equal to one ($SI = 1$), the effect is called an additive effect. If $SI > 1$ the combined process effect is much greater than the sum of the sole process effect, this is called a synergistic effect and if $SI < 1$ the effect is termed as antagonism (Li & Zhu 2016).

3. RESULTS AND DISCUSSION

3.1. Characteristics of pharmaceutical wastewater

Characterization of the pharmaceutical wastewater was done as per the APHA standard methods (Rice *et al.* 2012) and given in Table 3. The COD and BOD of the wastewater were 13,760 mg/L and 5,500 mg/L respectively with a BOD/COD ratio of 0.4.

Table 3 | Characteristics of pharmaceutical wastewater

Parameter	Concentration in mg/L (except pH)
pH	7.6–7.8
Chloride	174.9
Total solids	25,570
Total organic carbon	2,815
Biological oxygen demand	5,500
Chemical oxygen demand	13,760
BOD/COD	0.4

3.2. Analysis of variance (ANOVA)

The ANOVA in Table 4, provides the details of various coefficient values of process parameters and other details for COD removal from pharmaceutical wastewater with ultrasound cavitation treatment. The table shows that the model (Fischer test value) F-value is 9.42 and it shows the model value is significant as the *P*-value is less than 0.05, which implies that the model is significant. The significance of the *P*-value fixes the error probability of the regression co-efficient as significant. For optimized conditions of parameters quadratic and some of the interaction terms are significant, the parameter pH (A) is slightly significant, another parameter persulfate dosage (C) is significant, an interaction effect of amplitude intensity (B) and persulfate dosage (C), is also significant concerning model terms. According to the ANOVA, if the *P*-value of selected parameters is less than 0.05 (i.e. $P < 0.05$) then selected parameters are significant for the process efficiency, and when the *P*-value of parameters is greater than 0.05 ($P > 0.05$) then selected parameters are insignificant for the process efficiency. Table 5 gives the comparison of results between the predicted and experimental under different test conditions. The experimental results are comparable with the predicted with an error percentage variation of 0–30% under different parametrical conditions.

A mathematical representation of the independent variable to the dependent response by coded terms shown by this Equation (4):

$$\text{COD removal (mg/l)} = 30 + 2197.7A + 19.62B - 6.013C - 3.55AB + 0.88AC + 0.256BC - 240.2A^2 - 0.620B^2 - 0.0319 \tag{4}$$

From the above equation, it is noticed that there is a positive effect of linear terms on COD removal except persulfate dosage (C). With the change in linear terms, the COD removal changes according to its coefficients. The quadratic terms with a negative sign show a negative impact on COD removal. This means that an increase in the values of the quadratic coefficient will decrease the removal of the COD removal. Other interaction terms (AB, AC, BC) have mixed responses on the COD. The term with interaction (BC and AC) with persulfate dosage (C) will increase COD removal. Negative quadratic terms in Equation (2) indicate that the reduction in COD removal has been quadratic.

For the identification order of each process parameter and its effect while keeping the other process parameters constant indicated in perturbation plots, Figure 2 then factors sensitivity on COD removal, which is indicated by the nature of the curve for each process parameter and the factor with more slope has more noticeable effects on COD removal (Muthukumaran *et al.* 2017; Milano *et al.* 2018). The plot shows that factor B has a prominent effect on COD removal followed by A and C, similar to that reflected in the table for ANOVA.

Table 4 | ANOVA results for the obtained regression equations

Source	Sum of squares	df	Mean square	F-value	*P*-value	
Model	3.115E + 07	9	3.461E + 06	9.42	<0.050	Significant
A – pH	1.843E + 06	1	1.843E + 06	5.02	0.0652	Slightly significant
B – Amp (%)	1.191E + 05	1	1.191E + 05	0.3241	0.5938	
C – Dosage	3.942E + 06	1	3.942E + 06	10.73	<0.050	Significant
AB	4.096E + 05	1	4.096E + 05	1.11	0.3393	
AC	6.400E + 05	1	6.400E + 05	1.74	0.2441	
BC	5.345E + 06	1	5.345E + 06	14.55	<0.050	Significant
A^2	1.726E + 07	1	1.726E + 07	46.98	0.0010	
B^2	1.150E + 06	1	1.150E + 06	3.13	0.1371	
C^2	1.903E + 06	1	1.903E + 06	5.18	0.0719	
Residual	1.837E + 06	5	3.674E + 05			
Lack of fit	1.837E + 06	3	6.123E + 05			
Pure error	0.0000	2	0.0000			
Cor total	3.299E + 07	14	$R^2 = 94.43\%$			

Table 5 | ANOVA analysis results

Coded factors				COD removal	
Experimental runs	**pH**	**Amplitude**	**Persulfate dosage (mg/L)**	**Experimental (%) removal**	**Predicted (%) removal**
1	2	50	100	19%	22%
2	2	50	400	8%	6%
3	5	80	100	20%	20%
4	2	80	250	21%	19%
5	5	50	250	31%	32%
6	5	50	250	31%	32%
7	5	20	100	39%	36%
8	8	50	100	7%	9%
9	8	80	250	8%	7%
10	5	20	400	7%	8%
11	8	20	250	7%	10%
12	5	80	400	21%	26%
13	8	50	400	8%	5%
14	5	50	250	31%	32%
15	2	20	250	10%	12%

The individual effect of each parameter plot is shown in Figure 3(a)–3(c), which indicates the effect of parameter pH, amplitude intensity, and persulfate dosage, respectively. These plots are made by taking one factor as a variable and the other two factors fixed at constant levels and the experimental results are shown at three different levels compared to predicated plots that are discussed in the following section.

3.3. Effect of pH on COD removal

One of the important parameters that affects the process efficiency is pH. It plays a vital role in the generation of intermediate oxidants for the oxidation of various compounds it can also affect their condition during the cavitation process (Keenan & Sedlak 2008; Barik & Gogate 2017; Thanekar & Gogate 2019). Effects of pH on the degradation of organic pollutants were studied by varying pH from 2 to 8 with a fixed amplitude intensity of 50% and persulfate dosage of 250 mg/L and is shown in Figure 3(a). As shown in Figure 3(a), the maximum COD removal of 39% at pH 5 at a reaction time of 60 min was observed. Increasing pH from the lower level at pH 2 to mid-level pH 5, the removal efficiency also increased, further increasing the pH to 8 decreased the removal efficiency of COD. Maintaining the mid-level value of factors amplitude intensity of 50% and persulfate dosage of 250 mg/L at pH 2 and pH 8 corresponded to a predicated removal efficiency of 19 and 21% respectively.

The degradation of organic pollutants from the wastewater depends on the radicalization of the reactive compounds. The possible radicalization reactions of persulfate with respect to ultrasound as the activator are given below, in Equations (5)–(12). The significant radicals formed during the persulfate reaction are sulfate radicals and hydroxyl radicals. The radical compounds react with the organic compounds and are oxidized to by-products, carbon dioxide and water (see Li *et al.* 2013; Wang *et al.* 2015):

$$H_2O \xrightarrow{)))} \cdot OH + \cdot H \tag{5}$$

$$S_2O_8^{2-} +))) \rightarrow 2SO_4^{\cdot -} \tag{6}$$

$$SO_4^{\cdot -} + H_2O \rightarrow H^+ + SO_4^{2-} + HO^{\cdot} \tag{7}$$

$$S_2O_8^{2-} + HO^{\cdot} \rightarrow HSO_4^- + SO_4^{\cdot -} + \frac{1}{2}O_2 \tag{8}$$

$$S_2O_8^{2-} + HO^{\cdot} \rightarrow S_2O_8^{\cdot -} + OH^- \tag{9}$$

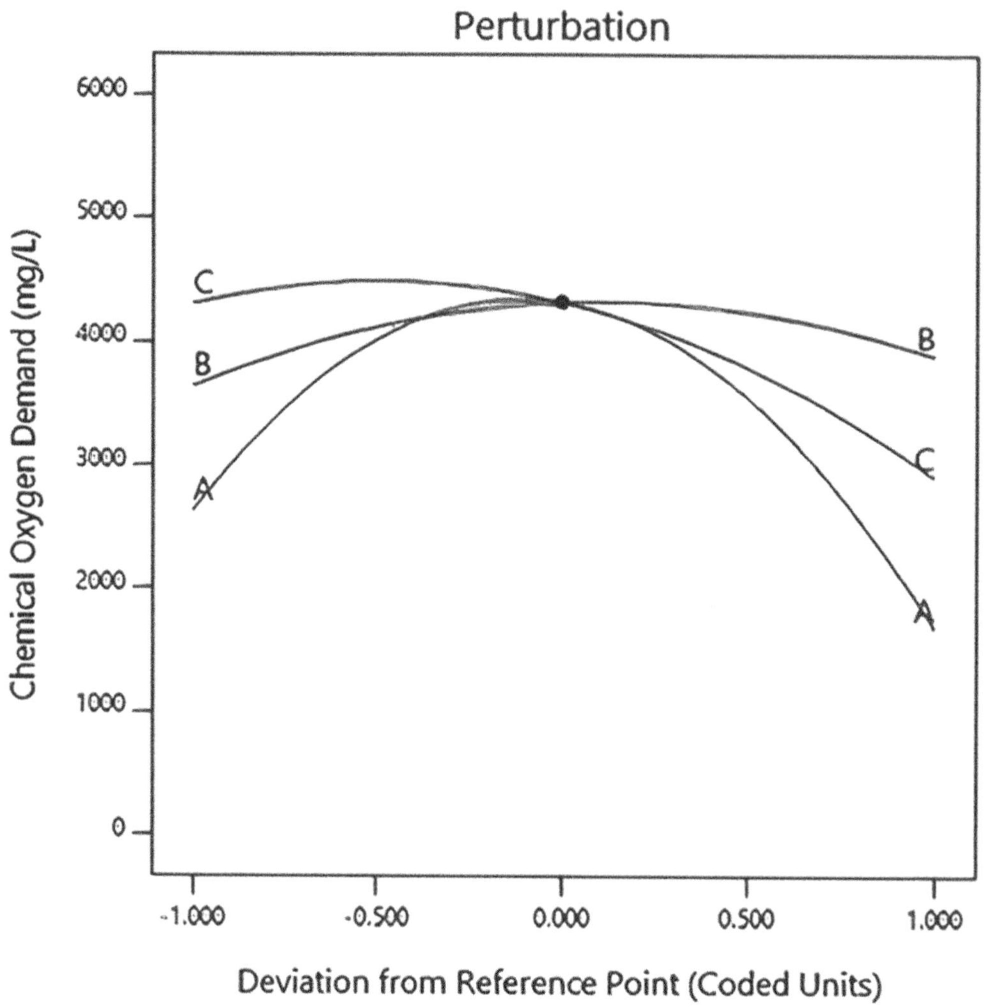

Figure 2 | Profile of perturbation plot showing significant parameters affecting COD removal with factors pH (curve A) 2–8, amplitude (curve B) 20–80%, persulfate dosage 100–400 mg/L (curve C) with middle level values. Temperature 50 ± 10 °C.

$$S_2O_8^{2-} + H^{\cdot} \rightarrow HSO_4^{-} + SO_4^{-\cdot} \tag{10}$$

$$S_2O_8^{2-} \xrightarrow{pyrolysis} 2SO_4^{-\cdot} \tag{11}$$

$$R + SO_4^{-\cdot} \rightarrow SO_4^{2-} + Products \tag{12}$$

Excessive loading of persulfate dosage above 250 mg/L has proven a scavenger for the process under the acidic and non-acidic conditions as given in Equations (13)–(15) (Peyton 1993; Wei *et al.* 2018).

$$S_2O_8^{2-} + SO_4^{-\cdot} \rightarrow S_2O_8^{-\cdot} + SO_4^{2-} \tag{13}$$

$$SO_4^{-\cdot} + SO_4^{-\cdot} \rightarrow S_2O_8^{2-} \tag{14}$$

$$S_2O_8^{2-} + 2e^- \rightarrow 2SO_4^{2-} \tag{15}$$

The observed results show near acidic conditions as more effective compared to alkaline conditions for the degradation of pollutants under a similar kind of analysis as shown in the published literature (Thanekar & Gogate 2019). Due to this, removal was low at a higher levels of pH 8. Mid-level pH 5 showed maximum removal efficiency compared with the highly acidic pH 2, which was due to a higher amount of cavitation yield and effective generation of sulfate radicals by

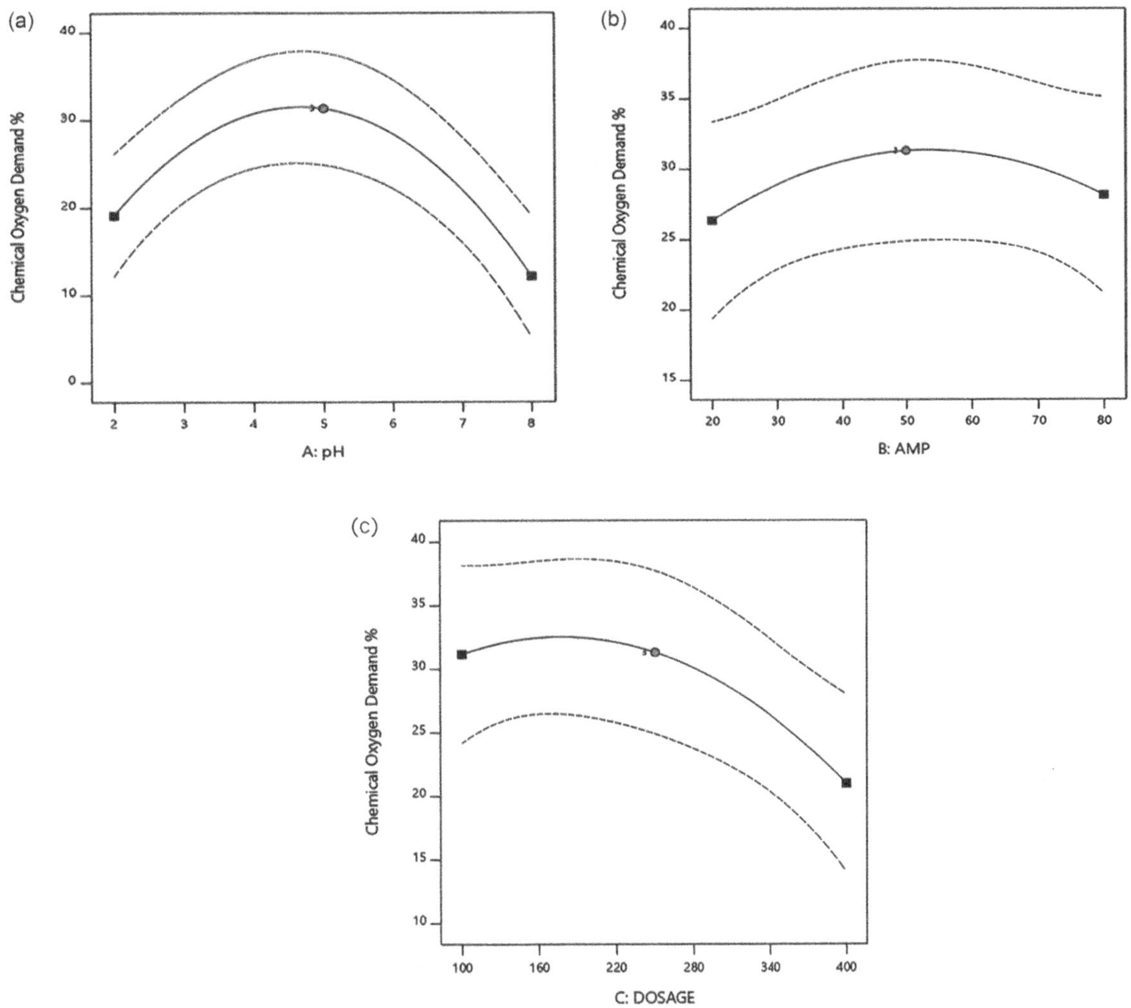

Figure 3 | Profile of an individual parameter plot with two reference curves based on ANOVA for process parameters. (a) pH from 2 to 8, with a fixed parameter amplitude of 50%, and persulfate dosage of 250 mg/L. (b) Amplitude from 20% (18 W) to 80% (78 W) with fixed parameter pH of 5, and persulfate dosage of 250 mg/L at mid-level. (c) Persulfate dosage 100–400 (mg/L) with fixed parameter of pH 5 and amplitude of 50%, on removal of COD with 95% of confidence interval band. Temperature 50 ± 10 °C.

persulfate oxidant (Romero *et al.* 2010; Sarath *et al.* 2016). At pH 2 more numbers of protons are present in the form of H_3O^+ which decreases effective decomposition of persulfate oxidant while increasing the pH, quantity of H_3O^+ ions decreases in solution which leads to effective generation of sulfate radicals from persulfate as shown in given Equations (16)–(17) (Romero *et al.* 2010; Sarath *et al.* 2016):

$$S_2O_8^{2-} + H^+ \rightarrow HS_2O_8^- \tag{16}$$

$$HS_2O_8^- + e^- \rightarrow H^+ + SO_4^{\cdot -} + SO_4^{2-} \tag{17}$$

3.4. Effect of amplitude on COD removal

Effect of power dissipation is one of the important influences in the ultrasound cavitation process, variation in amplitude generates more violent collapse of bubbles (Vega & Peñuela 2018). In this study, the effect of power dissipation on COD removal was studied by varying amplitude intensity from 20 to 80% (power dissipation 18–78 W) with fixed pH of 5 and persulfate dosage of 250 mg/L. The plot Figure 3(b) shows the variation in amplitude intensity at different levels and its effect on COD removal. The COD removal at 20, 50 and 80% of amplitude intensity is 27, 32 and 30% respectively. Maximum removal of 32% was observed at the mid-level of 50% (48 W), because Gaussian distribution probability of the cavitation bubble at the generation of cavitation is higher at lower amplitude (50%) and lower at higher amplitude (Feng *et al.* 2002). The amplitude intensity variation shows least significant by ANOVA as shown in Table 4. For higher power intensity at lower frequency, the decoupling effect occurs between sample solution and transducer. Thus bubble cloud formation occurs at the surface of the horn or transducer, resulting in a reduction in sound waves in the solution, hence at a lower frequency (20 kHz) the lower power cavitation becomes more effective (Sunartio *et al.* 2007).

3.5. Effect of persulfate oxidant on COD removal

Persulfate ($S_2O_8^{2-}$) is one of the strong oxidizing agents (Yang *et al.* 2019) for the degradation of organic matter. It becomes activated in the presence of ultrasound (Yang *et al.* 2019) as given in Equation (18) and produces sulfate radicals ($SO_4^{\cdot}$) with an oxidation potential of 2.40 V. The persulfate radical is a more stable radical compared to hydroxyl radicals (Romero *et al.* 2010). Persulfate oxidation has strong oxidation ability and efficient performance in a wide range of pH (Amor *et al.* 2019):

$$S_2O_8^{2-} + activator\ (Heat,\ US,\ Metal,\ Heat) \rightarrow 2SO_4^{\cdot} \tag{18}$$

In this study, the effect of combined ultrasound activated persulfate oxidation on COD removal was studied by varying the persulfate dosage from 100 mg/L to 250 mg/L with fixed factors pH 5 and amplitude intensity of 50%, as shown in Figure 3(c). Maximum COD removal of 32% was observed at a lower level of persulfate dosage of 100 mg/L. Predicted removal at a mid-level of 250 mg/L dosages and a higher level of 400 mg/L dosages of persulfate was 31 and 21% respectively. The increase in persulfate dosage from 100 to 400 mg/L decreases the removal efficiency of COD. It shows that an increase in persulfate concentration leads to the existence of excess persulfate in the system with which sulfate radicals in the system react and produce sulfate ions, and also extra sulfate radicals react themselves, as given in Equations (19) and (20) (Vu *et al.* 2004; Wei *et al.* 2018) . The additional persulfate loading under particularly acidic solutions are reported as a scavenger, thus the rate of reaction slightly decreases with an increase in dosage (Peyton 1993; Wei *et al.* 2018):

$$SO_4^{\cdot} + SO_4^{\cdot} \rightarrow S_2O_8^{2-} \tag{19}$$

$$SO_4^{\cdot} + S_2O_8^{2-} \rightarrow SO_4^{2-} + S_2O_8^{2-} \tag{20}$$

Figure 4 shows the comparison of COD removal with stand-alone ultrasound, persulfate process compared with the combined ultrasound and persulfate process under optimum conditions. The combined process is more effective with a rate constant of 0.0094 min^{-1} and removal efficiency of 39.5% compared with the stand-alone ultrasound cavitation process with 8% COD removal under the optimum conditions of pH 5, amplitude 20% provide and persulfate oxidation process with 5% COD removal under 100 mg/L dosage. A similar effect was observed by Yang *et al.* (2019) for the removal of pollutant from the wastewater. The combined effect leads to a higher amount of cavitation yield production and effective generation of sulfate radicals by persulfate, it also improves the degradability through degradation of polysaccharides ions to monomer ions by a combined ultrasound–persulfate oxidation process (Zhou & Ma 2006; Chen *et al.* 2021).

The synergistic index for the combined process has been computed based on the removal efficiency observed for the stand-alone process and for the combined process. When the wastewater was treated with ultrasound cavitation alone the maximum removal efficiency observed was 8% and that of persulfate oxidation was 5%. Whereas, when the persulfate process was activated with sonication the observed removal efficiency was almost 39.5%. Accordingly, the computed SI for the combined process was 3.038, which indicates there is a significant synergetic effect occurs when the ultrasound cavitation is combined with persulfate process as the reported synergetic index for the same process with different pollutants varies from 1.52 to 4.43 (Fedorov *et al.* 2022).

3.6. Effect of pH and amplitude on COD removal

The interaction effect of pH (A) and amplitude intensity (B) on COD removal is not significant based on ANOVA as shown in Table 4, its *P*-value is higher than 0.05 ($P > 0.05$). The combined effect of pH and amplitude on COD removal was studied by maintaining other parameters constant and the results are as shown in Figure 5(a). With the increasing amplitude at pH 2, removal of COD increased to 31% and, on the other hand, middle level pH 5 with amplitude up to 50% removal efficiency increased further as a result of the generation of effective cavitation and effective decomposition of persulfate

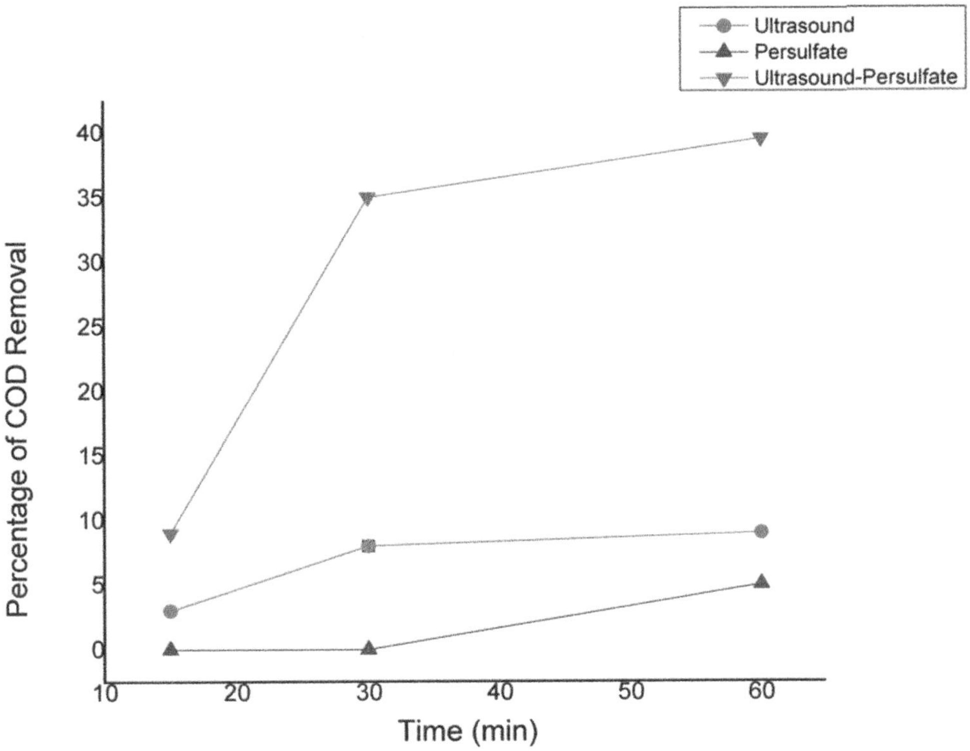

Figure 4 | Profile of percentage COD removal with individual treatment; ultrasound at pH 5, amplitude 20%; persulfate of 100 mg/L; temperature 50 ± 10 °C. Combined effect of ultrasound with persulfate at pH 5 and persulfate dosage of 100 mg/L, amplitude of 20%, temperature 50 ± 10 °C within 60 min reaction time.

(Sarath *et al.* 2016). By further increasing the amplitude to 80% at pH 5, this reduces removal due to reduction in the cavitation yield, thus increasing or decreasing amplitude beyond or below 50% under near acidic conditions has no significant impact on the removal efficiency (Thanekar & Gogate 2019).

3.7. Effect of amplitude and persulfate dosage on removal

The interaction effect of amplitude intensity (B) and persulfate dosage (C) on COD removal is significant as indicated by ANOVA outputs shown in Table 4, its *P*-value is lower than 0.05 ($P > 0.05$). The interaction effect of amplitude intensity and dosage of persulfate on COD removal is shown in Figure 5(b). For an amplitude intensity of 20% and persulfate dosages of 100 mg/L, the COD removal is significant due to the high availability of activation energy at low amplitude which was required for breakage of O-O bond and generation of sulfate radicals (Wacławek *et al.* 2017). Under higher amplitudes of 50 and 80% the persulfate amount of 250 and 400 mg/L became excessive for the system, leading to scavenger reactions, which reduced the removal (Wei *et al.* 2018) compared to 20% amplitude and 100 mg/L of dosage.

3.8. Effect of pH and persulfate dosage on COD removal

The interaction effect of pH (A) and persulfate dosage (B) on COD removal was not significant as indicated by ANOVA (Table 4), its *P*-value is higher than 0.05 ($P > 0.05$). The interaction effect of pH and dosage of persulfate on COD removal at 50% amplitude intensity is shown in Figure 5(c). Under an acidic pH of 2 and 5 with 100 mg/L dosages, there is more significant COD removal efficiency when compared to the removal achieved at 250 mg/L and 400 mg/L dosages. Excessive loading of persulfate leads to less utilization of persulfate ions under acidic conditions (Wei *et al.* 2018), persulfate loading above some extent acted as a proven scavenger under acidic conditions (Peyton 1993; Wei *et al.* 2018). Increasing dosage under near neutral condition of pH 8 did not have a significant effect on COD removal.

The comparison of removal efficiency of combined persulfate process with sonication with other similar advanced oxidation processes is given in Table 6. It is apparent that the removal efficiency of the present combined process is comparable with that of other similar processes. The present work was done with real wastewater containing a very high

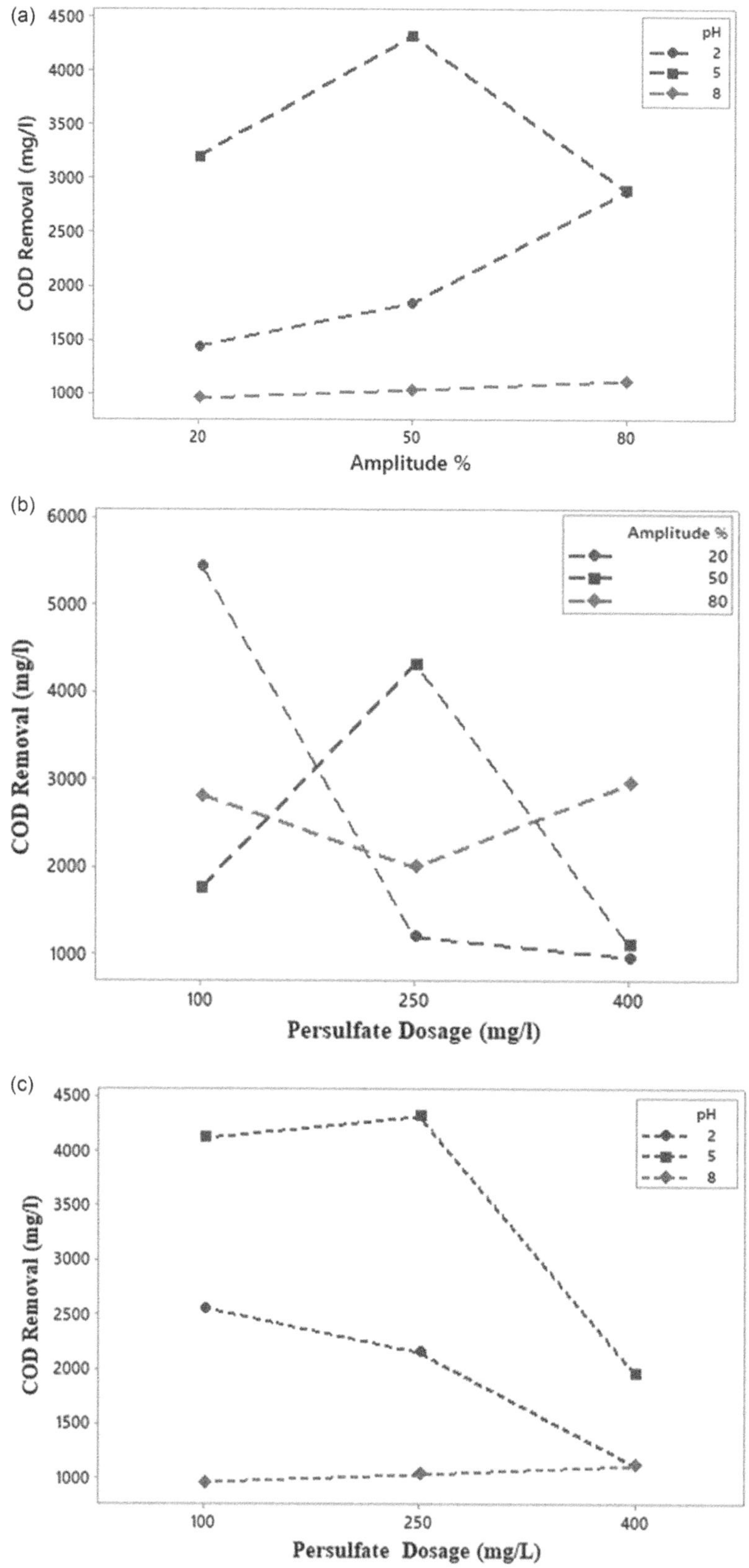

Figure 5 | Profile of the interaction effect plot for process parameters. (a) pH 2–8 to amplitude 20–80% (AB) maintaining persulfate dosage (mg/L) at middle level. (b) Amplitude 20–80% to persulfate dosage 100–400 (mg/L) (BC) maintaining pH at middle level. (c) pH 2–8 and persulfate dosage 100–400 (mg/L) (AC) maintaining amplitude % at middle level. Temperature 50 ± 10 °C.

Table 6 | Comparison of combined persulfate and ultrasound processes with other similar processes for the treatment of pharmaceutical wastewater

Treatment technology	Pollutant	Operating conditions	Efficiency	Reference
Photo Fenton	COD	pH – 3 COD – 8,370 ± 190 $[Fe^{2+}] = 0.05$ mol/L $[H_2O_2] = 0.25$ mol/L Reaction time – 120 min	58.4%	Changotra *et al.* (2019)
Electro-Fenton	COD	Batch experiment Current density – 17 mA/cm^2 Initial pH – 3 $Fe^{2+}/Fe^{3+}/Co^{2+}$ – 1:1:2 Oxygen flow rate – 0.5 L/min Reaction time – 2 h	93%	Quang *et al.* (2022)
Electro-Fenton (EF) + photocatalytic oxidation (PcO)	COD	EF Current density – 5 mA/cm^2 Reaction time – 1 h PcO $[TiO_2] = 1.5$ g/L Reaction time – 4 h	70.2%	Başaran Dindaş *et al.* (2020)
Aerobic biological treatment + Photo-Fenton	COD	pH –1 2.7 ± 0.25 COD – 8,370 ± 190	62.2%	Changotra *et al.* (2019)
Aerobic biological treatment + electro-fenton	COD	pH – 12.7 ± 0.25 COD – 8,370 ± 190	44.6%	Changotra *et al.* (2019)
Photocatalytic oxidation	COD	COD = 1,900 mg/L pH = 4.5, catalyst dosage – 2 g/l $[H_2O_2] = 0.32$ mM	72.7%	Berkün Olgun *et al.* (2021)
Ultrasound + persulfate	COD	COD 13,760 mg/L, pH 5, amplitude 20%, persulfate dosage 100 mg/L and reaction time 60 min	39.5%	Present study

COD of 13,760 mg/L and with a BOD/COD ratio less than 0.4. Also, it has been observed that the reaction time considered for the similar AOP process for the treatment pharmaceutical wastewater was very high compared to that of the present study.

In order to assess the effect of the oxidizing species in the present study, the radical scavenger tert-butyl alcohol was added. Tert-butyl is considered to be hydroxyl scavenger and has the effect of hydroxyl radicals in the electrolytic cell reaction (Li *et al.* 2017; Popat *et al.* 2019). At the optimum condition of pH 5, amplitude 20%, persulfate dosage of 100 mg/L,

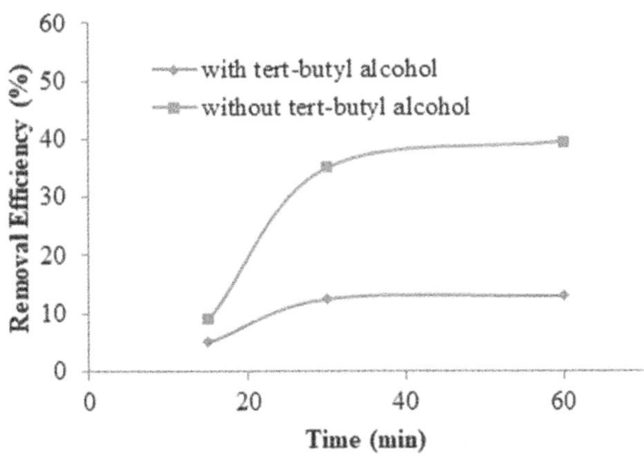

Figure 6 | Effect of tert-butyl alcohol in the combined ultrasound and persulfate process (optimum condition: pH 5, amplitude 20%, persulfate dosage 100 mg/L, temperature 50 ± 10 °C and reaction time 60 min).

1 mL of tert-butyl alcohol was added and the results are shown in Figure 6. From the figure it is very clear that the addition of tert-butyl alcohol did not cause any significant removal of COD, which shows that the hydroxyl and sulfate radicals have been scavenged. Whereas, without the addition of scavenger, significant removal of COD has been reported, which indicates the main oxidizing agents causing degradation of pharmaceutical waste water are hydroxyl radicals and persulfate radicals. A similar effect was observed by Popat *et al.* (2019) for the degradation of mixed industrial wastewater.

4. CONCLUSION

The synergetic index of the combined persulfate sonication process was computed for the treatment of pharmaceutical wastewater. The synergetic index was found to be 3.038, which is higher than 1, indicating the high synergetic effect of the combined process. It can also be concluded that ultrasound can be a better activator for the persulfate process. At the optimum condition of parameters pH 5, amplitude intensity of 20%, and persulfate dosage of 100 mg/L, the maximum COD removal of 5,440 mg/L (39.5%) was observed at 60 minutes of reaction time. The observed results concluded that process parameters like pH and persulfate oxidant dosage are very significant parameters for COD removal from pharmaceutical wastewater. The removal efficiency of the persulfate process, sonication and the combined persulfate sonication is in the following order $5 < 8 < 39.5\%$, respectively. The BOD/COD ratio of the treated wastewater increased from 0.4 to 0.53, thus the biodegradability of the treated wastewater had increased. The synergistic effect of ultrasound cavitation and persulfate oxidation was more effective for the degradation of organic matter from pharmaceutical wastewater compared with the effects of the stand-alone processes under optimum conditions. The major parameters that affect economic feasibility are chemical cost and energy consumption for sonication, which comes under the operational cost, and the cost of the sonicator and the reactor are the initial investments.

ACKNOWLEDGEMENTS

The authors would like to thank the Centre for Biofuels and Bioenergy Studies (CBBS) in PDEU for providing laboratory facilities for performing characterization and property estimation. The authors are thankful to the Director, Pandit Deendayal Energy University, Gandhinagar, India and Director, National Institute of Technology, Calicut, India for providing encouragement and kind permission for publishing the article.

CREDIT AUTHOR CONTRIBUTION STATEMENT

Karan Pandya: Methodology, Formal analysis, Writing – original draft, Writing – review and editing.
Thankappan Saraswathy Anantha Singh: Conceptualization, Methodology, Formal analysis, Supervision, Writing – original draft, Writing – review and editing.
Pravin Kodgire: Conceptualization, Methodology, Formal analysis, Supervision, Writing – original draft, Writing – review and editing.
Saji-Simon: Additional experimentation, Writing the revised article – review and editing.

ETHICAL APPROVAL

Not applicable.

AVAILABILITY OF DATA AND MATERIALS

Not applicable.

CONSENT TO PARTICIPATE

Not applicable.

CONSENT TO PUBLISH

Not applicable.

FUNDING

Not applicable.

DATA AVAILABILITY STATEMENT

All relevant data are included in the paper or its Supplementary Information.

CONFLICT OF INTEREST

The authors declare there is no conflict.

REFERENCES

Adishkumar, S., Sivajothi, S. & Rajesh Banu, J. 2012 Coupled solar photo-fenton process with aerobic sequential batch reactor for treatment of pharmaceutical wastewater. *Desalination and Water Treatment* **48** (1–3), 89–95. https://doi.org/10.1080/19443994.2012.698799.

Ahmadi, S., Igwegbe, C. A. & Rahdar, S. 2019 The application of thermally activated persulfate for degradation of Acid Blue 92 in aqueous solution. *International Journal of Industrial Chemistry* **10** (3), 249–260. https://doi.org/10.1007/s40090-019-0188-1.

Amor, C., Rodríguez-Chueca, J., Fernandes, J. L., Domínguez, J. R., Lucas, M. S. & Peres, J. A. 2019 Winery wastewater treatment by sulfate radical based-advanced oxidation processes (SR-AOP): thermally vs UV-assisted persulfate activation. *Process Safety and Environmental Protection*, 94–101. https://doi.org/10.1016/j.psep.2018.11.016.

Anipsitakis, G. P. & Dionysiou, D. D. 2003 Degradation of organic contaminants in water with sulfate radicals generated by the conjunction of peroxymonosulfate with cobalt. *Environmental Science and Technology* **37** (20), 4790–4797. https://doi.org/10.1021/es0263792.

Arvaniti, O. S., Bairamis, F., Konstantinou, I., Mantzavinos, D. & Frontistis, Z. 2020 Degradation of antihypertensive drug valsartan in water matrices by heat and heat/ultrasound activated persulfate: kinetics, synergy effect and transformation products. *Chemical Engineering Journal Advances* **4**, 100062. https://doi.org/10.1016/j.ceja.2020.100062.

Ayare, S. D. & Gogate, P. R. 2019 Sonocatalytic treatment of phosphonate containing industrial wastewater intensified using combined oxidation approaches. *Ultrasonics Sonochemistry* **51**, 69–76. https://doi.org/10.1016/j.ultsonch.2018.10.018.

Barik, A. J. & Gogate, P. R. 2017 Hybrid treatment strategies for 2,4,6-trichlorophenol degradation based on. *Ultrasonics – Sonochemistry*. https://doi.org/10.1016/j.ultsonch.2017.07.029.

Başaran Dindaş, G., Çalişkan, Y., Çelebi, E. E., Tekbaş, M., Bektaş, N. & Yatmaz, H. C. 2020 Treatment of pharmaceutical wastewater by combination of electrocoagulation, electro-Fenton and photocatalytic oxidation processes. *Journal of Environmental Chemical Engineering* **8** (3). https://doi.org/10.1016/j.jece.2020.103777.

Beckett, M. A. & Hua, I. 2001 Impact of ultrasonic frequency on aqueous sonoluminescence and sonochemistry. *Journal of Physical Chemistry A* **105** (15), 3796–3802. https://doi.org/10.1021/jp003226x.

Berkün Olgun, Ö., Palas, B., Atalay, S. & Ersöz, G. 2021 Photocatalytic oxidation and catalytic wet air oxidation of real pharmaceutical wastewater in the presence of Fe and $LaFeO_3$ doped activated carbon catalysts. *Chemical Engineering Research and Design* **171**, 421–432. https://doi.org/10.1016/j.cherd.2021.05.017.

Chandak, S., Ghosh, P. K. & Gogate, P. R. 2020 Treatment of real pharmaceutical wastewater using different processes based on ultrasound in combination with oxidants. *Process Safety and Environmental Protection* **137**, 149–157. https://doi.org/10.1016/j.psep.2020.02.025.

Changotra, R., Rajput, H. & Dhir, A. 2017 Natural soil mediated photo fenton-like processes in treatment of pharmaceuticals: batch and continuous approach. *Chemosphere* **188**, 345–353. https://doi.org/10.1016/j.chemosphere.2017.09.016.

Changotra, R., Rajput, H. & Dhir, A. 2019 Treatment of real pharmaceutical wastewater using combined approach of fenton applications and aerobic biological treatment. *Journal of Photochemistry and Photobiology A: Chemistry* **376**, 175–184. https://doi.org/10.1016/j.jphotochem.2019.02.029.

Chen, X., Sun-waterhouse, D., Yao, W., Li, X. & Zhao, M. 2021 Free radical-mediated degradation of polysaccharides: mechanism of free radical formation and degradation, influence factors and product properties. *Food Chemistry* **365**, 130524. https://doi.org/10.1016/j.foodchem.2021.130524.

Cortez, S., Teixeira, P., Oliveira, R. & Mota, M. 2010 Ozonation as polishing treatment of mature landfill leachate. *Journal of Hazardous Materials* **182** (1–3), 730–734. https://doi.org/10.1016/j.jhazmat.2010.06.095.

Crini, G. & Lichtfouse, E. 2019 Advantages and disadvantages of techniques used for wastewater treatment. *Environmental Chemistry Letters* **17** (1), 145–155. https://doi.org/10.1007/s10311-018-0785-9.

Dong, Z., Udepurkar, A. P. & Kuhn, S. 2020 Synergistic effects of the alternating application of low and high frequency ultrasound for particle synthesis in microreactors. *Ultrasonics Sonochemistry* **60**, 104800. https://doi.org/10.1016/j.ultsonch.2019.104800.

Dopar, M., Kusic, H. & Koprivanac, N. 2011 Treatment of simulated industrial wastewater by photo-Fenton process. part I: the optimization of process parameters using design of experiments (DOE). *Chemical Engineering Journal* **173** (2), 267–279. https://doi.org/10.1016/j.cej.2010.09.070.

Dular, M., Griessler-Bulc, T., Gutierrez-Aguirre, I., Heath, E., Kosjek, T., Krivograd Klemenčič, A., Oder, M., Petkovšek, M., Rački, N., Ravnikar, M., Šarc, A., Širok, B., Zupanc, M., Žitnik, M. & Kompare, B. 2016 Use of hydrodynamic cavitation in (waste)water treatment. *Ultrasonics Sonochemistry* **29**, 577–588. https://doi.org/10.1016/j.ultsonch.2015.10.010.

Estrada-Flórez, S. E., Serna-Galvis, E. A. & Torres-Palma, R. A. 2020 Photocatalytic vs. sonochemical removal of antibiotics in water: structure-degradability relationship, mineralization, antimicrobial activity, and matrix effects. *Journal of Environmental Chemical Engineering* **8** (5), 104359. https://doi.org/10.1016/j.jece.2020.104359.

Fedorov, K., Plata-Gryl, M., Khan, J. A. & Boczkaj, G. 2020 Ultrasound-assisted heterogeneous activation of persulfate and peroxymonosulfate by asphaltenes for the degradation of BTEX in water. *Journal of Hazardous Materials* **397**, 122804. https://doi.org/10.1016/j.jhazmat.2020.122804.

Fedorov, K., Dinesh, K., Sun, X., Darvishi Cheshmeh Soltani, R., Wang, Z., Sonawane, S. & Boczkaj, G. 2022 Synergistic effects of hybrid advanced oxidation processes (AOPs) based on hydrodynamic cavitation phenomenon – A review. *Chemical Engineering Journal* **432** https://doi.org/10.1016/j.cej.2021.134191.

Feng, R., Zhao, Y., Zhu, C. & Mason, T. J. 2002 Enhancement of ultrasonic cavitation yield by multi-frequency sonication. *Ultrasonics Sonochemistry* **9** (5), 231–236. https://doi.org/10.1016/S1350-4177(02)00083-4.

Ford, A. T. & Fong, P. P. 2016 The effects of antidepressants appear to be rapid and at environmentally relevant concentrations. *Environmental Toxicology and Chemistry* **35** (4), 794–798. https://doi.org/10.1002/etc.3087.

Gadipelly, C., Pérez-González, A., Yadav, G. D., Ortiz, I., Ibáñez, R., Rathod, V. K. & Marathe, K. V. 2014 Pharmaceutical industry wastewater: review of the technologies for water treatment and reuse. *Industrial and Engineering Chemistry Research* **53** (29), 11571–11592. https://doi.org/10.1021/ie501210j.

Gągol, M., Przyjazny, A. & Boczkaj, G. 2018 Wastewater treatment by means of advanced oxidation processes based on cavitation – A review. *Chemical Engineering Journal* **338**, 599–627. https://doi.org/10.1016/j.cej.2018.01.049.

Ghafoori, S., Mowla, A., Jahani, R., Mehrvar, M. & Chan, P. K. 2015 Sonophotolytic degradation of synthetic pharmaceutical wastewater: statistical experimental design and modeling. *Journal of Environmental Management* **150**, 128–137. https://doi.org/10.1016/j.jenvman.2014.11.011.

Hu, X., Wang, X., Ban, Y. & Ren, B. 2011 A comparative study of UV-Fenton, UV-H_2O_2 and Fenton reaction treatment of landfill leachate. *Environmental Technology* **32** (9), 945–951. https://doi.org/10.1080/09593330.2010.521953.

Huang, G. Y., Liu, Y. S., Chen, X. W., Liang, Y. Q., Liu, S. S., Yang, Y. Y., Hu, L. X., Shi, W. J., Tian, F., Zhao, J. L., Chen, J. & Ying, G. G. 2016 Feminization and masculinization of western mosquitofish (*Gambusia affinis*) observed in rivers impacted by municipal wastewaters. *Scientific Reports* **6**, 1–11. https://doi.org/10.1038/srep20884.

Ioannidi, A., Arvaniti, O. S., Nika, M. C., Aalizadeh, R., Thomaidis, N. S., Mantzavinos, D. & Frontistis, Z. 2022 Removal of drug losartan in environmental aquatic matrices by heat-activated persulfate: kinetics, transformation products and synergistic effects. *Chemosphere* **287** (P1), 131952. https://doi.org/10.1016/j.chemosphere.2021.131952.

Jeworski, M. & Heinzle, E. 2000 Combined chemical-biological treatment of wastewater containing refractory pollutants. *Biotechnology Annual Review* **6**, 163–196. https://doi.org/10.1016/S1387-2656(00)06022-1.

Jordens, J., Appermont, T., Gielen, B., Van Gerven, T. & Braeken, L. 2016 Sonofragmentation: effect of ultrasound frequency and power on particle breakage. *Crystal Growth and Design* **16** (11), 6167–6177. https://doi.org/10.1021/acs.cgd.6b00088.

Kansal, A., Siddiqui, N. A. & Gautam, A. 2013 Assessment of heavy metals and their interrelationships with some physicochemical parameters in eco-efficient rivers of Himalayan region. *Environmental Monitoring and Assessment* **185** (3), 2553–2563. https://doi.org/10.1007/s10661-012-2730-x.

Keenan, C. R. & Sedlak, D. L. 2008 Ligand-enhanced reactive oxidant generation by nanoparticulate zero-valent iron and oxygen. *Environmental Science and Technology* **42** (18), 6936–6941. https://doi.org/10.1021/es801438f.

Khorram, A. G. & Fallah, N. 2018 Treatment of textile dyeing factory wastewater by electrocoagulation with low sludge settling time: optimization of operating parameters by RSM. *Journal of Environmental Chemical Engineering* **6** (1), 635–642. https://doi.org/10.1016/j.jece.2017.12.054.

Leighton, T. G. 1995 Bubble population phenomena in acoustic cavitation. *Ultrasonics – Sonochemistry* **2** (2), 123–136. https://doi.org/10.1016/1350-4177(95)00021-W.

Li, B. & Zhu, J. 2016 Simultaneous degradation of 1,1,1-trichloroethane and solvent stabilizer 1,4-dioxane by a sono-activated persulfate process. *Chemical Engineering Journal* **284**, 750–763. https://doi.org/10.1016/j.cej.2015.08.153.

Li, B., Li, L., Lin, K., Zhang, W., Lu, S. & Luo, Q. 2013 Removal of 1,1,1-trichloroethane from aqueous solution by a sono-activated persulfate process. *Ultrasonics Sonochemistry* **20** (3), 855–863. https://doi.org/10.1016/j.ultsonch.2012.11.014.

Li, R., Kong, J., Liu, H., Chen, P., Liu, G., Li, F. & Lv, W. 2017 A sulfate radical based ferrous-peroxydisulfate oxidative system for indomethacin degradation in aqueous solutions. *RSC Advances* **7** (37), 22802–22809. https://doi.org/10.1039/c7ra03364h.

Liang, H., Sun, H., Patel, A., Shukla, P., Zhu, Z. H. & Wang, S. 2012 Excellent performance of mesoporous CO_3O_4/MnO_2 nanoparticles in heterogeneous activation of peroxymonosulfate for phenol degradation in aqueous solutions. *Applied Catalysis B: Environmental* **127**, 330–335. https://doi.org/10.1016/j.apcatb.2012.09.001.

Liu, H., Bruton, T. A., Li, W., Van Buren, J., Prasse, C., Doyle, F. M. & Sedlak, D. L. 2016 Oxidation of benzene by persulfate in the presence of Fe(III)- and Mn(IV)-containing oxides: stoichiometric efficiency and transformation products. *Environmental Science and Technology* **50** (2), 890–898. https://doi.org/10.1021/acs.est.5b04815.

Martínez, F., Molina, R., Rodríguez, I., Pariente, M. I., Segura, Y. & Melero, J. A. 2018 Techno-economical assessment of coupling fenton/biological processes for the treatment of a pharmaceutical wastewater. *Journal of Environmental Chemical Engineering* **6** (1), 485–494. https://doi.org/10.1016/j.jece.2017.12.008.

Milano, J., Ong, H. C., Masjuki, H. H., Silitonga, A. S., Chen, W. H., Kusumo, F., Dharma, S. & Sebayang, A. H. 2018 Optimization of biodiesel production by microwave irradiation-assisted transesterification for waste cooking oil-*Calophyllum inophyllum* oil via response surface methodology. *Energy Conversion and Management* **158**, 400–415. https://doi.org/10.1016/j.enconman.2017.12.027.

Mohapatra, D. P., Brar, S. K., Tyagi, R. D., Picard, P. & Surampalli, R. Y. 2014 Analysis and advanced oxidation treatment of a persistent pharmaceutical compound in wastewater and wastewater sludge-carbamazepine. *Science of the Total Environment* **470–471**, 58–75. https://doi.org/10.1016/j.scitotenv.2013.09.034.

Monteagudo, J. M., Durán, A., González, R. & Expósito, A. J. 2015 In situ chemical oxidation of carbamazepine solutions using persulfate simultaneously activated by heat energy, UV light, Fe^{2+} ions, and H_2O_2. *Applied Catalysis B: Environmental* **176–177**, 120–129. https://doi.org/10.1016/j.apcatb.2015.03.055.

Monteagudo, J. M., El-taliawy, H., Durán, A., Caro, G. & Bester, K. 2018 Sono-activated persulfate oxidation of diclofenac: degradation, kinetics, pathway and contribution of the different radicals involved. *Journal of Hazardous Materials* **357**, 457–465. https://doi.org/10.1016/j.jhazmat.2018.06.031.

Muthukumaran, C., Praniesh, R., Navamani, P., Swathi, R., Sharmila, G. & Manoj Kumar, N. 2017 Process optimization and kinetic modeling of biodiesel production using non-edible *Madhuca indica* oil. *Fuel* **195**, 217–225. https://doi.org/10.1016/j.fuel.2017.01.060.

Ng, K. K., Shi, X., Tang, M. K. Y. & Ng, H. Y. 2014 A novel application of anaerobic bio-entrapped membrane reactor for the treatment of chemical synthesis-based pharmaceutical wastewater. *Separation and Purification Technology* **132**, 634–643. https://doi.org/10.1016/j.seppur.2014.06.021.

Peyton, G. R. 1993 The free-radical chemistry of persulfate-based total organic carbon analyzers. *Marine Chemistry* **41** (1–3), 91–103. https://doi.org/10.1016/0304-4203(93)90108-Z.

Popat, A., Nidheesh, P. V., Anantha Singh, T. S. & Suresh Kumar, M. 2019 Mixed industrial wastewater treatment by combined electrochemical advanced oxidation and biological processes. *Chemosphere* **237**, 124419. https://doi.org/10.1016/j.chemosphere.2019.124419.

Quang, H. H. P., Dinh, N. T., Thi, T. N. T., Bao, L. T. N., Yuvakkumar, R. & Nguyen, V. H. 2022 Fe^{2+}, Fe^{3+}, Co^{2+} as highly efficient cocatalysts in the homogeneous electro-Fenton process for enhanced treatment of real pharmaceutical wastewater. *Journal of Water Process Engineering* **46**, 102635. https://doi.org/10.1016/j.jwpe.2022.102635.

Rayaroth, M. P., Aravind, U. K. & Aravindakumar, C. T. 2016 Degradation of pharmaceuticals by ultrasound-based advanced oxidation process. *Environmental Chemistry Letters* **14** (3), 259–290. https://doi.org/10.1007/s10311-016-0568-0.

Rayaroth, M. P., Aravindakumar, C. T., Shah, N. S. & Boczkaj, G. 2022 Advanced oxidation processes (AOPs) based wastewater treatment – unexpected nitration side reactions – a serious environmental issue: a review. *Chemical Engineering Journal* **430** (P4), 133002. https://doi.org/10.1016/j.cej.2021.133002.

Rice, E. W., Baird, R. B., Eaton, A. D. & Clesceri, L. A. (eds) 2012 Standard methods for the examination of water and wastewater, 22nd edn. American Public Health Association, Washington, DC, ISBN: 9780875530130.

Rokhina, E. V., Sillanpää, M., Nolte, M. C. M. & Virkutyte, J. 2008 Optimization of pulp mill effluent treatment with catalytic adsorbent using orthogonal second-order (Box-Behnken) experimental design. *Journal of Environmental Monitoring* **10** (11), 1304–1312. https://doi.org/10.1039/b811556g.

Romero, A., Santos, A., Vicente, F. & González, C. 2010 Diuron abatement using activated persulfate: effect of pH, Fe(II) and oxidant dosage. *Chemical Engineering Journal* **162** (1), 257–265. https://doi.org/10.1016/j.cej.2010.05.044.

Sarath, K., Gandhimathi, R., Ramesh, S. T. & Nidheesh, P. V. 2016 Removal of reactive magenta-MB from aqueous solution by persulfate-based advanced oxidation process. *Desalination and Water Treatment* **57** (25), 11872–11878. https://doi.org/10.1080/19443994.2015.1054886.

Serna-Galvis, E. A., Botero-Coy, A. M., Martínez-Pachón, D., Moncayo-Lasso, A., Ibáñez, M., Hernández, F. & Torres-Palma, R. A. 2019 Degradation of seventeen contaminants of emerging concern in municipal wastewater effluents by sonochemical advanced oxidation processes. *Water Research* **154**, 349–360. https://doi.org/10.1016/j.watres.2019.01.045.

Singh, R. & Prashant, R. 2017 A review on characterization and bioremediation of pharmaceutical industries 'wastewater: an Indian perspective. *Applied Water Science*, 1–12. https://doi.org/10.1007/s13201-014-0225-3.

Singh, S. K., Tang, W. Z. & Tachiev, G. 2013 Fenton treatment of landfill leachate under different COD loading factors. *Waste Management* **33** (10), 2116–2122. https://doi.org/10.1016/j.wasman.2013.06.019.

Soltani, R. D. C., Mashayekhi, M., Naderi, M., Boczkaj, G., Jorfi, S. & Safari, M. 2019 Sonocatalytic degradation of tetracycline antibiotic using zinc oxide nanostructures loaded on nano-cellulose from waste straw as nanosonocatalyst. *Ultrasonics Sonochemistry* **55**, 117–124. https://doi.org/10.1016/j.ultsonch.2019.03.009.

Spina, F., Anastasi, A., Prigione, V., Tigini, V. & Varese, G. C. 2012 Biological treatment of industrial wastewaters: a fungal approach. *Chemical Engineering Transactions* **27**, 175–180. https://doi.org/10.3303/CET1227030.

Sunartio, D., Ashokkumar, M. & Grieser, F. 2007 Study of the coalescence of acoustic bubbles as a function of frequency, power, and water-soluble additives. *Journal of the American Chemical Society* **129** (18), 6031–6036. https://doi.org/10.1021/ja068980w.

Tak, B. Y., Tak, B. S., Kim, Y. J., Park, Y. J., Yoon, Y. H. & Min, G. H. 2015 Optimization of color and COD removal from livestock wastewater by electrocoagulation process: application of Box-Behnken design (BBD). *Journal of Industrial and Engineering Chemistry* **28**, 307–315. https://doi.org/10.1016/j.jiec.2015.03.008.

Tan, C., Gao, N., Deng, Y., An, N. & Deng, J. 2012 Heat-activated persulfate oxidation of diuron in water. *Chemical Engineering Journal* **203**, 294–300. https://doi.org/10.1016/j.cej.2012.07.005.

Thanekar, P. & Gogate, P. R. 2019 Combined hydrodynamic cavitation based processes as an efficient treatment option for real industrial effluent. *Ultrasonics Sonochemistry* **53**, 202–213. https://doi.org/10.1016/j.ultsonch.2019.01.007.

Tiwari, B., Drogui, P. & Tyagi, R. D. 2020 Removal of emerging micro-pollutants from pharmaceutical industry wastewater. In: *Current Developments in Biotechnology and Bioengineering*. Elsevier B.V. https://doi.org/10.1016/b978-0-12-819594-9.00018-8.

Tran, N., Drogui, P. & Brar, S. K. 2015 Sonochemical techniques to degrade pharmaceutical organic pollutants. *Environmental Chemistry Letters* **13** (3), 251–268. https://doi.org/10.1007/s10311-015-0512-8.

Vega, L. P. & Peñuela, G. A. 2018 High frequency sonochemical degradation of benzophenone-3 in water. *Journal of Environmental Engineering (United States)* **144** (8), 1–7. https://doi.org/10.1061/(ASCE)EE.1943-7870.0001406.

Verma, A. K., Dash, R. R. & Bhunia, P. 2012 A review on chemical coagulation/flocculation technologies for removal of colour from textile wastewaters. *Journal of Environmental Management* **93** (1), 154–168. https://doi.org/10.1016/j.jenvman.2011.09.012.

Vu, X. Y., Bao, Z. C. & Barker, J. R. 2004 Free radical reactions involving Cl•, Cl^{2-}•, and SO^{4-}• in the 248 nm photolysis of aqueous solutions containing S2O82- and Cl-. *Journal of Physical Chemistry A* **108** (2), 295–308. https://doi.org/10.1021/jp036211i.

Wacławek, S., Lutze, H. V., Grübel, K., Padil, V. V. T., Černík, M. & Dionysiou, D. D. 2017 Chemistry of persulfates in water and wastewater treatment: a review. *Chemical Engineering Journal* **330**, 44–62. https://doi.org/10.1016/j.cej.2017.07.132.

Wang, S. & Zhou, N. 2016 Removal of carbamazepine from aqueous solution using sono-activated persulfate process. *Ultrasonics Sonochemistry* **29**, 156–162. https://doi.org/10.1016/j.ultsonch.2015.09.008.

Wang, S., Zhou, N., Wu, S., Zhang, Q. & Yang, Z. 2015 Modeling the oxidation kinetics of sono-activated persulfate's process on the degradation of humic acid. *Ultrasonics Sonochemistry* **23**, 128–134. https://doi.org/10.1016/j.ultsonch.2014.10.026.

Wei, Z., Villamena, F. A. & Weavers, L. K. 2017 Kinetics and mechanism of ultrasonic activation of persulfate: an in situ EPR spin trapping study. *Environmental Science and Technology* **51** (6), 3410–3417. https://doi.org/10.1021/acs.est.6b05392.

Wei, L., Chen, W., Li, Q., Gu, Z. & Zhang, A. 2018 Wastewater in heat-activated persulfate system. *RSC Advances* **8**, 20603–20611. https://doi.org/10.1039/c8ra01995a.

Xie, P., Ma, J., Liu, W., Zou, J., Yue, S., Li, X., Wiesner, M. R. & Fang, J. 2015 Removal of 2-MIB and geosmin using UV/persulfate: contributions of hydroxyl and sulfate radicals. *Water Research* **69**, 223–233. https://doi.org/10.1016/j.watres.2014.11.029.

Yang, Q., Zhong, Y., Zhong, H., Li, X., Du, W., Li, X., Chen, R. & Zeng, G. 2015 A novel pretreatment process of mature landfill leachate with ultrasonic activated persulfate: optimization using integrated Taguchi method and response surface methodology. *Process Safety and Environmental Protection* **98**, 268–275. https://doi.org/10.1016/j.psep.2015.08.009.

Yang, L., Xue, J., He, L., Wu, L., Ma, Y., Chen, H., Li, H., Peng, P. & Zhang, Z. 2019 Review on ultrasound assisted persulfate degradation of organic contaminants in wastewater: influences, mechanisms and prospective. *Chemical Engineering Journal* **378**. https://doi.org/10.1016/j.cej.2019.122146.

Yuan, R., Jiang, Z., Wang, Z., Gao, S., Liu, Z., Li, M. & Boczkaj, G. 2020 Hierarchical MnO_2 nanoflowers blooming on 3D nickel foam: a novel micro-macro catalyst for peroxymonosulfate activation. *Journal of Colloid and Interface Science* **571**, 142–154. https://doi.org/10.1016/j.jcis.2020.03.041.

Zhang, M. F., Qin, Y. H., Ma, J. Y., Yang, L., Wu, Z. K., Wang, T. L., Wang, W. G. & Wang, C. W. 2016 Depolymerization of microcrystalline cellulose by the combination of ultrasound and fenton reagent. *Ultrasonics Sonochemistry* **31**, 404–408. https://doi.org/10.1016/j.ultsonch.2016.01.027.

Zhou, C. & Ma, H. 2006 Ultrasonic degradation of polysaccharide from a red algae (*Porphyra yezoensis*). *Journal of Agricultural and Food Chemistry* **54**, 2223–2228. https://doi.org/10.1021/jf052763h.

Zou, X., Zhou, T., Mao, J. & Wu, X. 2014 Synergistic degradation of antibiotic sulfadiazine in a heterogeneous ultrasound-enhanced Fe0/persulfate fenton-like system. *Chemical Engineering Journal* **257**, 36–44. https://doi.org/10.1016/j.cej.2014.07.048.

First received 21 May 2022; accepted in revised form 16 September 2022. Available online 3 October 2022

doi: 10.2166/wst.2023.059

Ciprofloxacin degradation by catalytic activation of monopersulfate using Mn–Fe oxides: performance and mineralization

Gelavizh Barzegar[a], Mohammad Sabaghan[a], Omid Azadbakht[a], Ehsan Aghayani[b], Mostafa Mahdavianpour[b], Abudukeremu Kadier[c,d], Saeid Fallahizadeh[e] and Farshid Ghanbari[b,*]

[a] Behbahan Faculty of Medical Sciences, Behbahan, Iran
[b] Research Center for Environmental Contaminants (RCEC), Abadan University of Medical Sciences, Abadan, Iran
[c] Laboratory of Environmental Science and Technology, The Xinjiang Technical Institute of Physics and Chemistry, Key Laboratory of Functional Materials and Devices for Special Environments, Chinese Academy of Sciences, Urumqi 830011, China
[d] Center of Materials Science and Optoelectronics Engineering, University of Chinese Academy of Sciences, Beijing 100049, China
[e] Social Determinants of Health Research Center, Yasuj University of Medical Sciences, Yasuj, Iran
*Corresponding author. E-mail: ghanbari.env@gmail.com

FG, 0000-0001-5599-3490

ABSTRACT

In this work, Mn_2O_3/Fe_2O_3 (MFO) was synthesized and used to activate monopersulfate (MPS) for the degradation of ciprofloxacin (CIP). The effect of several parameters was studied on CIP degradation. Under the optimum conditions (pH = 6.3 (natural pH), MFO = 300 mg/L and MPS = 2 mM), around 92% of CIP was decomposed. Nitrite, phosphate and bicarbonate ions had a strong inhibitory effect on the MFO/MPS process while the effect of chloride and nitrate ions was neutral. The catalytic activity of MFO was also studied by other chemical oxidants such as peroxydisulfate, periodate, hydrogen peroxide, percarbonate and peracetic acid. Scavenging tests showed that the role of sulfate radicals is more than hydroxyl radicals. MFO exhibited high catalytic activity in four recycling with insignificant leaching of Mn and Fe. During CIP oxidation, 45.5% carbon mineralization occurred and antibacterial activity of treated CIP solution was reduced. Finally, MFO/MPS was applied on actual wastewater (hospital effluent) and the results showed that MFO/MPS can be considered as a practical method for the treatment of contaminated water with emerging pollutants.

Key words: advanced oxidation processes (AOPs), ciprofloxacin, monopersulfate, toxicity, transition metals

HIGHLIGHTS

- Mn_2O_3/Fe_2O_3 (MFO) was simply synthesized to activate MPS for CIP degradation.
- Catalytic activity of MFO was evaluated in the presence of several oxidants.
- Reusability and stability of MFO was studied in four cycles.
- Carbon and fluorine mineralization and antibacterial activity were evaluated.

GRAPHICAL ABSTRACT

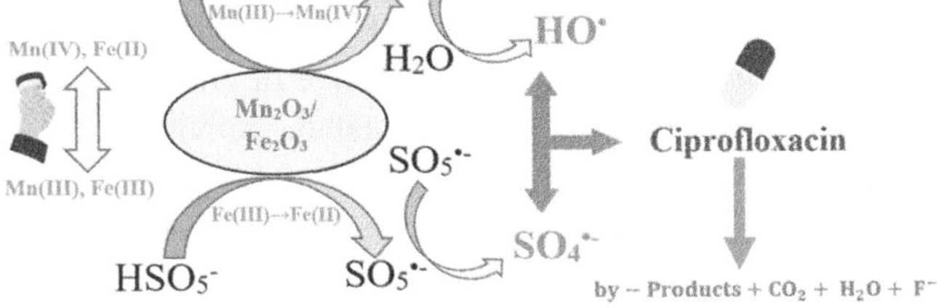

1. INTRODUCTION

The discovery of antibiotics is a great revolution since their usage greatly decreased the number of deaths from infection. Since 1928 to now, thousands of antibiotics are known. Fluoroquinolone antibiotics are antimicrobial agents which are used for the treatment of humans and veterinarians. Among the fluoroquinolone antibiotics, ciprofloxacin (CIP) is one of the most broadly utilized antibiotics in the world. CIP can be emitted into the environment in form of metabolites. The presence of wide spectrum of drugs and antibiotics like CIP in water bodies, even at low concentrations, results in serious threats to the ecosystem and human health by inducing proliferation of bacterial drug resistance (An *et al.* 2010; Khavari Kashani *et al.* 2022). Due to its high solubility under various pH conditions, CIP may be identified in wastewater treatment plants (Liu *et al.* 2017). It has been reported that a high concentration of CIP (up to 31,000 μg/L) was detected in the effluent of drug production factories (Ghanbari *et al.* 2020).

In last decades, several advanced treatments have been employed for the decontamination of emerging pollutants such as CIP. Most of classical treatment processes such as coagulation process and membrane technologies are not able to complete degradation of antibiotics. Moreover, these processes produce a concentrated effluent containing antibiotics which need a further remediation process (Chi *et al.* 2022). Advanced oxidation processes (AOPs) are a new method for complete oxidation or the detoxification of emerging and refractory pollutants in which they produce high reactive species for the degradation of organic pollutants (Eslami *et al.* 2021). Classically, hydroxyl radical ($HO^•$) is the main high reactive species which is produced by various methods (Pandis *et al.* 2022). Recently, other reactive species such as sulfate radical ($SO_4^{•-}$) (Madihi-Bidgoli *et al.* 2021), chlorine radical ($Cl^•$) (Yaghoot-Nezhad *et al.* 2023), iodate radical ($IO_3^•$) (Li *et al.* 2022) and organic radical ($R^•$) (Zhu & Liu 2022) have been used as a new AOP for the treatment of water and wastewater (Ghanbari *et al.* 2022). There are many chemical oxidants for AOPs application including hydrogen peroxide (HP), percarbonate (PC), peroxydisulfate (PDS), monopersulfate (MPS), chlorine, periodate (PI) and peracetic acid (PAA) (Eslami *et al.* 2020; Ghanbari *et al.* 2022). Sulfate radicals are more selective than hydroxyl radicals; moreover, they have a much longer lifetime compared to hydroxyl radicals (30–40 μs *vs.* 20 ns). The redox potential (E^0) of sulfate radicals is higher than hydroxyl radicals (2.5–3.1 V *vs.* 1.9–2.7 V) (Wacławek *et al.* 2017; Yousefi *et al.* 2017). Monopersulfate (HSO_5^-) is a precursor for the production of sulfate radicals and hydroxyl radicals. Several methods have been proposed for MPS activation to generate free radicals. UV irradiation, ultrasound (US) irradiation, transition metals, carbon catalyst and heat (Hassani *et al.* 2023b). Thanks to the reusability of transition metals and easy electron transfer, several studies have focused on the synthesis and application of transition metal-based heterogeneous catalysts for MPS activation in recent years (Liu *et al.* 2017; Madihi-Bidgoli *et al.* 2021; Chi *et al.* 2022). Cobalt has been indicated as the best activator for MPS to generate sulfate radicals. However, cobalt ion is a threat to water bodies due to high toxicity at low concentrations (Xiao *et al.* 2020). Iron and manganese are the best alternatives for MPS activation due to low toxicity and high performance. Various Mn and Fe oxides along with several supports have been fabricated for MPS activation in the last decades (Liu *et al.* 2017; Madihi-Bidgoli *et al.* 2021). However, there are several unknown configurations of composite with high activity for catalyzing chemical oxidants in AOPs application. Mn_2O_3/Fe_2O_3 as a new composite has been used in different fields including the reduction of NO (Zhou *et al.* 2022), adsorption of H_2S (Kim *et al.* 2021) and anode for lithium ion batteries (Wang *et al.* 2021). Recently, Mn_2O_3/Fe_2O_3 has been also applied as catalyst for AOPs like photocatalysis (Ghaffari *et al.* 2020) and Fenton oxidation (Baek *et al.* 2022). Mn_2O_3/Fe_2O_3 has been rarely used for the activation of new oxidants to generate free radicals (Chen *et al.* 2020). Therefore, further in-depth research is needed to fully describe the catalytic activity of Mn_2O_3/Fe_2O_3 for the generation of free radicals.

In current study, Mn_2O_3/Fe_2O_3 (MFO) composite was simply synthesized and applied for MPS activation. CIP was selected as an emerging pollutant for the evaluation of the performance of MFO/MPS. The influence of changing MPS dosage, pH, catalyst loading and water constituents were studied on CIP decomposition. The catalytic activity of MFO was also assessed in the presence of various chemical oxidants (HP, PAA, PC, PDS and PI). Mineralization degree of CIP was studied based on carbon and fluorine element. Reusability and stability of MFO were investigated and the MFO/MPS was studied on real wastewater. Finally, antibacterial activity was conducted on the treated CIP solution.

2. MATERIALS AND METHODS

2.1. Chemical and reagents

Monopersulfate compound was purchased from Sigma-Aldrich based on Oxone salt. Ciprofloxacin (CIP, $C_{17}H_{18}FN_3O_3$, > 98%), sulfuric acid (H_2SO_4, 98%), sodium hydroxide (NaOH, 99%), sodium chloride (NaCl > 99%), sodium nitrate

($NaNO_3 > 99\%$), sodium nitrite ($NaNO_2 > 99\%$), furfuryl alcohol (FFA) and sodium bicarbonate ($NaHCO_3 > 99\%$) were purchased from Sigma Aldrich company. Ferric nitrate ($Fe(NO_3)3 > 98\%$) and manganese nitrate ($Mn(NO_3)_2 > 99\%$) and cetyltrimethylammonium bromide (CTAB > 99%) were supplied from Merck Inc. An actual hospital effluent was collected from a hospital in Abadan city in Iran and its characterizations are presented in Table S1.

2.2. The synthesis of MFO

The composite was synthesized based on a surfactant-mediated co-precipitation method (Ghaffari *et al.* 2020). 6.15 g of ferric nitrate and 2.5 g of manganese nitrate were dissolved in 200 mL of deionized water and heated up to 60 °C for 10 min. Then, 1.0 g CTAB was introduced to the solution and stirred for 45 min. NaOH solution (4%) was dropwisely added to the solution until pH 12 was obtained. After stirring for 1 h, the precipitate was filtrated and washed for several times with deionized water and dried at 105 °C for 24 h. Finally, the dried precipitates were finely grounded before calcination at 600 °C for 5 h. The morphology and structure of MFO composite were analyzed by field-emission scanning electron microscopy (FESEM) using MIRA3 TESCAN. X-ray powder diffraction (XRD) analysis was utilized for the identification of crystalline phase of MFO by a D8 Bruker advanced diffractometer with Cu–Kα radiation. To determine the organic and inorganic bonds of the composite, Fourier transform-infrared spectroscopy (FTIR) was used by a Vertex 70 spectrophotometer (Bruker, Germany). The surface area, size and volume of the pores of MFO composite were measured by the Brunaeur, Emmett and Teller (BET) method (BEL, BELSORP MINI II, Japan).

2.3. The experiment of MFO/MPS for CIP degradation

For the evaluation of CIP degradation efficiency using MPS activated by MFO, a batch-type experiment was conducted. In brief, a known amount of MPS (0.5–2.5 mM) was introduced in 200 mL of CIP solution (50 μM). After that, a specified dosage of MFO was immediately added to CIP solution. The degradation reaction was initiated by shaking the solution (200 rpm). At present reaction times, the sample was collected from the CIP solution and filtered to remove MFO. The remaining concentration of CIP was determined by high-performance liquid chromatography (HPLC, Waters), equipped with a C18 column (25 cm × 4.6 mm). The mobile phase was methanol and water (65:35) with a flow rate 1 mL/min (Khavari Kashani *et al.* 2022). Total organic carbon (TOC) was measured via a Shimadzu 5050 TOC analyzer. Atomic absorption spectroscopy (Perkin-Elmer) was utilized for measuring iron and manganese.

2.4. Antibacterial evaluation

The antibacterial activity of CIP and treated solution by MFO/MPS were assessed using *Escherichia coli* in the disk diffusion susceptibility test as a semi-quantitative indicator for antibacterial activity of CIP degradation. To determine antibacterial activity, the diameter of the inhibition zone formed on an agar plate was measured. Agar plates were used for the injection of 20 μL samples into holes of 6 mm diameter and then were incubated at 37 °C for 1 day. These tests were conducted in triplicate.

3. RESULTS AND DISCUSSION

3.1. Characteristics of MFO

This section is presented in Supplementary Information (Fig. S1–3 and Text S1).

3.2. CIP degradation by various processes

Figure 1(a) shows the CIP degradation by different conditions under a fixed situation. As can be seen, oxidative removal of CIP by MPS was slight since the oxidation power of MPS is weak. The capacity of MFO adsorption for CIP removal was also poor in a way that around 12% of CIP was removed during 60 min reaction time. To clarify the role of each component of MFO, MPS/Fe_2O_3 and MPS/Mn_2O_3 were also evaluated. MPS/Fe_2O_3 could degrade 66% of CIP while a better performance was observed in MPS/Mn_2O_3 with 81% CIP removal efficiency. Coupling these components as MFO increased catalytic activity and CIP degradation consequently. More than 92% of CIP was eliminated in MPS/MFO process indicating that the composite has a better function compared to Fe_2O_3 and Mn_2O_3. The simultaneous presence of Fe_2O_3 and Mn_2O_3 for MPS activation resulted in more production of free radicals. Figure 1(b) shows the rate constant of the pseudo-first order model for CIP degradation by various processes. The rate constants of MPS and MFO were insignificant (~0.002 1/min) while MPS/MFO had the highest rate constant (0.0447 1/min) which was 2.5 and 1.5 times more than the rate constants of MPS/Fe_2O_3 and MPS/Mn_2O_3, respectively.

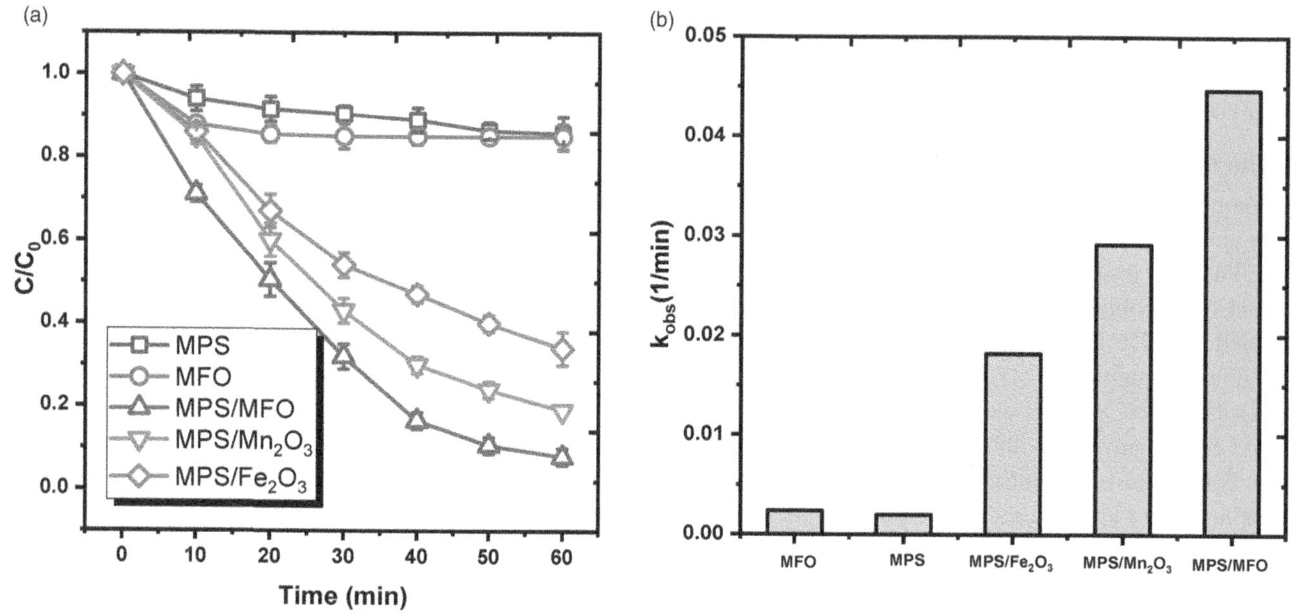

Figure 1 | (a) The CIP degradation using various processes. (b) The rate constant of pseudo-first order model for CIP degradation (MPS = 2 mM, catalyst = 300 mg/L and pH = 6.3).

To confirm the performance of MFO activity, MPS was monitored after 60 min reaction and the results are presented in Fig. S4. One can see that self-decomposition of MPS was 8.9% while a dramatic increase was seen in the presence of catalysts. Mn_2O_3 and Fe_2O_3 decomposed 53.5 and 47.7% of MPS indicating that Mn_2O_3 has a higher catalytic activity compared to Fe_2O_3. MFO nanocomposite could increasingly decompose MPS (69.9%) confirming the results of CIP degradation by various catalysts. Simply put, it can be concluded that more MPS activation resulted in higher CIP degradation.

3.3. The effect of operating parameters

The solution pH is a critical parameter in sulfate radicals-based AOPs. The impact of solution pH was investigated in various values. Figure 2(a) depicts the CIP degradation in various conditions. As can be seen, a wide range of pH (4–8) indicated an

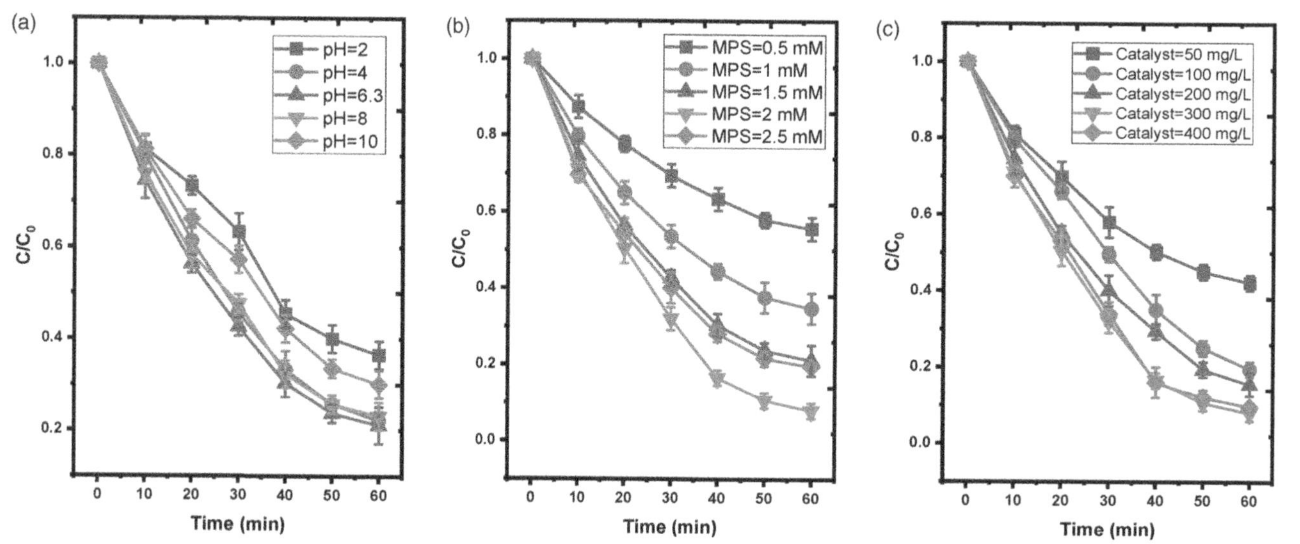

Figure 2 | (a) Effect of pH (MPS = 1.5 M and MFO = 300 mg/L). (b) The effect of MPS (pH = 6.3 and MFO = 300 mg/L). (c) The effect of MFO loading on CIP degradation (MPS = 2 mM and pH = 6.3).

acceptable result for CIP degradation while a suppression was seen in acidic (pH = 2) and alkaline (pH = 10) conditions. Several studies have reported that sulfate radicals-based AOPs can be operated in a wide range of pH (Madihi-Bidgoli *et al.* 2021; Han *et al.* 2022a). However, acidic conditions may affect the free radical generated ($HO^{\bullet}$ and $SO_4^{\bullet-}$) through the protonation of free radicals based on the following equations (Isari *et al.* 2021).

$$HO^{\bullet} + H^{+} + e^{-} \rightarrow H_2O \tag{1}$$

$$SO_4^{\bullet-} + H^{+} + e^{-} \rightarrow HSO_4^{-} \tag{2}$$

At alkaline conditions (pH = 10), an inhibitory effect was observed in which the CIP degradation rate was reduced during 60 min reaction time. The *pka* of HSO_5^- is 9.4, hence, SO_5^{2-} is dominated at pH = 10 in which SO_5^{2-} species resulted in a strong repulsion between catalyst and HSO_5^-/SO_5^{2-}. On the other hand, it has been reported that MPS is decomposed in the presence of OH^- according to the following equation (Madihi-Bidgoli *et al.* 2021). Therefore, the pathway of MPS decomposition would not lead to the production of free radicals.

$$HSO_5^- + SO_5^{2-} + OH^- \rightarrow 2SO_4^{2-} + O_2 + H_2O \tag{3}$$

The MPS concentration can determine the amount of free radicals produced during catalytic degradation of CIP. As shown in Figure 2(b), the degradation efficiency of CIP enhanced from 45 to 92% when the MPS concentration raised from 0.5 to 2 mM at pH = 6.3. This is related to more generation of free radicals resulting from MPS decomposition. However, a further increase in MPS concentration (2–2.5 mM) led to the same results. A chain of reactions took place in excess MPS in which additional MPS (HSO_5^-/SO_5^{2-}) may react with sulfate radicals and hydroxyl radicals and produce $SO_5^{\bullet-}$ with weak reactivity ($E^0 = 1.1$ V) (Equations (4)–(6)) (Chen *et al.* 2023; Sun *et al.* 2023). Besides, more generation of free radicals may have a negative effect through their recombination to form persulfate anions and HP (Equations (7)–(9)). The high reaction rate of recombining free radicals ($>10^8\ M^{-1}\ s^{-1}$) decreased the degradation process in SR-AOPs (Madihi-Bidgoli *et al.* 2021).

$$HSO_5^- + SO_4^{\bullet-} \rightarrow SO_5^{\bullet-} + SO_4^{2-} + H^+ \quad k < 1 \times 10^5\ M^{-1}s^{-1} \tag{4}$$

$$HSO_5^- + HO^{\bullet} \rightarrow SO_5^{\bullet-} + H_2O \quad k = 1.7 \times 10^7\ M^{-1}s^{-1} \tag{5}$$

$$SO_5^{2-} + HO^{\bullet} \rightarrow SO_5^{\bullet-} + HO^- \quad k = 2.1 \times 10^9\ M^{-1}s^{-1} \tag{6}$$

$$HO^{\bullet} + HO^{\bullet} \rightarrow H_2O_2 \quad k = 5.5 \times 10^9\ M^{-1}s^{-1} \tag{7}$$

$$SO_4^{\bullet-} + SO_4^{\bullet-} \rightarrow S_2O_8^{2-} \quad k = 7 \times 10^8\ M^{-1}s^{-1} \tag{8}$$

$$SO_4^{\bullet-} + HO^{\bullet} \rightarrow HSO_5^- \quad k = 1 \times 10^{10}\ M^{-1}s^{-1} \tag{9}$$

The impact of MFO loading (50–400 mg/L) was studied on CIP degradation and the results are presented in Figure 2(c). Increase in MFO loading improved CIP degradation significantly up to 300 mg/L dosage. After that, there was no change in the function of the process. Generally, an enhancement of MFO dosage increased the active site of Mn and Fe for enhanced MPS activation. However, further increments of MFO dosage did not improve the CIP degradation. The maximum generation of sulfate and hydroxyl radicals is obtained at an optimal ratio of MPS and MFO dosages. Therefore, the MFO dosage of 300 mg/L was selected as the optimum dosage for this process. It should be noted that obtaining optimum dosage of MFO and MPS is also an economical aspect to reduce material waste on a large scale.

3.4. The effect of scavengers and anions

Scavenging experiments are simple tests for determining the contribution of sulfate and hydroxyl radicals in MPS-based AOPs. According to the rate constants of ethyl alcohol (EtOH) with sulfate and hydroxyl radicals, EtOH is a strong scavenger for both sulfate and hydroxyl radicals while tert-Butyl alcohol (TBA) can scavenge hydroxyl radicals significantly and its rate constant with sulfate radicals is 1000 fold lower than that of hydroxyl radicals. Hence, the difference between the CIP degradation in the presence of EtOH and TBA indicates the contribution of the sulfate radicals in the CIP degradation. Moreover, free fatty acid (FFA) is utilized as a scavenger of $^1O_2$ (Wacławek *et al.* 2017; Isari *et al.* 2021). Figure 3(a) shows the CIP

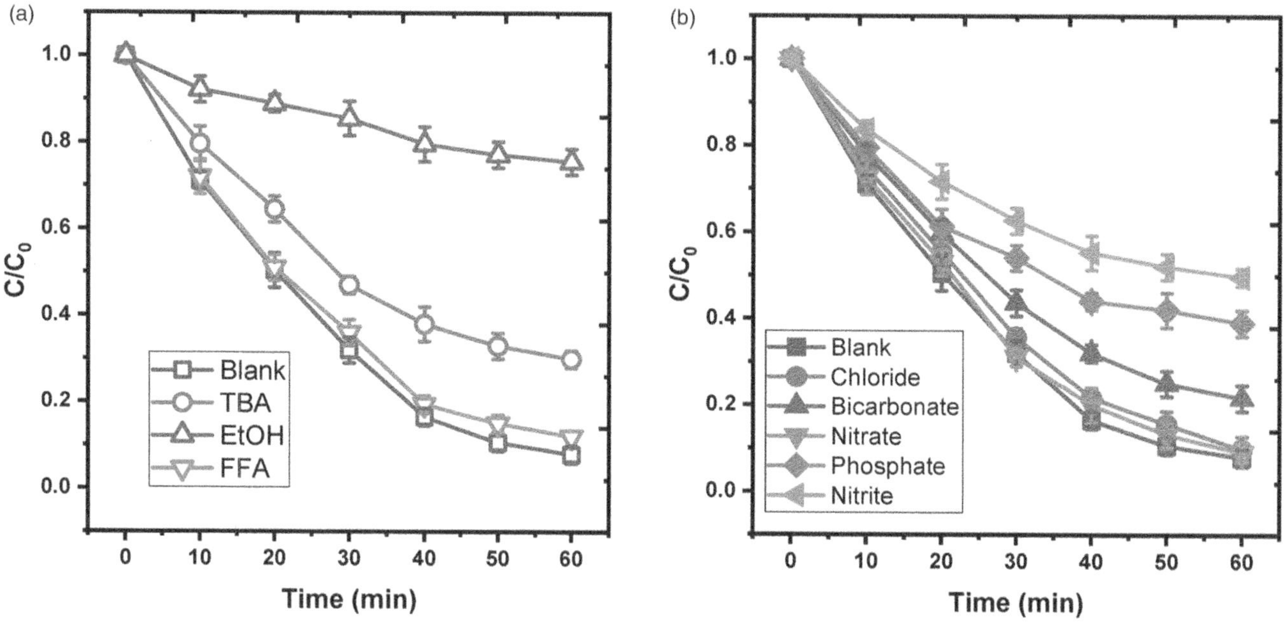

Figure 3 | (a) The effect of TBA, FFA and EtOH on CIP degradation. (b) The effect of anions on CIP degradation (pH = 6.3, MPS = 2 mM, MFO = 300 mg/L, TBA = EtOH = 0.1 M and anion = 4 mM).

degradation in the presence of EtOH and TBA. As can be seen, EtOH significantly inhibited the CIP degradation indicating that sulfate and hydroxyl radicals were the main oxidative agents for the oxidation of CIP in MPS/MFO process. In the presence of TBA, the CIP degradation was modestly reduced indicating that hydroxyl radicals had a lesser role than sulfate radicals in oxidizing CIP molecules. In the case of FFA, a negligible effect on CIP degradation was observed indicating that singlet oxygen did not contribute to CIP degradation.

When AOPs and SR-AOPs apply for a real condition, anions may affect the yield of the treatment process. In this work, the effect of several anions was studied on CIP degradation by MPS/MFO (Figure 3(b)). Chloride and nitrate ions had a neutral effect on CIP degradation. Chloride ions have been known as a scavenger for hydroxyl and sulfate radicals with high reactivity ($\sim 10^9\ M^{-1}\ s^{-1}$) (Equations (10)–(12)) (Wang & Wang 2021). Chlorine radical produced has high reactivity with organic compounds and is able to oxidize CIP. Moreover, it can be directly reacted with MPS to produce HOCl and Cl_2 which are strong oxidants (Equations (13) and (14)) (Jaafarzadeh *et al.* 2018).

$$HO^{\bullet} + Cl^{-} \rightarrow ClOH^{\bullet -} \quad k = 4.3 \times 10^{9}\ M^{-1}\ s^{-1} \tag{10}$$

$$ClOH^{\bullet -} + H^{+} \rightarrow Cl^{\bullet} + H_2O \quad k = 2.1 \times 10^{10}\ M^{-1}\ s^{-1} \tag{11}$$

$$Cl^{-} + SO_4^{\bullet -} \rightarrow Cl^{\bullet} + SO_4^{2-} \quad k = 3 \times 10^{8} M^{-1}\ s^{-1} \tag{12}$$

$$Cl^{-} + HSO_5^{-} \rightarrow HOCl + SO_4^{2-} \tag{13}$$

$$2Cl^{-} + HSO_5^{-} + H^{+} \rightarrow SO_4^{2-} + Cl_2 + H_2O \tag{14}$$

Bicarbonate ions inhibited the CIP degradation in MPS/MFO process. Bicarbonate ions are well known as powerful scavengers for free radicals and they are converted to carbonate radicals ($E^0 = 1.59$ V) as weak oxidants. It should be noted that bicarbonate has a higher reaction rate with hydroxyl radicals ($\sim 10^8\ M^{-1}\ s^{-1}$) compared to sulfate radicals ($\sim 10^6\ M^{-1}\ s^{-1}$) (Wacławek *et al.* 2017; Wang & Wang 2021). On the other hand, bicarbonate can affect MPS concentration in a way that MPS reacts with bicarbonate instead of the catalyst surface and produce peroxymonocarbonate (HCO_4^{-}) (Madihi-Bidgoli *et al.* 2021).

$$HSO_5^{-} + HCO_3^{-} \rightarrow HSO_4^{-} + HCO_4^{-} \tag{15}$$

Phosphate ions inhibited the CIP degradation rate considerably. Phosphate ion is well known as an attaching agent in heterogeneous catalysts. Phosphate ions can poison –OH sites of the catalyst surface since phosphate ions can rapidly occupy the -OH sites of MFO to form an inner-sphere complex. Since –OH is a reactive site on MFO for MPS activation, catalytic activity of MFO was dropped in the presence of phosphate ions as a competitor for MPS. Accordingly, it can be confirmed that -OH sites are the reactive sites of MFO for MPS activation and have a key role in the surface of MFO for the formation of MPS-Fe(III) or MPS-Mn(III) complex (Han *et al.* 2022b).

According to Figure 3(b), nitrate ions did not impact the CIP elimination. Although nitrate ions can quench sulfate radicals, the rate of scavenging sulfate radicals is relatively slight compared to other anions ($\sim 10^4\ M^{-1}\ s^{-1}$). It has been reported that nitrate ion has a neutral impact on the performance of SR-AOPs (Hassani *et al.* 2023a). Hence, there is no concern about the presence of nitrate in the water matrix for SR-AOPs in the degradation of organic pollutants.

Nitrite ions had the maximum inhibitory effect on the CIP degradation. Nitrite ion is a strong reductant which can scavenge both free radicals (Equations (16) and (17)) and MPS (Equation (18)). Hence, the presence of nitrite ions negatively affects sulfate radicals and their precursor (MPS) (Wacławek *et al.* 2017).

$$SO_4^{\bullet -} + NO_2^- \rightarrow SO_4^{2-} + NO_2^{\bullet} \quad k = 9 \times 10^8\ M^{-1}\ s^{-1} \tag{16}$$

$$HO^{\bullet} + NO_2^- \rightarrow HO^- + NO_2^{\bullet} \quad k = 2.5 - 14 \times 10^9\ M^{-1}\ s^{-1} \tag{17}$$

$$HSO_5^- + NO_2^- \rightarrow HSO_4^- + NO_3^- \tag{18}$$

3.5. Catalytic activity of MFO for other chemical oxidants

The function of MFO was investigated for the activation of several chemical oxidants (PDS, PI, HP, PAA and PC). Figure 4(a) shows the CIP degradation by activated chemical oxidants by MFO under the same conditions. As can be seen, MPS, PDS and PI have a similar trend for the CIP degradation (more than 90% CIP removal efficiency). PDS similar to MPS generates sulfate radicals in the presence of a catalyst. Hence, it is expected that the results obtained to be as same as the results of MPS. On the other hand, PI can generate $IO_3^{\bullet}$, $HO^{\bullet}$ and O(3P) (Equations (19)–(21)) which are strong reactive species for the

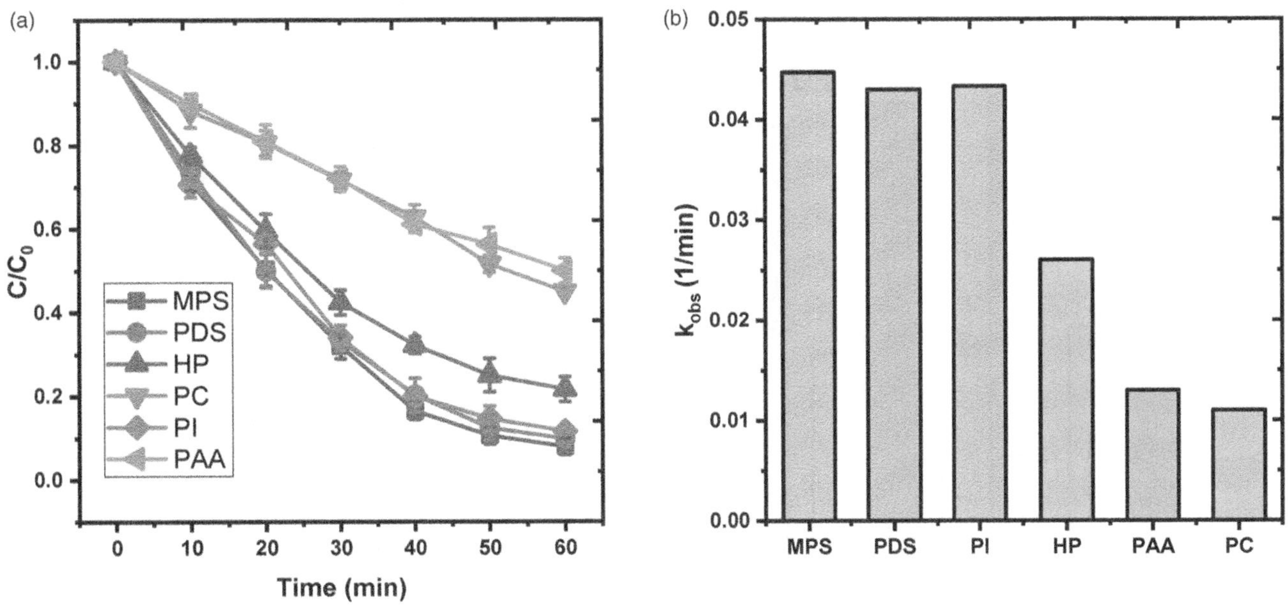

Figure 4 | Catalytic activity of MFO in the presence of various oxidants (a) CIP degradation, (b) the rate constant of the pseudo-first order model (MFO = 300 mg/L, oxidant = 2 mM and pH = 6.3).

destruction of organic compounds (Djaballah *et al.* 2021; Li *et al.* 2022).

$$IO_4^- + UV \rightarrow IO_3^\bullet + O^{\bullet -} \quad k = (0.5 - 4.16) \times 10^{-3}\ s^{-1} \tag{19}$$

$$IO_4^- + UV \rightarrow IO_3^\bullet + O(3P) \quad k = (1 - 2) \times 10^{-4}\ s^{-1} \tag{20}$$

$$O^{\bullet -} + H_2O \rightarrow HO^\bullet + OH^- \quad k = 1 \times 10^8\ M^{-1}s^{-1} \quad pKa = 11.9 \tag{21}$$

Catalytic decomposition of PAA, HP and PC results in organic radicals/hydroxyl radicals, hydroxyl radicals and hydroxyl radicals/carbonate radicals, respectively. The results showed that activating these oxidants by MFO were relatively ineffective for the degradation of CIP. Low reactivity of organic radicals (PAA/MFO) can be reasonable for lower degradation of organic compounds. On the other hand, the presence of carbonate ions in the structure of PC may reduce the performance of free radicals generated in activated PC by MFO. Overall, the results of this section and the previous section showed that sulfate radicals were more effective than hydroxyl radicals in the oxidation of CIP. Accordingly, the rate constants of pseudo-first-order model is presented in Figure 4(b). As can be seen, the rate constants of MPS, PDS and PI were close to each other (~0.04 1/min) while a significant reduction was observed in HP, PAA and PC. Furthermore, the solution pH was monitored during CIP oxidation and the results are presented in Fig. S5. All oxidants reduced the solution pH except for PC in which the presence of carbonate sodium increased the solution pH. The role of sulfate radicals in the degradation of CIP can be a rational reason for this performance. However, it should be noted that MFO exhibited an effective catalyst for the activation of several chemical oxidants, especially persulfate anions.

3.6. Mineralization, reusability and stability

Carbon mineralization is an index to demonstrate the complete degradation of CIP. Figure 5(a) shows the mineralization of CIP in four consecutive cycles of MFO application in MPS/MFO process. In fresh use of MFO, 45.5% of CIP was mineralized during 60 min reaction time indicating a high reactivity of free radicals generated. Reusability of the catalyst is a critical parameter for scaling up the process. The CIP degradation and TOC removal were investigated in four cycles and the results are presented in Figure 5(a). As shown in this figure, MFO kept the catalytic activity for four cycles in which CIP degradation and TOC removal from fresh use to the fourth cycle was changed from 92.1% and 45.5% to 85.6% to 34.4%, respectively. With

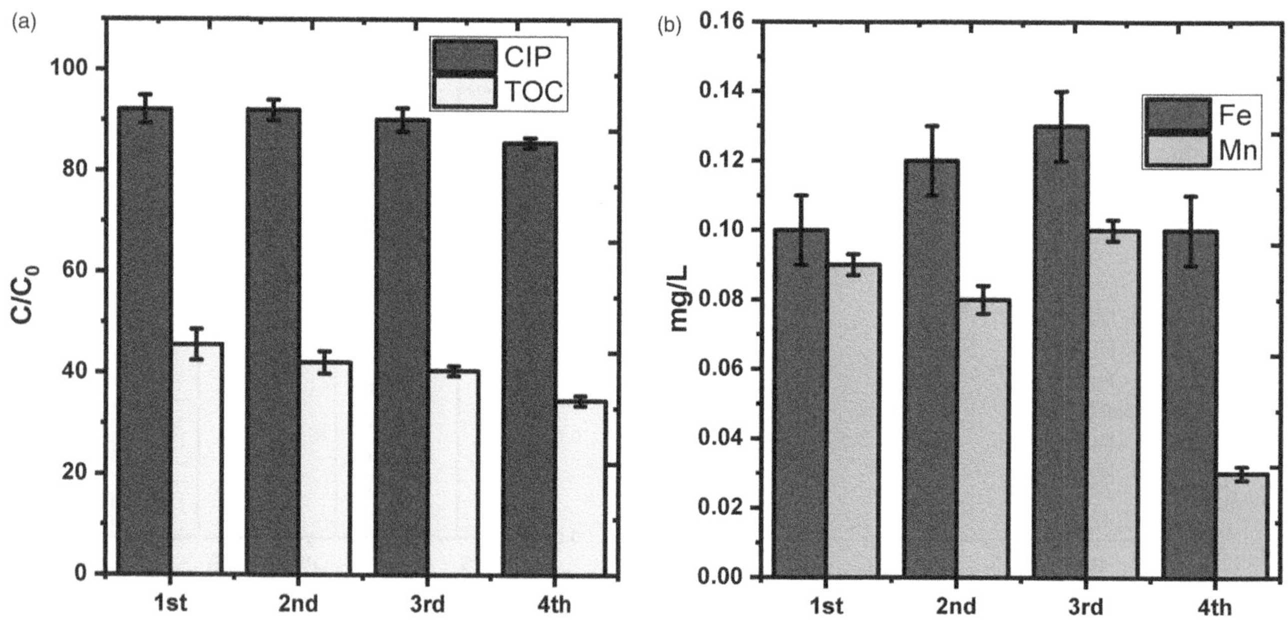

Figure 5 | (a) CIP degradation and TOC removal in four cycles of MFO usage, (b) Mn and Fe leaching in four cycles of MFO usage (MFO = 300 mg/L, MPS = 2 mM and pH = 6.3).

around 10% reduction in efficiency, these results showed that MFO can be acceptably applied for the fourth cycle in catalytic degradation of CIP.

Stability of MFO was studied by measuring Mn and Fe leaching in the solution. Figure 5(b) shows the amount of Mn and Fe leached in the solution during four cycles of MFO usage. Maximum leaching of Mn and Fe were 0.1 and 0.13 mg/L, respectively. These values are negligible indicating that MFO is a stable catalyst for MPS activation. Moreover, it is worthwhile to mention that these values are below than discharge standards of effluent to receiving waters (0.3 mg/L) (Ghanbari *et al.* 2020).

XRD and FTIR analyses of reused MFO after the fourth usage were conducted. As shown in Fig. S6a, there was no change in the intensity of peaks in XRD pattern for MFO. On the other hand, some changes were observed in FTIR spectrum (Fig. S6b). Two stretches were observed in 1,575.6 and 1,418.2 cm^{-1} which corresponded to C = C bonds resulting from the adsorption of CIP and its by-products. Moreover, a peak in 1,013.1 cm^{-1} appeared which might be related to S–O bond driving from the interaction of MPS with MFO.

To confirm the role of Mn and Fe leached in homogeneous behavior of MPS activation, an experiment was conducted as MPS/(Mn (0.1 mg/L) + Fe (0.13 mg/L) and the results are presented in Fig. S7. As can be seen, there is no significant MPS activation by Mn and Fe leached in the solution indicating that MPS activation has been conducted on the surface of MFO. Overall, MFO indicated an effective, durable and recyclable catalyst for SR-AOPs.

3.7. The antibiotic activity of the treated CIP solution and fluorine release

The antibiotic activity of the CIP solution and treated solution by MPS/MFO process was evaluated. CIP induces further DNA replication failure in bacteria cells through topoisomerase IV and DNA gyrase. The basic structure of CIP is quinolone group, while the peripheral moiety plays an additional role in antibacterial activity. The carboxyl and carbonyl groups are critical for the cleavage and destruction of DNA and are necessary for the decrease of bacteria resistance (Ou *et al.* 2016; Hu *et al.* 2020). Moreover, the fluorine at CIP structure improves the antibacterial activity of CIP. The inhibition zone diameters were applied as a semi-quantitative test for the antibacterial activity (Figure 6(a)). As can be seen, CIP solution had a strong inhibition zone diameter of ~ 23 mm while with the increase in reaction time this value significantly reduced and reached ~6 mm. This reduction can be related to the transformations of the carboxyl, the piperazine ring, the cyclopropyl and fluorine bond which are made by free radicals to reduce the antibacterial activity of CIP (De Bel *et al.* 2009).

To confirm the reduction of antibacterial activity of CIP, fluorine released was monitored during CIP degradation (Figure 6(b)). According to the concentration of CIP (16.5 mg/L), the maximum fluorine released can be 0.95 mg/L while

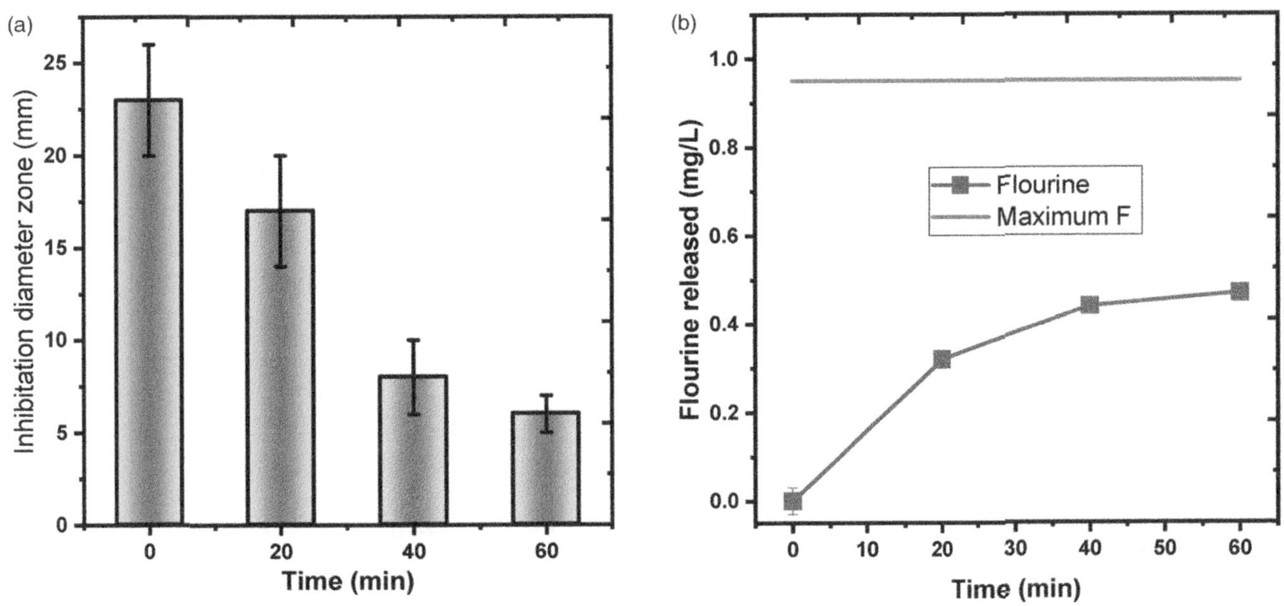

Figure 6 | (a) Bacterial growth inhibition zone diameter for CIP degradation by MPS/MFO during 60 min reaction time, (b) fluorine release from CIP degradation by MPS/MFO (MFO = 300 mg/L, MPS = 2 mM and pH = 6.3).

complete mineralization occurs. In this study, during 60 min reaction time fluorine release increased by the increase of reaction time in a way that 0.47 mg/L fluorine was released at 60 min reaction time. This value was about 50% fluorine mineralization indicating that the fluorine bond was effectively destructed in MPS/MFO process.

3.8. MPS/MFO process for other pollutants and an actual wastewater

The performance of MPS/MFO was evaluated on other pollutants such as bisphenol A (BPA) and acid yellow 36 (AY36) and the results are presented in Figure 7(a). As can be seen, AY36 was completely degraded within 60 min reaction time. There is a possibility in which free radicals ($HO^{\bullet}$ and $SO_4^{\bullet-}$) rapidly attacked the chromophore of the dye. On the other hand, CIP and BPA degradation were, respectively, 92% and 95% with a similar trend. These results showed that MPS/MFO is able to degrade a variety of pollutants.

MPS/MFO was operated for the treatment of actual wastewater (hospital effluent) containing CIP and the results are presented in Figure 7(b). As shown in this figure, CIP degradation was enhanced by increasing MPS dosage. CIP degradation efficiencies were 42.3, 57.2, 73.2 and 75.5% for MPS dosages of 1, 2, 3 and 4 mM, respectively. Compared to the aqueous solution with a maximum efficiency of 92% with 2 mM, to reach the maximum efficiency in hospital effluent a further MPS dosage is necessary. In this way, 4 mM was used for 75% CIP degradation in a real condition which was attributed to the presence of various organic and inorganic compounds. According to Table S1, the presence of 110 mg/L COD indicates that a part of the free radicals generated was consumed by other organic compounds. Moreover, the existence of anions and cations (total dissolved solids = 1,170 mg/L) in a real condition can be another reason for an additional need for MPS dosage. Similar results were obtained for TOC value in which with an increase of MPS dosage, TOC removal enhanced from 17.8 to 33.6% when MPS dosage increased from 1 to 4 mM. It is worthwhile to note that the real conditions in terms of the solution pH and water constituents are very different from aqueous solutions. Hence, to scale up the AOPs, this issue should be considered before implementation.

MPS/MFO process was compared to other AOPs based on previous works and the results are presented in Table 1. As can be seen, MPS/MFO process showed an acceptable function compared to other processes. However, there are some advantages and disadvantages of SR-AOPs. High performance, rapid reaction time, operation in a wide range of pH, safe storage of MPS on a large scale and contribution of both $HO^{\bullet}$ and $SO_4^{\bullet-}$ are the advantages of SR-AOPs while their limitations include residual sulfate ions in the effluent, an increase of cations (K^+) and higher cost compared to HP.

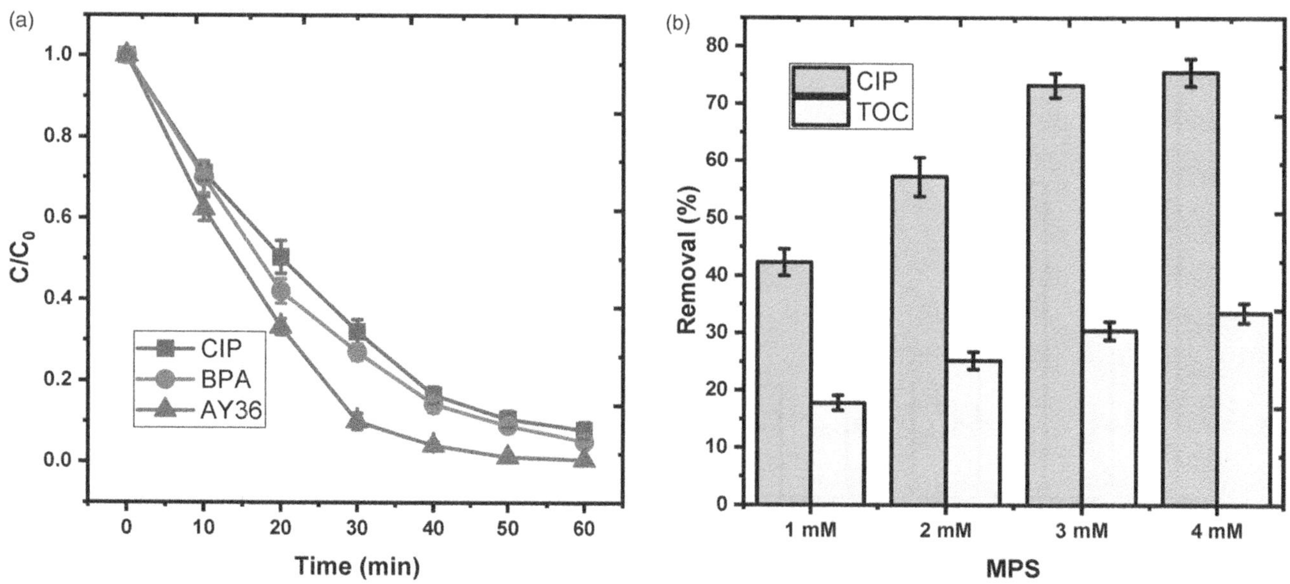

Figure 7 | (a) The performance of MPS/MFO for other pollutants (MPS = 2 mM, MFO = 300 mg/L and pH = 6.3), (b) CIP and TOC removals from an actual hospital effluent by MPS/MFO by various MPS dosages (MFO = 300 mg/L and pH = 6.3).

Table 1 | Comparison of CIP degradation by different AOPs

AOP	Conditions	CIP (mg/L)	Removal (%)	Ref.
TiO_2/Montmorillonite/US	pH = 6, catalyst = 0.2 g/L, US power = 650 W and time = 120 min	10	65	Hassani *et al.* (2017)
Bi_2WO_6/Fe_3O_4/visible light	Catalyst = 0.1 g/L and time = 120 min	15	90	Song *et al.* (2020)
PDS//UV	pH = 7, PDS = 15 mM, UV power = 650 3 and time = 40 min	10	94	Ma *et al.* (2022)
Electro-Peroxone	pH = 3, O_3 = 33.1 mg/L, applied current = 180 mA and time = 40 min	20	100	Khavari Kashani *et al.* (2022)
UV/Chlorine	pH = 7, ClO^- = 0.1 mM, UV fluence rate = 0.4 mW/cm^2 and time = 10	3.33	98	Deng *et al.* (2019)
PDS/Ferrous ions	pH = 6, PDS = 0.6 mM, Fe(II) = 0.6 mM and time = 240 min	10	95.6	Ji *et al.* (2014)
PI/CuO	pH = 6, PI = 0.5 mM, CuO = 0.5 g/L and time = 30 min	20	92	Zhang *et al.* (2023)
MPS/MFO	pH = 6.3 (natural pH), MPS = 2 mM, MF = 0.3 g/L and time = 30 min	16.55	92	Current work

3.9. The mechanism of MPS activation via MFO

Based on the results obtained above and a focus on literature, the mechanism on MPS activation with the MFO composites was illustrated in Figure 8. The MPS activation by MFO composite was initiated by the reaction of Fe(III) by MPS to generate peroxymonosulfate radicals ($SO_5^{\bullet -}$) and Fe(II) through the electron transfer mechanism (Equation (22)). On the other hand, Mn(III) reacts with MPS and generates sulfate radicals ($SO_4^{\bullet -}$) and Mn(IV) (Equation (23)). Mn(IV) and Fe(II) may activate MPS according to Equations (24) and (25). Moreover, the regeneration of Mn(III) and Fe(III) may be obtained from the reaction of Mn(IV) and Fe(II) to pervade the free radical generation (Equation (26)). In addition, the recombination of $SO_5^{\bullet -}$ can result in the generation of $SO_4^{\bullet -}$ (Equation (27)). As mentioned above, hydroxyl radicals contributed to the degradation of CIP. In this way, $SO_4^{\bullet -}$ may react with a water molecule and hydroxyl radicals ($HO^{\bullet}$) are formed (Equation (28)). Hence, sulfate radicals and hydroxyl radicals destructed CIP molecule (Equation (29)).

$$\equiv Fe(III) + HSO_5^- \rightarrow \equiv Fe(II) + SO_5^{\bullet -} + H^+ \quad (22)$$

$$\equiv Mn(III) + HSO_5^- \rightarrow \equiv Mn(IV) + SO_4^{\bullet -} + OH^- \quad (23)$$

$$\equiv Mn(IV) + HSO_5^- \rightarrow \equiv Mn(III) + SO_5^{\bullet -} + H^+ \quad (24)$$

$$\equiv Fe(II) + HSO_5^- \rightarrow \equiv Fe(III) + SO_4^{\bullet -} + OH^- \quad (25)$$

$$\equiv Mn(III) + \equiv Fe(III) \rightarrow \equiv Mn(IV) + \equiv Fe(II) \quad (26)$$

$$SO_5^{\bullet -} + SO_5^{\bullet -} \rightarrow 2SO_4^{\bullet -} + O_2 \quad k = 1 \times 10^8\ M^{-1}\ s^{-1} \quad (27)$$

$$SO_4^{\bullet -} + H_2O \rightarrow HSO_4^- + HO^{\bullet} \quad k = 660\ s^{-1} \quad (28)$$

$$SO_4^{\bullet -} + HO^{\bullet} + CIP \rightarrow by - products + CO_2 + H_2O + F^- \quad (29)$$

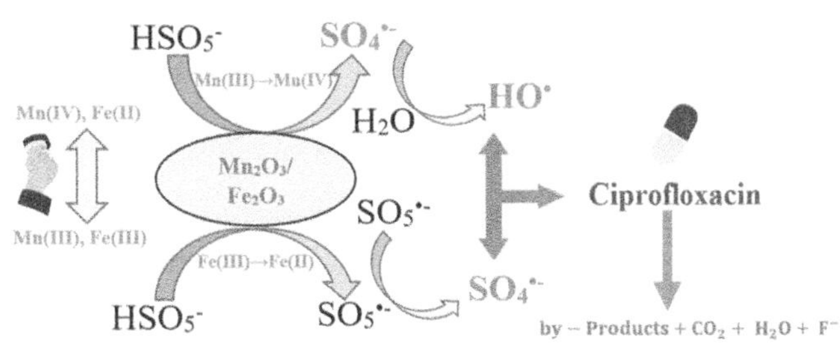

Figure 8 | The mechanism of MPS activation by MFO for CIP degradation.

4. CONCLUSIONS

In this work, MFO was simply synthesized and its catalytic activity was evaluated for the activation of MPS to degrade an emerging pollutant. In this way, MFO/MPS could remove 92% of CIP under optimal conditions (pH = 6.3, MFO = 300 mg/L and MPS = 2 mM). MFO/MPS could be operated in a wide range of pH (4–8). Moreover, MFO showed that it can be a suitable catalyst for other oxidants such as PDS and PI. Chloride and nitrate ions had no inhibitory effect on the performance of MFO/MPS process while phosphate, bicarbonate and nitrite ions significantly suppressed the removal efficiency through the scavenging free radicals and occupying of active sites in the catalyst surface. MFO was an effective, stable and recoverable catalyst keeping the catalytic activity for four cycles. MFO/MPS could reduce the toxicity of CIP solution considerably. Moreover, the function of MFO/MPS on a real water matrix was also investigated and the results were convenient. This work provided essential knowledge for the practical utilization of MPS activation via heterogeneous catalyst (MFO) for the degradation of emerging pollutants in water and wastewater.

ACKNOWLEDGEMENTS

This research project has been financially supported by Behbahan Faculty of Medical Sciences (Iran) under contract number of IR.BHN.REC.1398.029.

DATA AVAILABILITY STATEMENT

All relevant data are included in the paper or its Supplementary Information.

CONFLICT OF INTEREST

The authors declare there is no conflict.

REFERENCES

An, T., Yang, H., Li, G., Song, W., Cooper, W. J. & Nie, X. 2010 Kinetics and mechanism of advanced oxidation processes (AOPs) in degradation of ciprofloxacin in water. *Applied Catalysis B: Environmental* **94** (3), 288–294.

Baek, S., Ghaffari, Y. & Bae, J. 2022 Synthesis of Fe_2O_3/Mn_2O_3 nanocomposites and impregnated porous silicates for dye removal: insights into treatment mechanisms. *Catalysts* **12** (9), 1045.

Chen, G., Nengzi, L.-C., Gao, Y., Zhu, G., Gou, J. & Cheng, X. 2020 Degradation of tartrazine by peroxymonosulfate through magnetic Fe_2O_3/Mn_2O_3 composites activation. *Chinese Chemical Letters* **31** (10), 2730–2736.

Chen, W., Huang, J., Shen, Y., Zhu, K., Lei, L., He, H. & Ai, Y. 2023 Fe-N co-doped coral-Like hollow carbon shell toward boosting peroxymonosulfate activation for efficient degradation of tetracycline: singlet oxygen-dominated non-radical pathway. *Journal of Environmental Sciences* **126**, 470–482.

Chi, Y., Wang, P., Lin, M., Lin, C., Gao, M., Zhao, C. & Wu, X. 2022 Manganese oxides activated peroxymonosulfate for ciprofloxacin removal: effect of oxygen vacancies and chemical states. *Chemosphere* **299**, 134437.

De Bel, E., Dewulf, J., Witte, B. D., Van Langenhove, H. & Janssen, C. 2009 Influence of pH on the sonolysis of ciprofloxacin: biodegradability, ecotoxicity and antibiotic activity of Its degradation products. *Chemosphere* **77** (2), 291–295.

Deng, J., Wu, G., Yuan, S., Zhan, X., Wang, W. & Hu, Z.-H. 2019 Ciprofloxacin degradation in UV/chlorine advanced oxidation process: influencing factors, mechanisms and degradation pathways. *Journal of Photochemistry and Photobiology A: Chemistry* **371**, 151–158.

Djaballah, M. L., Merouani, S., Bendjama, H. & Hamdaoui, O. 2021 Development of a free radical-based kinetics model for the oxidative degradation of chlorazol black in aqueous solution using periodate photoactivated process. *Journal of Photochemistry and Photobiology A: Chemistry* **408**, 113102.

Eslami, A., Mehdipour, F., Lin, K.-Y. A., Sharifi Maleksari, H., Mirzaei, F. & Ghanbari, F. 2020 Sono-photo activation of percarbonate for the degradation of organic dye: the effect of water matrix and identification of by-products. *Journal of Water Process Engineering* **33**, 100998.

Eslami, A., Khavari Kashani, M. R., Khodadadi, A., Varank, G., Kadier, A., Ma, P.-C., Madihi-Bidgoli, S. & Ghanbari, F. 2021 Sono-peroxi-coagulation (SPC) as an effective treatment for pulp and paper wastewater: focus on pH effect, biodegradability, and toxicity. *Journal of Water Process Engineering* **44**, 102330.

Ghaffari, Y., Gupta, N. K., Bae, J. & Kim, K. S. 2020 One-step fabrication of Fe_2O_3/Mn_2O_3 nanocomposite for rapid photodegradation of organic dyes at neutral pH. *Journal of Molecular Liquids* **315**, 113691.

Ghanbari, F., Zirrahi, F., Olfati, D., Gohari, F. & Hassani, A. 2020 TiO_2 nanoparticles removal by electrocoagulation using iron electrodes: catalytic activity of electrochemical sludge for the degradation of emerging pollutant. *Journal of Molecular Liquids* **310**, 113217.

Ghanbari, F., Giannakis, S. & Samoili, S. 2022 Chapter 10 persulfate application for landfill leachate treatment: current status and challenges. In: Zhu, M., Bian, Z. & Zhao, C. (eds) *Persulfate-based Oxidation Processes in Environmental Remediation*. The Royal Society of Chemistry, London, UK, pp. 252–288.

Han, X., Zhang, W., Li, S., Cheng, C., Zhou, L., Jia, Q. & Xiu, G. 2022a Efficient activation of peroxymonosulfate by MnS/Fe-MOF hybrid catalyst for sulfadiazine degradation: synergistic effects and mechanism. *Separation and Purification Technology* **287**, 120509.

Han, Y., Li, Z., Zhang, M., Han, F., Liu, Z. & Zhou, W. 2022b Isomorphic substitution of goethite by cobalt: effects on the pathway and performance of peroxymonosulfate activation. *Chemical Engineering Journal* **450**, 138460.

Hassani, A., Khataee, A., Karaca, S., Karaca, C. & Gholami, P. 2017 Sonocatalytic degradation of ciprofloxacin using synthesized TiO_2 nanoparticles on montmorillonite. *Ultrasonics Sonochemistry* **35**, 251–262.

Hassani, A., Eghbali, P., Mahdipour, F., Wacławek, S., Lin, K.-Y. A. & Ghanbari, F. 2023a Insights into the synergistic role of photocatalytic activation of peroxymonosulfate by UVA-LED irradiation over $CoFe_2O_4$-rGO nanocomposite towards effective bisphenol A degradation: performance, mineralization, and activation mechanism. *Chemical Engineering Journal* **453**, 139556.

Hassani, A., Scaria, J., Ghanbari, F. & Nidheesh, P. V. 2023b Sulfate radicals-based advanced oxidation processes for the degradation of pharmaceuticals and personal care products: a review on relevant activation mechanisms, performance, and perspectives. *Environmental Research* **217**, 114789.

Hu, X., Hu, X., Peng, Q., Zhou, L., Tan, X., Jiang, L., Tang, C., Wang, H., Liu, S., Wang, Y. & Ning, Z. 2020 Mechanisms underlying the photocatalytic degradation pathway of ciprofloxacin with heterogeneous TiO_2. *Chemical Engineering Journal* **380**, 122366.

Isari, A. A., Moradi, S., Rezaei, S. S., Ghanbari, F., Dehghanifard, E. & Kakavandi, B. 2021 Peroxymonosulfate catalyzed by core/shell magnetic ZnO photocatalyst towards malathion degradation: enhancing synergy, catalytic performance and mechanism. *Separation and Purification Technology* **275**, 119163.

Jaafarzadeh, N., Ghanbari, F. & Zahedi, A. 2018 Coupling electrooxidation and oxone for degradation of 2,4-dichlorophenoxyacetic acid (2,4-D) from aqueous solutions. *Journal of Water Process Engineering* **22**, 203–209.

Ji, Y., Ferronato, C., Salvador, A., Yang, X. & Chovelon, J.-M. 2014 Degradation of ciprofloxacin and sulfamethoxazole by ferrous-activated persulfate: implications for remediation of groundwater contaminated by antibiotics. *Science of the Total Environment* **472**, 800–808.

Khavari Kashani, M. R., Kiani, R., Hassani, A., Kadier, A., Madihi-Bidgoli, S., Lin, K.-Y. A. & Ghanbari, F. 2022 Electro-peroxone application for ciprofloxacin degradation in aqueous solution using sacrificial iron anode: a new hybrid process. *Separation and Purification Technology* **292**, 121026.

Kim, S., Gupta, N. K., Bae, J. & Kim, K. S. 2021 Fabrication of coral-Like Mn_2O_3/Fe_2O_3 nanocomposite for room temperature removal of hydrogen sulfide. *Journal of Environmental Chemical Engineering* **9** (3), 105216.

Li, R., Wang, J., Wu, H., Zhu, Z. & Guo, H. 2022 Periodate activation for degradation of organic contaminants: processes, performance and mechanism. *Separation and Purification Technology* **292**, 120928.

Liu, M., Zhang, L., Xi, B.-d., Yu, S., Hu, X. & Hou, L.-a. 2017 Degradation of ciprofloxacin by TiO_2/Fe_2O_3/zeolite catalyst-activated persulfate under visible LED light irradiation. *RSC Advances* **7** (81), 51512–51520.

Ma, Y., Wang, Z., Yang, W., Chen, C., Li, J., He, R. & Liu, S. 2022 Insights into the radical and nonradical oxidation degradation of ciprofloxacin in peroxodisulfate activation by ultraviolet light. *Journal of Water Process Engineering* **49**, 103184.

Madihi-Bidgoli, S., Asadnezhad, S., Yaghoot-Nezhad, A. & Hassani, A. 2021 Azurobine degradation using Fe2O3@multi-walled carbon nanotube activated peroxymonosulfate (PMS) under UVA-LED irradiation: performance, mechanism and environmental application. *Journal of Environmental Chemical Engineering* **9** (6), 106660.

Ou, H.-s., Ye, J.-s., Ma, S., Wei, C.-h., Gao, N.-y. & He, J.-z. 2016 Degradation of ciprofloxacin by UV and UV/H_2O_2 via multiple-wavelength ultraviolet light-emitting diodes: effectiveness, intermediates and antibacterial activity. *Chemical Engineering Journal* **289**, 391–401.

Pandis, P. K., Kalogirou, C., Kanellou, E., Vaitsis, C., Savvidou, M. G., Sourkouni, G., Zorpas, A. A. & Argirusis, C. 2022 Key points of advanced oxidation processes (AOPs) for wastewater, organic pollutants and pharmaceutical waste treatment: a mini review. *ChemEngineering* **6** (1), 8.

Song, S. Y., Chen, H. D., Li, C. X., Shi, D. S., Ying, Y., Han, Y. B., Xu, J. C., Hong, B., Jin, H. X., Jin, D. F., Peng, X. L., Ge, H. L. & Wang, X. Q. 2020 Magnetic Bi_2WO_6 nanocomposites: synthesis, magnetic response and their visible-light-driven photocatalytic performance for ciprofloxacin. *Chemical Physics* **530**, 110614.

Sun, J., Xin, X., Sun, S., Du, Z., Yao, Z., Wang, M. & Jia, R. 2023 Experimental and theoretical investigation on degradation of dimethyl trisulfide by ultraviolet/peroxymonosulfate: reaction mechanism and influencing factors. *Journal of Environmental Sciences* **127**, 824–832.

Wacławek, S., Lutze, H. V., Grübel, K., Padil, V. V. T., Černík, M. & Dionysiou, D. D. 2017 Chemistry of persulfates in water and wastewater treatment: a review. *Chemical Engineering Journal* **330**, 44–62.

Wang, J. & Wang, S. 2021 Effect of inorganic anions on the performance of advanced oxidation processes for degradation of organic contaminants. *Chemical Engineering Journal* **411**, 128392.

Wang, F., Li, T., Fang, Y., Wang, Z. & Zhu, J. 2021 Heterogeneous structured Mn_2O_3/Fe_2O_3 composite as anode material for high performance lithium ion batteries. *Journal of Alloys and Compounds* **857**, 157531.

Xiao, S., Cheng, M., Zhong, H., Liu, Z., Liu, Y., Yang, X. & Liang, Q. 2020 Iron-mediated activation of persulfate and peroxymonosulfate in Both homogeneous and heterogeneous ways: a review. *Chemical Engineering Journal* **384**, 123265.

Yaghoot-Nezhad, A., Wacławek, S., Madihi-Bidgoli, S., Hassani, A., Lin, K.-Y. A. & Ghanbari, F. 2023 Heterogeneous photocatalytic activation of electrogenerated chlorine for the production of reactive oxygen and chlorine species: a new approach for Bisphenol A degradation in saline wastewater. *Journal of Hazardous Materials* **445**, 130626.

Yousefi, M., Ghanbari, F., Zazouli, M. A. & Madihi-Bidgoli, S. 2017 Brilliant Blue FCF degradation by persulfate/zero valent iron: the effects of influencing parameters and anions. *Desalination and Water Treatment* **70**, 364–371.

Zhang, X., Kamali, M., Xue, Y., Li, S., Costa, M. E. V., Cabooter, D. & Dewil, R. 2023 Periodate activation with copper oxide nanomaterials for the degradation of ciprofloxacin – a new insight into the efficiency and mechanisms. *Journal of Cleaner Production* **383**, 135412.

Zhou, H., Qian, L., Cheng, T., Du, B., Xia, F. & Zhu, C. 2022 Effect of MnOx/α-Fe_2O_3 prepared from goethite on selective catalytic reduction of NO with NH_3. *Journal of Chemistry* **2022**, 5049161.

Zhu, T. & Liu, B. 2022 Mechanism study on the effect of peracetic acid (PAA), UV/PAA and ultrasonic/PAA oxidation on ultrafiltration performance during algae-laden water treatment. *Water Research* **220**, 118705.

First received 15 August 2022; accepted in revised form 16 February 2023. Available online 1 March 2023

doi: 10.2166/wst.2023.069

Applications of biochar in sulfate radical-based advanced oxidation processes for the removal of pharmaceuticals and personal care products

Sourabh M. Badiger[a,b] and P. V. Nidheesh[a,b,*]

[a] CSIR-National Environmental Engineering Research Institute, Nagpur 440020, India
[b] Academy of Scientific and Innovative Research (AcSIR), Ghaziabad 201002, India
*Corresponding author. E-mail: pv_nidheesh@neeri.res.in

ABSTRACT

Recently, biochar (BC) has been increasingly used as a catalyst for the degradation of 'emerging pollutants' (EPs). Pharmaceuticals and personal care products (PPCPs), which come under 'EPs', can be harmful to the aquatic ecosystem despite being present in very low concentrations (ng/L–μg/L). Advanced oxidation processes (AOPs), which produce sulfate radical (SR-AOPs), show a great potential to degrade PPCPs effectively from wastewater. It is mainly due to the higher stability, long half-lives and better non-selectivity of $SO_4^{\bullet -}$ compared with AOPs with •OH generation. Furthermore, research focus is now given on AOPs coupled with BC-supported catalyst to enhance the degradation of PPCPs because of quicker generation of radicals (•OH, $SO_4^{\bullet -}$) by the activation of persulfate (PS) and peroxymonosulfate (PMS). This article sheds light on the catalytic ability of BC after its physical and chemical modifications such as acid/alkali treatment and metal doping. The role of persistent free radicals (PFRs) in the BC for effective removal of PPCPs has been elaborated. Its potential applications in synthetic as well as real wastewater have also been discussed.

Key words: advanced oxidation processes, biochar, pharmaceuticals and personal care products, sulfate radicals

HIGHLIGHTS

- Advanced oxidation processes coupled with sulfate radical (SR-AOPs) for degradation of PPCPs have been introduced.
- Implementation of biochar as a catalyst in SR-AOPs has been explained.
- The mechanism and pathways of biochar-added catalyst for PPCP degradation have been summarized.
- A potential application of biochar catalyst with SR-AOPs in real wastewater and sewage water has been discussed.

LIST OF ABBREVIATIONS

ACT	Acetaminophen
AOP	Advanced oxidation process
API	Active pharmaceutical ingredients
BC	Biochar
BTA	Benzotriazole
CBC	Corn stalk biochar
CBZ	Carbamazepine
CEX	Cephalexin
CIP	Ciprofloxacin
EDC	Endocrine disrupting chemicals
EfOM	Effluent organic matter
EP	Emerging pollutants
EPFR	Environmentally persistent free radical
FTIR	Fourier transform infrared spectroscopy
IBU	Ibuprofen
NOR	Norfloxacin
OFG	Oxygen containing functional group
OFX	Ofloxacin
PDS	Peroxydisulfate

PFR	Persistent free radical
PMS	Peroxymonosulfate
POP	Persistent organic pollutant
PPCP	Pharmaceuticals and personal care product
PS	Persulfate
ROS	Reactive oxygen species
SFG	Surface functional group
SMX	Sulfamethoxazole
SPS	Sodium persulfate
SR-AOP	Sulfate radical-based advanced oxidation process
STP	Sewage treatment plant
WWTP	Wastewater treatment plant
XPS	X-ray photoelectron spectroscopy

1. INTRODUCTION TO PPCPS

Contamination of water bodies by 'emerging pollutants' (EPs) has become a huge problem in the recent years (Gogoi *et al.* 2018; Rathi *et al.* 2021; Scaria *et al.* 2021a, 2021b). One such emerging problem is the contamination by pharmaceuticals and personal care products (PPCPs) which has created a major problem by degrading the water quality and endangering the aquatic ecosystem (Dey *et al.* 2019; Patel *et al.* 2019; Ngqwala & Muchesa 2020; Mohan *et al.* 2021). PPCPs also constitute active pharmaceuticals ingredients (APIs) of prescribed, non-prescribed and illegal drugs for human, plant and veterinary use (Dey *et al.* 2019).

Major pharmaceutical contamination in water bodies arise mainly due to unused medicines from hospital waste and human excretion via urine and feces. This is because the drug and their metabolites are partially consumed in the human body (Taylor & Senac 2014; Patel *et al.* 2020; Scaria *et al.* 2021a, 2021b). PPCPs in the form of APIs have also been shown to generate from a small molecule into at least 20,000 varieties of drug products (Evgenidou *et al.* 2015).

Conventional wastewater treatment systems that involve physical, chemical and biological removal processes are not efficient to treat PPCPs completely due to their persistent nature. Advanced oxidation processes (AOPs) are steadily emerging as a potential chemical technology that primarily relies on partial or complete oxidation of several recalcitrant compounds by the generation of free radical species ($^{\bullet}OH$ and $SO_4^{\bullet -}$), which result in the breakdown of pollutants (both organic and inorganic) into more fundamental and non-toxic molecules (Kanakaraju *et al.* 2018; Babu *et al.* 2019; Nidheesh *et al.* 2022a). Various AOP methods such as ozonation, Fenton, Photo-Fenton and UV/H_2O_2 processes have been previously applied as suitable alternative chemical processes for tertiary or post-tertiary water and wastewater treatment in degrading PPCPs (Anjali & Shanthakumar 2019; Krishnan *et al.* 2021; Scaria & Nidheesh 2022c). Furthermore, AOPs in combination with porous adsorbents (activated carbon (AC), graphene) have been a popular technology to be used as catalysts, which increase the degrading efficiency of several PPCPs in aquatic systems by enhancing the generation of free radical species in the system (Nidheesh 2017; Rubeena *et al.* 2018; Nidheesh *et al.* 2021; Scaria *et al.* 2022).

Among these materials, biochar (BC), a carbon-centered material obtained from plants such as bamboo, wheat straw (He *et al.* 2021) and animal biomasses like swine bone (Zhou *et al.* 2021) and shrimp shell (Yu *et al.* 2020), has gained importance for making supported catalysts in AOPs. There is a broad interest in the applications of BC in AOPs due to its combined ability as an adsorbent as well as a catalyst (Nidheesh *et al.* 2021; Nakarmi *et al.* 2022). BC has been applied to soil in agriculture to increase its nutrient content, thereby increasing the fertility of soil for enhancing crop yields, as an adsorbent material for removal of organic pollutants and also studied as a potential CO_2 mitigation material (Mohan *et al.* 2014; Ding *et al.* 2016; Lee *et al.* 2017). This is attributed to its physicochemical properties such as a high surface area, porosity, cation exchange capacity and abundant surface functional groups (SFGs) (Ni *et al.* 2020; Tomczyk *et al.* 2020). These groups can be functionalized suitably by altering the pyrolysis temperature and the feedstock source.

There is a growing trend in the applications of BC as a catalyst in AOPs for effective degradation of toxic-pollutants, including endocrine-disrupting chemicals (EDCs) and persistent organic pollutants (POPs). Studies have shown that BC contains persistent free radicals (PFRs), alternatively termed as 'Environmentally Persistent Free Radicals' (EPFRs), which help to enhance free radical generation in AOPs resulting in the degradation of several PPCPs (Nidheesh *et al.* 2021; Kumar *et al.* 2022; Scaria & Nidheesh 2022a).

Recently, BC as a catalyst in combination with AOPs that particularly generate $SO_4^{\bullet -}$ (SR-AOPs) is becoming a popular method for the degradation of PPCPs from wastewater (Guerra-Rodríguez *et al.* 2018; Delgado *et al.* 2020; Ushani *et al.* 2020). Preference for $SO_4^{\bullet -}$ over the conventional $^{\bullet}OH$ is that $SO_4^{\bullet -}$ has a higher oxidation potential (2.5–3.1 V), longer half-life (30–40 μs), can support a broader range of pH (2.0–8.0) and has better non-selectivity for degrading organic contaminants compared with $^{\bullet}OH$ (Nidheesh & Rajan 2016; Ushani *et al.* 2020; Kohantorabi *et al.* 2021; Gujar *et al.* 2022; Scaria & Nidheesh 2022b). This review focuses on summarizing the role of BC as a supported catalyst in SR-AOPs in removal of PPCPs. The catalytic ability of the BC will be stressed and its possible utilization and major challenges in real wastewater treatment for treatment of PPCPs will be discussed.

1.1. PPCPs and their effects on human health and the aquatic ecosystem

PPCPs, due to their increasing applications as medicines, have received much attention for their contamination in aquatic ecosystems because of their persistent nature even at very low concentrations (ng/L–μg/L). PPCPs mainly constitute antibiotics, anti-inflammatories, analgesics, anticonvulsants, antidepressants, lipid regulators, ß-blockers, artificial sweeteners and personal care products such as sunscreens, body lotions, sprays and musks (Wang & Wang 2016). Exposure of PPCPs to humans can occur mainly through ingesting of packaged/canned food, breathing contaminated air and interacting with contaminated soil whereas the effluent streams from the WWTPs are the main exposure to the aquatic ecosystem. There are no concrete experimental studies for the direct toxic effects of exposure of PPCPs to humans (Wilkinson *et al.* 2016). However, in a research work by Hind M. Ewadh and group, exposure of PPCPs (ibuprofen (IBU), ketoprofen) on the cells of human embryonic kidney (HEK 293) were studied. About 90% reduction of cell proliferations in the kidney were observed due to the increased exposures to high concentrations of IBU (750 μg/L) and ketoprofen (75 μg/L), indicating the extensive sensitivity of cells in human organs toward increased chemical dosages (Ewadh *et al.* 2018).

The presence of trace amounts (ng/L) of PPCPs greatly induces toxic effects on ecological systems by harming their psychological and hormonal responses. Joon-Woo Kim and group studied the toxicity effects of pharmaceutical chemicals such as IBU and carbamazepine (CBZ) on freshwater crustaceans (*Thamnocephalus platyurus*) and fish (*Oryzias latipes*) and observed that the 24- and 96-h median lethal concentration (LC_{50}) levels of IBU (19.59 mg/L) and CBZ (81.92 mg/L) affected the central nervous systems, thereby decreasing the overall neuronal activity of the species (Kim *et al.* 2009).

Similarly, antibiotics such as ofloxacin, lincomycin and ciprofloxacin (CIP) are known to be toxic to freshwater algae. Low concentrations (ng/L) of oestrogen 17-alpha-ethinyloestradiol have been shown to cause abnormalities in fish livers and affect sexual characteristics in male fish (Ngqwala & Muchesa 2020). It is reported that the major exposure of PPCPs to the aquatic flora and fauna are through the effluents discharged into the water bodies (runoff, rivers and floods) (Balakrishna *et al.* 2020).

1.2. Sources and fate of PPCP contamination

Figure 1 shows the possible pathways of pharmaceuticals in water bodies. Contamination of PPCPs in water bodies is happening mainly through two types of sources, point and non-point sources. Point sources (also called effluent-based point sources) generally refer to the origin of PPCP intrusions that are STPs, WWTPs, domestic waste, hospital waste and industrial wastes. PPCPs also penetrate the subsurface soil and contaminate the groundwater through landfill leachate. Non-point sources (also called diffused-based sources), which are generally difficult to identify, constitute soil erosion/runoff and unidentified leakage in wastewater systems. PPCPs have also been occurred in sewage. A study by Chang and colleagues collected sewage samples from hospitals, nurseries and slaughter house as well as from the Three Gorges reservoir to detect several antibiotics. Results indicated that the chemical ofloxacin (OFX) showed the highest levels in sewage generated from hospitals, indicating that the residual antibiotics could discharge into the environment and contaminate the water bodies due to inefficient treatment processes in STPs (Chang *et al.* 2010). Even though sewage is found as the primary source of PPCPs in China, WWTPs receiving hospital waste have more effects than STPs (Balakrishna *et al.* 2020).

Current PPCPs removal processes in WWTPs and STPs are adsorption and biodegradation. Biodegradation is not effective to mineralize or breakdown PPCPs completely due to the persistency nature of PPCPs, and its toxicity to microorganisms (Wang & Wang 2016). Moreover, the decreased removal efficiency of chemicals such as diclofenac and carbamazepine in the activated sludge process of WWTPs and STPs was also evident with only 30% of carbamazepine and 10% of diclofenac removal observed in the WWTPs processes (Zhang *et al.* 2008). This signifies the inefficiency of the conventional wastewater treatment systems to degrade PPCPs effectively.

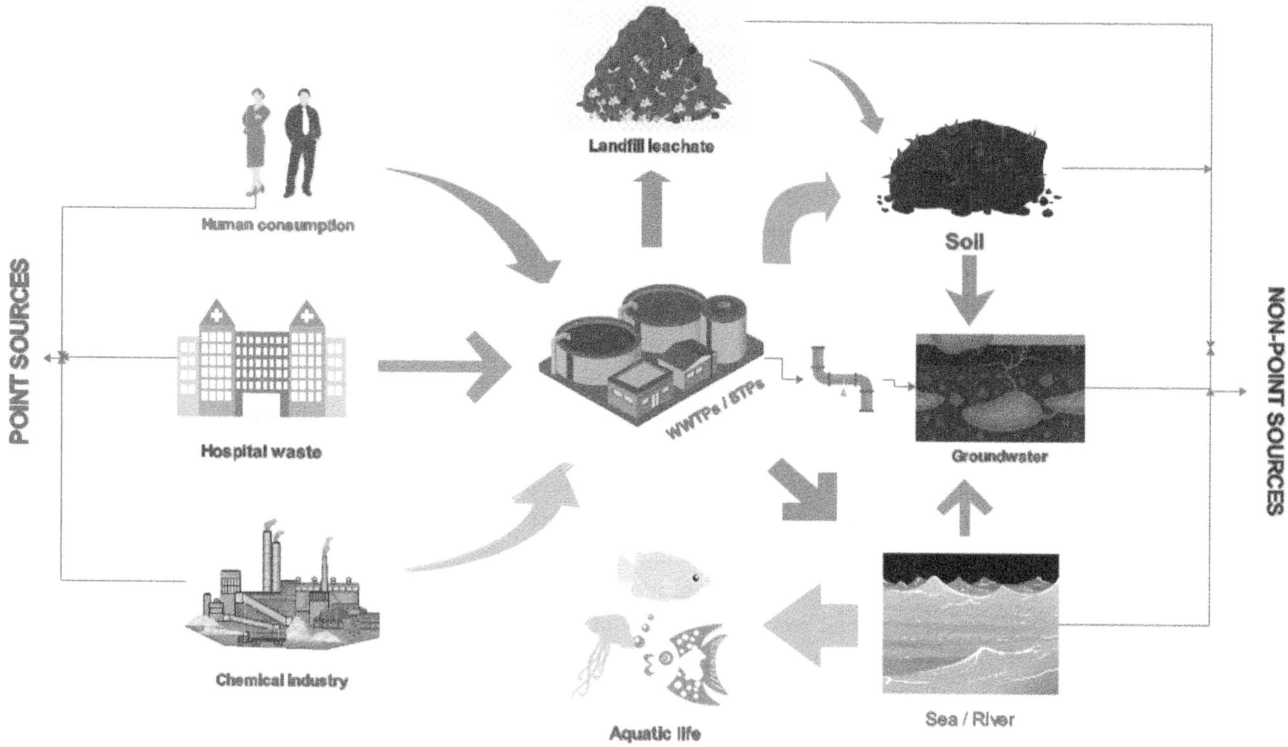

Figure 1 | Contamination route of PPCPs into the aquatic ecosystem.

Furthermore, only a particular category of PPCPs called fluoroquinolones were effectively removed in STPs through biodegradation and adsorption (Lin *et al.* 2010; Lin & Gan 2011; Balakrishna *et al.* 2020). Other categories of antibiotics such as macrolides, sulfonamides, penicillin and imidazole showed no effect in treatment. A total of 71 pharmaceuticals in the districts of Shanghai were also monitored in a study by Liu and group. About 10 PPCPs were detected in very high concentrations, which was attributed to the wastewater treatment plants in the vicinity (Liu *et al.* 2019). Table 1 shows the removal rates of different PPCPs from STPs and WWTPs treatment processes.

The occurrences of PPCPs due to their persistent nature have also been observed in sediments, soil strata, rivers and lakes. PPCPs such as triclosan can adsorb onto sediments and may release back into the water environment (Liu & Wong 2013). Anti-inflammatory drugs such as diclofenac and IBU result in incomplete sorption onto surface soils. Moreover, the incomplete sorption permits its leaching into the groundwater during irrigation or surface spreading of treated wastewater, which increase the risk of groundwater contamination (Lin & Gan 2011).

Studies have mentioned that reclaimed wastewater used for agricultural purposes and biosolids along with wastewater treatment byproducts in agricultural areas are two of the major sources of PPCP contamination in soils. Most commonly detected compounds in sludge and biosolids are carbamazepine, diclofenac and IBU. All these compounds are known to have a lower sorption potential which restricts its adsorption onto organic matter. Concentration levels of antibiotics such as clarithromycin (235 ng/L), erythromycin (320.5 ng/L), metronidazole (1,195.5 ng/L) and SMX (326 ng/L) were predominantly found in the lakes of Jarma, Manzanares, Guadarrama, Henares and Tagus in Spain (Valcárcel *et al.* 2011). Similarly, rivers such as Pearl, Yellow, Hai and Liao in China have detected antibiotic concentrations of norfloxacin (NOR) (5,770 ng/L), ofloxacin (OFX) (1,290 ng/L) and CIP (653 ng/L) (Peng *et al.* 2008, 2009, 2011).

Higher levels of PPCPs have also been found in sludge. A study by Huber and group observed high concentrations of cetrimonium salts (680 μg/L) in the sludge of the Icelandic wastewater treatment plant and concluded that sludge could act as sinks for PPCP accumulation (Huber *et al.* 2016). Moreover, higher concentration levels of norfloxacin (6,490 ng/g) and CIP (6,050 ng/g) were detected in sediment samples collected from an urban drain (Ashfaq *et al.* 2019). Table 2 shows a summary table depicting the concentration levels of some PPCPs in different environments.

Table 1 | Removal rates of some PPCPs in STPs and WWTPs processes

Sr.no	Pollutant	Initial concentration	Removal efficiency (%)	Treatment process	Reference
1	Fluoxetine	50 ng/L	23.1	Grit channels + primary clarifies + conventional activated sludge	Blair *et al.* (2015)
2	Atenolol	255 ng/L	47.1	Grit tanks + primary sedimentation + bioreactor + clarifiers	Roberts *et al.* (2016)
3	Gemfibrozi	190 ng/L	50.8	Grit channels + primary clarifies + conventional activated sludge	Zhang *et al.* (2013)
4	Benzafibrate	50 μg/kg dry weight sludge	30	Anaerobic sludge digestion	Narumiya *et al.* (2013)
5	Trimethoprim	570 ng/L	64.1	Grit channels + primary clarifies + conventional activated sludge	Blair *et al.* (2015)
6	Diclofenac	20–70 mg/L	10–60	Primary treatment + Orbal oxidation ditch + UV	Sun *et al.* (2014)
7	Sulfadiazine	20 ng/L	64.1	Grit channels + primary clarifies + conventional activated sludge	Blair *et al.* (2015)
8	Ofloxacin	10 μg/kg dry weight sludge	45	Anaerobic sludge digestion	Narumiya *et al.* (2013)
9	Sulfamethoxazole (SMX)	7,400 ng/L	35.8	Grit channels + primary clarifies + conventional activated sludge	Blair *et al.* (2015)
10	Propanolol	151 ng/L	49.9	Grit tanks + primary sedimentation + bioreactor + clarifiers	Roberts *et al.* (2016)
11	Carbamazepine	589 ng/L	16.3	Grit tanks + primary sedimentation + bioreactor + clarifiers	Roberts *et al.* (2016)
12	Erythromycin	15 μg/kg dry weight sludge	45	Anaerobic sludge digestion	Narumiya *et al.* (2013)
13	Ampicillin	160 ng/L	1.30	Grit channels + primary clarifies + conventional activated sludge	Blair *et al.* (2015)
14	Azithromycin	1,300 ng/L	47.90	Grit channels + primary clarifies + conventional activated sludge	Blair *et al.* (2015)
15	Triclosan	300 ng/L	55.30	Grit channels + primary clarifies + conventional activated sludge	Blair *et al.* (2015)

PPCPs have also been found in plant and animal tissues, indicating their bioaccumulative properties and their occurrence are said to be influenced greatly by the efficiency of the wastewater treatment plants. Plants absorb the minerals and nutrients from the soil through the uptake process. N4-acetyl-sulfamethoxazole (96 ng/g), sulfamethoxazole (SMX) (24 ng/g) and triclosan (18 ng/g) had been predominantly detected in the crops grown from an agricultural system in China irrigated by reclaimed wastewater (Reyes *et al.* 2021). A global scenario on the concentration of different PPCPs in STPs and WWTPs at different locations has also been listed in Supplementary Table S1.

2. SULFATE RADICAL-BASED ADVANCED OXIDATION PROCESSES (SR-AOPS)

SR-AOPs have been increasingly studied as a potential process for water and wastewater treatment. The keen interest of using sulfate radicals is due to the limitations of hydroxyl radicals such as non-selectiveness which makes it vulnerable toward scavenging by matrix constituents such as common anions and effluent organic matter (EfOM). Kinetic studies on the generation of sulfate radical and hydroxyl radical to degrade organic pollutants depicted that sulfate radical-generated AOPs were 10 times faster in degrading some selected antibiotics (Brienza & Katsoyiannis 2017). $SO_4^{\bullet-}$ is generated in water medium by the homolytic or heterolytic cleavage breaking of the (hydro)peroxide bond (O–O) of chemicals such as persulfate (PS) (also known as peroxydisulfate (PDS)) and peroxymonosulfate (PMS) (Nidheesh *et al.* 2022b). Methods to generate

Table 2 | Concentrations of PPCPs in different environmental matrices

Sr.no	Pollutant	Matrix	Sampling location	Concentrations	Reference
1	Norfloxacin	Surface waters Sediments Sludge	Laizhou Bay, Shandong Baiyangdian Lake Xiamen City, WWTP	7.5–102.7 (ng/L) 49.4–1,140 (ng/g) 20,600 (ng/g)	Zhang *et al.* (2012) Li *et al.* (2012) Wang *et al.* (2018)
2	Ofloxacin	Surface waters Sediments	Victoria Harbor, Hong Kong Chengtaizi Drainage River	8.1–634 (ng/L) 197–3,876 (ng/g)	Minh *et al.* (2009) Gao *et al.* (2012)
3	Ciprofloxacin	Surface waters Sediments Biosolids	Qiantang River, Zhejiang Chengtaizi Drainage River WWTP, USA and Canada	9.3–11 (ng/L) 64.2–1,515 (ng/g) 3,438 (ng/g)	Tong *et al.* (2011) Gao *et al.* (2012) Brown *et al.* (2019)
4	Tetracycline	Surface waters Sediments	Jialingjiang River, Chongqing Jiulongjiang River	<5 (ng/L) 6.9–7,614 (ng/g)	Chang *et al.* (2010) Zhang *et al.* (2011)
5	Doxycycline	Surface waters Sediments	Huangpujiang River, Shanghai Tiaoxi River	ND–46.9 (ng/L) 6–15.6 (ng/g)	Jiang *et al.* (2011) Chen *et al.* (2011)
6	Amoxicillin	Surface waters	Victoria Harbor, Hong Kong	ND–76 (ng/L)	Minh *et al.* (2009)
7	Penicillin V	Surface waters	Qiantang River, Zhejiang	ND–450 (ng/L)	Chen *et al.* (2012)
8	Sulfamethoxazole	Surface waters Sediments	Pearl River, Guangdong Jiulongjiang River	20–350 (ng/L) 1.2–3.4 (ng/g)	Peng *et al.* (2011) Zhang *et al.* (2011)
9	Diclofenac	Surface waters	Pearl River, Guangdong	17.6–150 (ng/L)	Zhao *et al.* (2010b)
10	Naproxen	Surface waters Raw wastewater	Pearl River, Guangdong WWTP, California	20.9–125 (ng/L) 210 (μg/L)	Zhao *et al.* (2010b) Yu *et al.* (2013)
11	Ibuprofen	Surface waters	Qiantang River, Zhejiang	60–70 (ng/L)	Chen *et al.* (2012)
12	Sulfadiazine	Surface waters Sediments	Qin River, Guangxi Dagu Sewage Discharge Channel	0.65–4.8 (ng/L) 622–12,300 (ng/g)	Zheng *et al.* (2012) Hu *et al.* (2012)
13	Erythromycin	Surface waters Sediments	Jialingjiang River, Chongqing Dagu Drainage River	12–23 (ng/L) ND–56.8 (ng/g)	Chang *et al.* (2010) Gao *et al.* (2012)
14	Triclosan	Sediments Biosolids	Zhujiang RiverBEZ WWTP, USA and Canada	12.2–196 (ng/g) 9,322 (ng/g)	Zhao *et al.* (2010a) Brown *et al.* (2019)
15	Carbamazepine	Surface waters	Pearl River, Guangdong	15.6–43.1 (ng/L)	Zhao *et al.* (2010b)

ND, not detected.

sulfate radicals are via energy supplied through sonochemical, photochemical, thermal or by carbonaceous materials (AC, graphene, biochar), and transition metals (iron, copper, etc.) as heterogeneous catalysts. In BC, the PFRs help in the electron transfer process in the PS and PMS system by acting as an electron donor to generate $SO_4^{\bullet -}$ (Equation (1)) and (Equation (2)).

$$S_2O_8^{2-} + e_{BC}^- \rightarrow SO_4^{\bullet -} + SO_4^{2-} \tag{1}$$

$$HSO_5^- + e_{BC}^- \rightarrow SO_4^{\bullet -} + OH^- \tag{2}$$

SR-AOPs have also been applied for groundwater and soil remediation, specifically because sulfate radicals have higher stability and can travel through subsurface gradients in the soil efficiently compared with hydroxyl radicals (Duan *et al.* 2020). SR-AOPs in the presence of nitrite to remove phenolic substrates from groundwater and wastewater have been investigated as a potential *in-situ* chemical oxidation treatment. Thermally operated PS oxidation of phenol in the presence of NO_2^- can lead to the formation of nitrophenol byproducts such as 2-nitrophenol (2-NP), 4-nitrophenol (4-NP), 2,4-dinitrophenol (2,4-DNP) and 2,6-dinitrophenol (2,6-DNP). The degradation of phenols was attributed to the generation of $NO_2^{\bullet}$ from the scavenging of $SO_4^{\bullet -}$ and NO_2^-. It was also concluded from the studies that even though generation of $NO_2^{\bullet}$ results in the degradation of phenols, the formation of nitrophenol byproducts during implementation in SR-AOPs needs control measures as such nitrated aromatic compounds can be carcinogenic, mutagenic and genotoxic to the aquatic organisms (Ji *et al.* 2017).

SR-AOPs in removing pharmaceuticals such as bisphenol-A by thermally activated PS show its credibility to degrade not only refractory organics but also PPCPs from wastewater. The performance of SR-AOPs greatly increases with the addition of a catalyst by enhancing the generation of sulfate radicals. Metal-based catalysts such as metal ions (cobalt, silver and manganese) are known to donate electrons to generate sulfate radicals in PS/PMS systems.

Carbonaceous materials such as graphene, AC and BC are said to contribute to $SO_4^{\bullet-}$ generation via the formation of π-electrons upon interacting with SFGs (–COOH, –OH, C = O) (Nidheesh *et al.* 2021; Kumar *et al.* 2022). The catalytic ability of BC-supported catalyst in SR-AOPs and its functionalization to improve surface properties for the degradation of PPCPs have been discussed in subsequent sections.

3. CATALYTIC ABILITY OF BC

BC has been widely accepted as an adsorbent to majorly transfer pollutants from one medium to another (Ahmad *et al.* 2014; Ghaffar *et al.* 2018). This exceptional behavior of BC is attributed to its morphological structure which comprises aromatic carbon and condensed aromatic structures (Mohan *et al.* 2014; Tomczyk *et al.* 2020). BC as a potential heterogeneous catalyst has been demonstrated in biodiesel production, removing tar in bio-oil and syngas and serving as far as electrode material in fuel cells (Lee *et al.* 2017). Moreover, its participation as a catalyst has gathered wide attention recently because of the presence of abundant oxygen-containing functional groups (OFGs) which comprise of carbonyl (C = O), carboxyl (–COOH) and hydroxyl (–OH) moieties that contribute to the degradation of target pollutants. The presence of OFGs again strongly depends on the pyrolysis temperature, residence time and the feedstock of the material used (Faheem *et al.* 2020). These OFGs, upon photo-based irradiation, generate reactive oxygen species (ROS) ($^{\bullet}OH$, $SO_4^{\bullet-}$, $^1O_2$ and $O_2^{\bullet-}$) through an electron transfer mechanism (Luo *et al.* 2021). This results in the interaction of OFGs on the surface of the BC with the ROS.

From the pyrolysis of biomass, large groups of phenol or quinone moieties are produced from the lignin in biomass for transferring electrons to transition metals, which leads to the formation of PFRs in BC (Wang *et al.* 2019c). The existence of PFRs in BC is the main factor for the degradation of complex species and has attracted interest in exploring its catalytic abilities. PFRs in BC are generated from the incomplete carbonization of organic substances resulting in the formation of 'dangling bonds'. These bonds form complex structures with metal oxides to form stabilized radicals such as semiquinones, cyclopentadienyls and phenoxyls (Luo *et al.* 2021).

PFRs are categorized into two types, namely: oxygen-centered and carbon-centered radicals. Oxygen-centered radicals mainly consist of semiquinolone free radicals and are present up to a pyrolysis temperature of 300 °C. When the temperature further reaches 400 °C, carbon-centered radicals are formed with adjacent oxygen. Carbon-centered radicals usually have better electron-donating capacities and are responsible for radical ($^{\bullet}OH$ and $SO_4^{\bullet-}$) generation in SR-AOPs. Optimum generation of PFRs in BC from coconut shell, eucalyptus leaves and walnut shell were observed at a pyrolysis temperature of 500 °C and result in optimum generation of sulfate radical in AOPs (Zhang *et al.* 2021). The formation of PFRs in BC at different stages of pyrolysis of biomass has been shown in Table 3.

Biomass consists of cellulose, hemicellulose and lignin. The decomposition of cellulose and hemicellulose in the biomass occurs at a pyrolysis temperature above 300 °C, leaving behind lignin residues. Hence, the formation of PFRs in the BC is influenced with the presence of lignin content (Kibet *et al.* 2012; Luo *et al.* 2021; Zhang *et al.* 2022).

In summary, the PFRs help in the generation of $SO_4^{\bullet-}$ by the activation of PS/PDS and PMS in SR-AOPs. The performance of the catalytic ability of the BC is strongly influenced by its pyrolysis temperature and the feedstock material used. The OFGs on the BC surface activates the PS and PMS by generating ROS due to electron transfer between BC surface and PS in the

Table 3 | Different thermal stages of biochar synthesis in relation to $SO_4^{\bullet-}$ activation (Odinga *et al.* 2020)

Sr.no	Pyrolysis temperatures of biochar (°C)	PFR production in biochars	Reasons
1	0–300	Little to no radical production	No alteration in overall structure of biomass with minute mass loss due to loss of water elimination, bond breakage (not preferred for $SO_4^{\bullet-}$ activation)
2	300–500	Oxygenated and carbon-centered radical generation	Conversion of oxygen-centered radicals into carbon-centered radicals (most preferable for $SO_4^{\bullet-}$ activation)
3	500–700	Gradual reduction of PFRs	Production of graphene-like sheet structures

aqueous degradation of the PPCPs. Hence, it is required to maintain adequate pyrolysis temperature for the optimum generation of PFRs in the BC.

4. RADICAL AND NON-RADICAL PATHWAYS OF DEGRADATION

ROS generated from the activation of PS/PDS or PMS result in two pollutant degradation pathways: radical ($^{\bullet}OH$ and $SO_4^{\bullet -}$) and non-radical ($^1O_2$). Degradation of pollutants via non-radical pathway involves singlet oxygen ($^1O_2$), electron transfer and bridging of carbon catalyst for facilitating electron transfer from contaminants to PS/PDS or PMS. Studies have also referred to BC as an 'electron shuttle' which depicts the transfer of electrons as a result, directly affecting the degradation pathway (Klüpfel *et al.* 2014; Faheem *et al.* 2020). Studies on the non-radical pathway carry several merits such as complete utilization of the oxidizing capability of PMS and segregation of organic/inorganic contaminants in the selected matrix (Kohantorabi *et al.* 2021). Zhou *et al.* investigated the catalytic performance of bone-derived BC pyrolyzed at 1,300 °C for degrading acetaminophen (ACT). The presence of the ketone (C = O) group facilitated the electron transfer mechanism by generating singlet oxygen ($^1O_2$) by the non-radical pathway which then acted as active sites for effective degradation of ACT. Second, carboxyl (–COOH) and hydroxyl (–OH) groups proved beneficial for generating $^{\bullet}OH$ and $SO_4^{\bullet -}$ via the radical pathway (Zhou *et al.* 2020, 2021).

In additional studies, a comparison between two heterogeneous Fe-based catalysts (bentonite-combined Fe–Ni and BC-combined Fe composite) for the degradation of sulphapyridine (SPY) and oxytetracycline (OTC) demonstrated that the BC/Fe catalyst initiated singlet oxygen ($^1O_2$) generation, which resulted in increased degradation rate of SPY and OTC (Li *et al.* 2020c). The contribution of radical and non-radical pathways was investigated in a study by Zou *et al.* where metal-free BC catalyst made from corn cob was investigated for removing five PPCPs from pharmaceutical wastewater. The higher contribution for the PPCPs degradation was shown by the radical pathway which was governed by the carbon framework (*sp*^{2} hybridized carbon) of the BC to produce free radicals. In contrast, activation into $^1O_2$ by the ketone group (C = O) was responsible for the non-radical pathway, both of which were achieved at a pyrolysis temperature of 800 °C (Zou *et al.* 2020).

Surprisingly, contradictory results regarding the formation of the oxidized groups (C = O, –COOH, –OH) were also observed in accordance with the activation of BC for sodium persulfate (SPS) treatment through the non-radical pathway (Ntzoufra *et al.* 2021). Fourier transform infrared spectroscopy (FTIR) observations showed the formation of sulfonic groups ($-SO_3H$) on the surface of the BC. This was due to the oxidation property of $S_2O_8^{2-}$. It was also observed that the peak intensities in the FTIR spectrum showed that the presence of –C = O groups gradually increased upon PS oxidation, pointing out the oxidation of the –C–C and –C = O groups.

Additionally, X-ray photoelectron spectroscopy (XPS) analysis of the BCs also showed that higher binding energies were ascribed to the oxidized C groups (–C = O or O–C = O). The amount of non-oxidized C species in the BC was higher before PS oxidation, whereas after treatment, formation of oxidized C species (C = O) was more predominant. This was also further validated by Raman spectroscopy where the peak (at 1,233 cm^{-1}) was more intense with the absence of oxygen-containing groups. In the results reported, this peak which represents *sp*^{3} hybridized ($-CH_3$ group) structures diminished upon PS oxidation indicating the enrichment by oxygen. The results concluded that although BC was a stable catalyst for the oxidation of organic contaminants, the effectiveness of PS treatment for longer periods could result in the oxidation of functional groups on the surface of the BC which is a main deactivating factor for its catalytic ability (Ntzoufra *et al.* 2021).

This phenomenon was also observed in BC for PMS activation. The formation of –C–O and –C = O groups could act as an interference for the active sites in the formation of singlet oxygen ($^1O_2$) by the non-radical pathway (Kohantorabi *et al.* 2021). All in all, it is evident that the radical pathway of degradation in PPCPs is strongly influenced by the electron transfer mechanism of the BC surface. The non-radical mechanism of degradation is dominated by the singlet oxygen ($^1O_2$) and its interaction with the functional groups on the BC surface. On the other hand, the radical pathway results in the generation of $^{\bullet}OH$ and $SO_4^{\bullet -}$. Figure 2 describes a general overview of the functionalization methods of the BC that cater to the formation of radicals in radical and non-radical mechanism in the degradation of PPCPs.

5. ENGINEERED BC FOR PPCP REMOVAL

Functional BCs are becoming an upcoming sustainable material for the remediation of pollutants, particularly for removing antibiotics from wastewater systems through the adsorption mechanism (Muter *et al.* 2019; Krasucka *et al.* 2021).

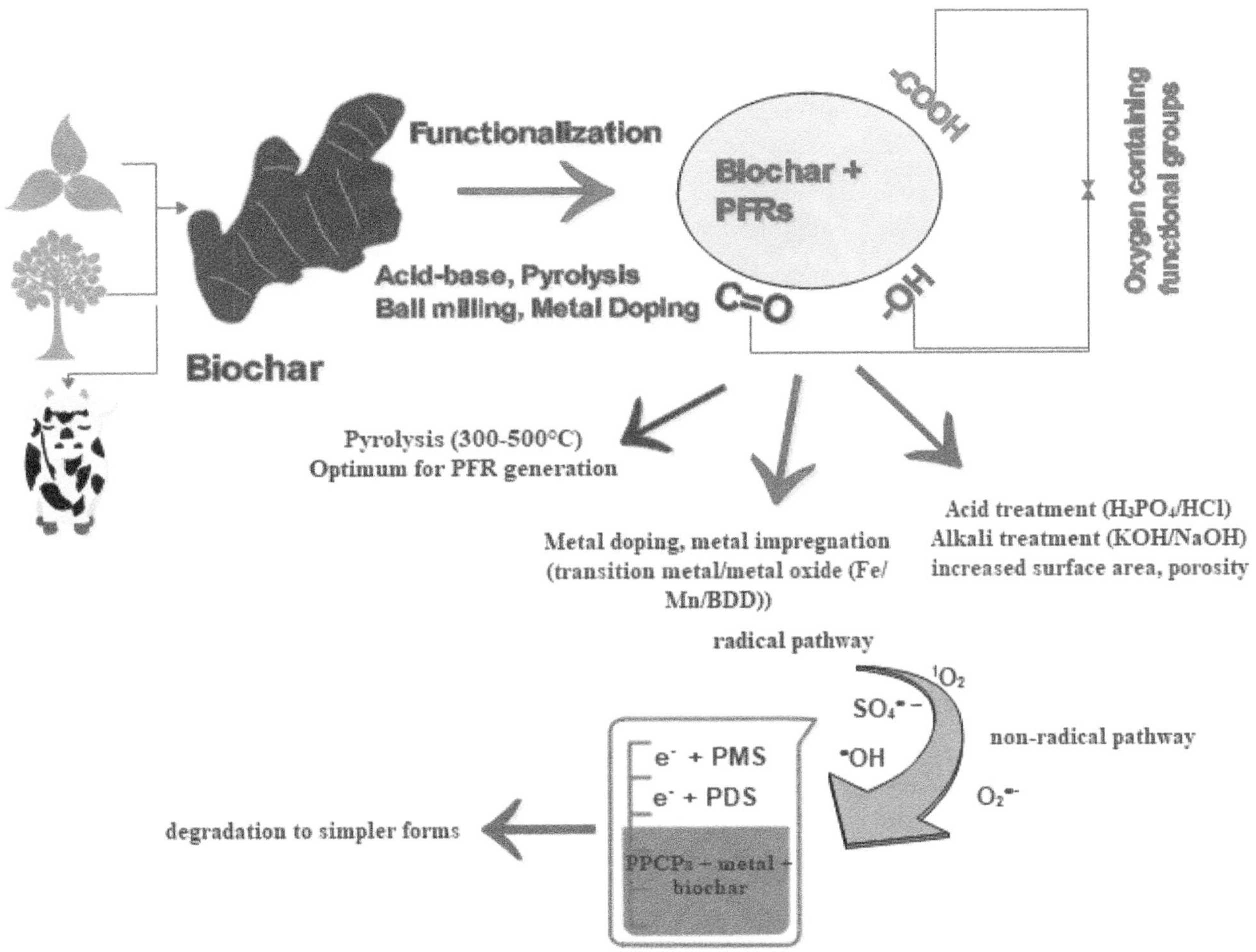

Figure 2 | General overview of functionalization of BC (PFRs generation), influence of oxygen-containing functional groups and persulfate activation process for PPCP degradation.

Engineered BC was initially applied for soil improvement; for example, BC of high porous nature is preferred to enhance the crop yield. Biochar, due to its high organic carbon, nitrogen and ash content can increase the soil organic carbon (SOC), which is a health indicator of soil fertility (Kazemi Shariat Panahi *et al.* 2020). Pristine BC has a lower density and smaller particle size, making it difficult to separate from water and limiting its applications (Tan *et al.* 2016; Wang *et al.* 2017).

Surface properties of BC can be greatly increased by acid-base treatment, metal loading, high temperature treatment and heteroatom doping. There are several examples of increasing surface area of BC by impregnation with acid and alkali. Bone-derived BC treated with H_2SO_4 resulted in a surface area increase up to 83%. This was due to the removal of inorganic compositions such as $CaCO_3$ and hydroxyapatite of the bone BC that resulted in the increased number of active sites (Zhang *et al.* 2022). The treatment with acid affects the carbonaceous phase of the biochar by solubilization of minerals such as hemicellulose and partially cellulose, leaving only the insoluble lignin part of the BC that results in the increase of active sites in the BC. Similarly, BC during pyrolysis treated with basic solutions such as KOH and NaOH is responsible for solubilizing the lignin part in the BC, leaving behind the hemicellulosic mineral in the cellulose residue (Jin *et al.* 2014; Ntaflou & Vakros 2020). Ball milling is another technique which refers to the grinding of the BC samples into a nano-material size, which increases the pore structure and specific surface area (SSA) of the BC (Cheng *et al.* 2021). BC derived from ball milling at a pyrolysis temperature of 525 ± 1 °C was converted into nanobiochar and investigated for removing CBZ from water. About 95% of CBZ was removed in the process (Naghdi *et al.* 2017). A pie chart on the different functionalization methods of BC has been shown in Figure 3.

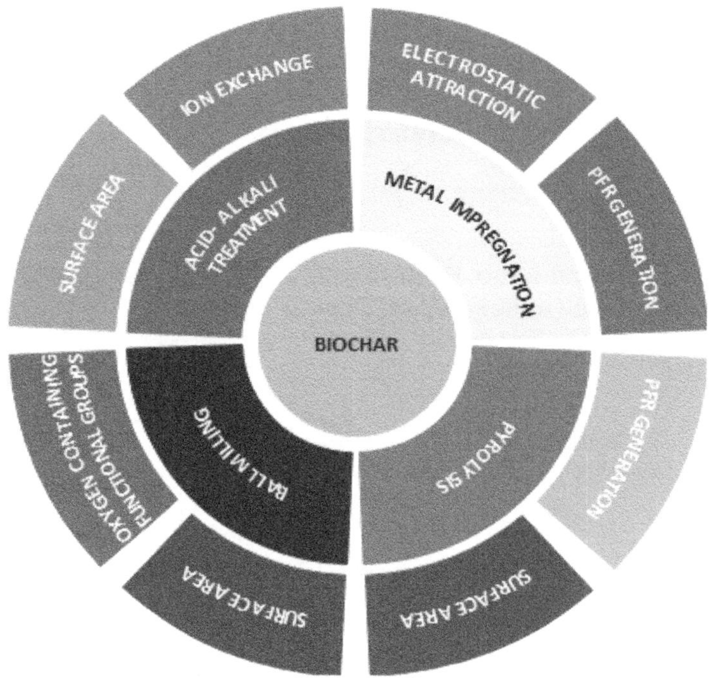

Figure 3 | Functionalization methods for biochar.

Engineered BCs have been investigated for removing pharmaceuticals such as diclofenac, naxoprofen and triclosan from water medium. Higher adsorption rates were achieved for the selected PPCPs on the engineered BC. Low pyrolysis temperature favored increased adsorption of PPCPs due to the pore filling mechanism and hydrophobicity of the chemicals (Czech *et al.* 2021).

Another study on woodchip-derived BC pyrolyzed at a temperature of 725 °C for adsorbing PPCPs such as xylazine, metoprolol, azithromycin, ketoprofen, diclofenac, IBU, sulfamethoxazole (SMX) and simvastatin indicated that the adsorption followed by incubation could provide better results for some selected pharmaceuticals. However, this was majorly dependent on the sorption capability of the compounds (Muter *et al.* 2019). Modification of BC through addition of a methanol solvent can alter the SFGs or oxygen-containing groups of the BC to adsorb tetracycline (TCE) (Jing *et al.* 2014). Engineered BCs are typically used as a tool to enhance the adsorption capabilities which are governed by the selected contaminant and the production method. Results suggest that BC-based sorbents, pristine BC produced at different conditions, modified BCs and BC composites are effective in the removal of certain antibiotics (Krasucka *et al.* 2021). The effectiveness depends on the physicochemical properties of both antibiotic and adsorbent. The binding of antibiotic onto the BC surface is dependent on the surface functional group, pH and the carbon content of the BC as well as the sorption parameters of the target pollutants (Log K_{OW}). The phenomena of pore filling mechanisms have also been stressed (Krasucka *et al.* 2021).

Engineered BC have been well established as adsorbents that increase the removal rate of many PPCPs. The modification of BC can alter its morphological structure, thereby producing functional groups on the BC surface. Different modification methods increase the micropores of the BC which in turn increases the active sites of the BC for better interaction toward target pollutants. Hence, proper modification of the BC surface can result in better generation of PFRs in the BC which can enhance the radical generation in SR-AOPs.

6. APPLICATIONS OF BC CATALYST IN SR-AOPS

6.1. Synthetic water applications

SR-AOPs by PS activation have wide applications in the degradation of antibiotics from synthetic wastewater such as tetracycline (TCE) (Zhong *et al.* 2020), CIP (Gao *et al.* 2020), SMX (Avramiotis *et al.* 2021) and cephalexin (CEX) (Song *et al.* 2021). All the studies included the identification of reaction pathways and mechanisms through quenching experiments and electron paramagnetic resonance (EPR) analysis. Zhong and group studied the removal of TCE using BC co-doped with metals. The degradation of TCE by the nitrogen and copper co-doped BC (N–Cu/BC) succeeded even after reusing with efficiencies reported up to 80 and 65% on the 2nd and 3rd times indicating the increased catalytic ability of the metal-doped BC. EPR studies also confirmed the generation of $SO_4^{\bullet-}$ and $^{\bullet}OH$. The $^{\bullet}OH$ played a crucial role in the degradation of pollutants. Moreover, the non-radical pathway was attributed to the electron transfer between TCE and PS at the surface of the BC catalyst (Zhong *et al.* 2020).

SR-AOPs to generate sulfate radicals have also been increasingly applied to treat synthetic groundwater matrices. Selectivity of PS/PDS is attributed to its low price, high stability and easier transport in the subsurface environments compared with PMS (Duan *et al.* 2020). BC derived from maize stock and cob pyrolyzed at different temperatures (300 and 600 °C) were examined for the degradation of TCE from groundwater, superoxide and singlet oxygen along with ketone (C = O) groups were the main ROS and OFGs influencing the degradation rate of TCE. However, it was also noted that the presence of other anions in the groundwater matrix (chloride (Cl^-), nitrate (NO_3^-)) acted as scavengers in restricting the generation of sulfate radical thereby reducing the degradation rate of TCE. Inhibiting of ROS generation ability was also observed due to the reaction between Fe species with phosphate anions (Li *et al.* 2020b).

BC/Fe-activated PS treatment was applied by (Wang *et al.* 2019b) for the degradation of SMX. A BC-induced Fe(III) reduction system efficiently degraded SMX due to the presence of PFRs in BC which facilitated electron transfer between the BC surface and Fe(III). Fe(II) was generated for activation of PS. Tert-butanol was responsible to inhibit the degradation of SMX by hindering the electron transfer between BC and Fe(III). Iron and its oxide are widely used for their nontoxicity, low cost and eco-friendly nature (Wang *et al.* 2019a, 2019b, 2019c). Pristine BC derived from coconut shell has also been investigated for the removal of SMX from an aqueous solution. About 85% removal was achieved which was attributed to the generation of ROS ($SO_4^{\bullet-}$, $^{\bullet}OH$) and $^1O_2$ by PMS activation through the non-radical pathway (Hung *et al.* 2022). Bo Wang and group did a combined study of norfloxacin (NOR) removal in deionized water using corn stalk biochar (CBC) at a pyrolyzed at 500 °C. The combination of CBC, PS and CBC/PS systems were applied for the degradation of NOR. The rate of NOR increased with the increase of CBC dosage. This was effective in catalyzing PS consequently leading to increased generation of $SO_4^{\bullet-}$ ultimately leading to the acceleration of the reaction (Wang *et al.* 2019a, 2019b, 2019c). Different pharmaceuticals and their removal using BC as a catalyst in SR-AOPs are listed in Table 4.

6.2. Real water applications

SR-AOPs by activation of PS/PDS have been focused due to the stability of $SO_4^{\bullet-}$ at ambient conditions followed by strong water solubility of PS. PMS requires higher energy to produce radicals due to a larger dissociation energy (377 kJ mol^{-1}) to break the peroxide O–O bond length (1.437 Å) compared with PDS (1.497 Å and 97 kJ mol^{-1}) (Flanagan *et al.* 1984; Nidheesh *et al.* 2022b). Degradation of SMX and its human metabolite N4-acetyl-sulfamethoxazole (NSMX) in urine by PDS combined with BC derived from cotton straw biomass under oxygen-free conditions at 350 °C was studied (Zhang *et al.* 2020). The results concluded that the BC-aided technology sufficiently degraded SMX and NSMX in the urine matrix with minor interventions due to formations of HCO_3^- and NH_4^+. Synergistic effects of $^{\bullet}OH$ generation and singlet oxygen species were the dominant mechanisms in the degradation (Zhang *et al.* 2020).

In the study for CIP degradation, sulfide and nanoscale zero-valent iron-doped BC (S-nZVI/BC) observed a similar influence of $SO_4^{\bullet-}$ and $^{\bullet}OH$ as the primary ROS for the process (Gao *et al.* 2020). However, in the case of SMX removal by rice husk-derived BC, a comparison between spiked samples with bicarbonates and real wastewater indicated that ionic species (Cl^-, Br^-) in real wastewater inhibit the performance of the BC catalyst (41% against 51% in 90 min) (Avramiotis *et al.* 2021). Li and group conducted a similar study by applying a composite of BC-supported by copper oxide (CuO) for removal of pharmaceuticals such as ciprofloxacin and atrazine from highly saline wastewater by PMS activation. It was observed that ciprofloxacin was completely removed (100%) followed by atrazine (78.27%). A comparison between synthetic salt water system and actual saline wastewater showed no significant changes in the removal rates of the antibiotics. The non-radical pathway was the dominant pollutant degradation mechanism. The influence of OFGs was also responsible for the generation of $^1O_2$ (Li *et al.* 2020a).

Phenolic byproducts are also a result of industrial effluent discharge of pharmaceuticals and other detrimental sources. BC modified by lanthanum chloride effectively degraded phenol by PS activation with efficiencies reaching about 97%. Phenol concentration, pH, catalyst dosage as well as PS addition governed the activation of PS. It was observed that the optimum removal was achieved at a pH of 4.0. This was attributed to the surface charge of the BC catalyst as the +ve charge of BC

Table 4 | Treatment of PPCPs in water by BC catalyst in SR-AOPs

Sr.no	Pollutant	Feedstock	Catalyst	Operating optimum condition	Removal efficiency (%)	Reference
1	Tetracycline	Maize stock and cob	Maize stock and cob-derived biochar	1 g/L biochar catalyst and 5 mM PMS concentration	100	Li *et al.* (2020b)
		Glucose and urea	Nitrogen and copper co-doped biochar (N–Cu/Biochar)	200 mg/L biochar catalyst and 0.5 mM PS concentration	100	Zhong *et al.* (2020)
		Soybean residue	Ball milling-assisted KOH-derived biochar (MKBC)	0.2 g/L catalyst dosage and 1 mM PDS concentration	93.1	Li *et al.* (2021)
2	Tetracycline hydrochloride	Filament biomass of Ramie (*Boehmeria nivea* (L.) Gaud)	Nitrogen-doped biochar from Ramie biomass (NRBF)	100 mg/L biochar catalyst dosage and 1 mM PDS concentration	92.05	Ye *et al.* (2021)
3	Sulfamethoxazole	Coconut shell	Metal-free pristine coconut shell-derived biochar (CSBC)	150 mg/L biochar catalyst and 0.05 mM PMS concentration	85	Hung *et al.* (2022)
		Rice husk	Rice husk biochar	500 mg/L biochar catalyst and 500 mg/L PS concentration	>92	Avramiotis *et al.* (2021)
		Spent coffee grounds	BC derived from spent coffee grounds	50–200 mg/L biochar catalyst and 100–1,000 mg/L PS concentration	100	Lykoudi *et al.* (2020)
		Red mud-sewage sludge	BC derived from red mud-sewage sludge (RSDBC)	0.5–1.0 g/L catalyst dosage and 0.2 mM PMS concentration	95.7	Wang *et al.* (2020)
4	Benzene	Watermelon rind	BC-supported iron sulfide composite (Fe_xS_y@biochar)	1 g/L Fe_xS_y@biochar and 3 mM PS concentration	99.8	Zhu *et al.* (2022)
5	Norfloxacin (NOR)	Corn stalk	Corn stalk-derived biochar (CBC)	3.2 g/L catalyst dosage at PS/NOR molar ratio of 240:1	99.57	Wang *et al.* (2019a, 2019b, 2019c)
6	Cephalexin (CFX)	Loofah	Magnetic loofah biochar (Fe_2O_3@LBC)	0.4 g/L catalyst dosage and 0.1 g/L PS concentration	73.9	Song *et al.* (2021)
7	2,4-Dichlorophenol	Egg shell	ES-biochar	0.167 g/L catalyst dosage and up to 2 g/L PS concentration	>90	Liu *et al.* (2020)
8	Triclosan	Sewage sludge	Sludge-derived biochar	1.0 g/L catalyst dosage and 0.8 mM PMS concentration	98.9	Wang & Wang (2019)
9	*p*-Nitrophenol	Sunflower stalk	Sunflower stalk-derived biochar	10 g/L catalyst dosage and 10 mM PS concentration	83.83	Sun *et al.* (2019)
10	Bisphenol-A	Deciduous leaves of *Platanus orientalis*	nanoFe_3O_4-biochar	2.0 g/L catalyst dosage and 5 mM PMS concentration	100	Cui *et al.* (2021)
11	1,4-Dioxane	Pine needle	Pine needle-derived biochar	1.0 g/L catalyst dosage and 0.8 mM PMS concentration	85.2	Ouyang *et al.* (2019)
12	Aniline	Rice straw	Rice straw-derived biochar (RSBC)	0.6 g/L catalyst dosage and 90 mg/L PS concentration	94	Wu *et al.* (2018)

surface attracts the −ve charge phenolate anion leading to subsequent removal of phenol (Razmi *et al.* 2019). Bai *et al.* (2022) studied the effect of chitosan-based BC for PS activation to degrade phenol from saline wastewater. The non-radical pathway using $^1O_2$ was the predominant pollutant degradation mechanism followed by radicals ($SO_4^{\bullet-}$ and $^{\bullet}OH$). The presence of NaCl added to produce a saline media had an inhibitory effect on the phenol degradation and increased with additional dosage (Bai *et al.* 2022). Bo Wang and group also studied the effect of aquatic systems on the degradation of norfloxacin (NOR) by PS activation using BC obtained from corn stalk. Groundwater and surface water samples from Liaoning province and Hunhe river showed that PS activation by BC as catalyst was negligibly affected and catered successful degradation of NOR. The pH conditions of the aquatic sources indicated that a slightly acidic medium favored the degradation of NOR due to the formation of $SO_4^{\bullet-}$ and $^{\bullet}OH$ but also depended on the PS dosage (Wang *et al.* 2019a, 2019b, 2019c).

In a similar study, the applicability of egg shell-derived biochar at a pyrolysis temperature of 600 °C (ES-biochar) by PS activation for removal of 2,4-dichlorophenol in real water sample of Taozi Lake in China was investigated (Liu *et al.* 2020). Up to 88% efficiency was achieved for about 2 h, although the removal rate was slower compared with the results obtained by aqueous solutions. This was attributed to the presence of other organic compounds in the lake water.

Overall, the superiority of the BC-supported catalyst has been well established as far as controlled environments are concerned. The effect of BC as a catalyst material shows positive results in completely degrading PPCPs from the aqueous medium. The above studies show that the amount of catalyst as well as PS and PMS dosage is largely dependent on the target pollutant to be removed. However, the suitability of BC-supported catalyst in SR-AOPs in real applications are fairly limited. One reason is that SR-AOPs are associated with high generation of sulfate ion which can breach the regulation limits (500 mg/L). Moreover, adequate analytical techniques such as EPR, free radical quenching experiments and density functional theory (DFT) calculations are needed to identify the radicals generated and the degradation pathways of pollutant degradation (Duan *et al.* 2020). Acidic pH environments are shown to be favorable in most studies as the acidic nature helps in the quicker generation of $SO_4^{\bullet-}$ in the system.

7. ADVANTAGES OF BC CATALYST OVER METAL-BASED CATALYST IN SR-AOPS

Although metal-based catalysts are well known for activation of PDS and PMS in SR-AOPs, there are certain limitations which can pose a challenge and can be avoided by using BC-based catalysts:

(1) Application of transition metals or their oxides are generally reduced by excessive leaching of metal ions which can increase the overall pH of the aqueous medium (Kohantorabi *et al.* 2021). BC-based catalyst are found much effective for the degradation pollutants with lesser metal leaching (Delgado *et al.* 2020; Zhao *et al.* 2021).
(2) The metal ions can form secondary toxic intermediates upon reacting with existing ions in the contaminated water and pose a challenge for the aquatic health. BC-based catalysts are said to reduce the agglomeration of transition metals by ensuring their uniform distribution and can accelerate the transfer of electrons due to the presence of free radical species in the BC matrix (Yu *et al.* 2019; Wang *et al.* 2021).
(3) The synergistic advantages of modified biochar both as an adsorbent and a catalyst material can be accessed *in-situ* for the removal of antibiotics from water medium. A study reported the dual applications of corn stalk biochar pyrolyzed at 700 °C to remove TCE from water medium. Adsorption of TCE was credited to the increased surface area of the BC whereas the reduction of Cu(I) on the BC resulting in generation of $SO_4^{\bullet-}$ and $^{\bullet}OH$ by using the PS activation method was responsible for the increased catalytic ability of the BC for degrading TCE (Chen *et al.* 2020).

8. CHALLENGES IN THE TREATMENT

Although there are many studies conducted on BC as a catalyst in SR-AOPs, the applications are limited to laboratory-scale analysis. The implementation of BC as a catalyst in SR-AOPs for real systems/scale has potential challenges and their possible mitigation can be summarized in the following points:

(1) Adsorptive abilities of BC are more pronounced for its application in removal of organic pollutants due to its large surface area, SFGs and porosity. Applying BC catalyst for SR-AOPs can initiate dominance of adsorption over its catalytic ability for activation of radicals. Moreover, BC produced at high pyrolysis temperature reduces the presence of PFRs required for activation of PS but favors better surface properties of the BC to adsorb pollutants. On the contrary, BC produced from low pyrolysis temperatures facilitate high amount of PFRs but caters to reduced surface properties for adsorption. This

can be somewhat minimized by producing BC at adequate temperatures (between 300 and 500 °C) to enhance the combined abilities of adsorption as well as a catalyst.

(2) Another challenge can be the subsequent poisoning or deactivation of the surface of the BC catalysts due to the adsorption of intermediates or byproducts during the degradation process. This can potentially reduce the efficiency of BC due to the accumulation of toxic intermediates into the active sites of the BC surface and can create a challenge for producing tailored BC (adequate surface properties as well as sufficient PFR generation) for producing $SO_4^{\bullet -}$.

(3) Implementation of SR-AOPs in real wastewater containing ions such as chloride (Cl^-) and bromide (Br^-) are said to form unwanted toxic intermediates like chlorate (ClO_3^-) and bromate (BrO_3^-) by the generation of $SO_4^{\bullet -}$. This can be prevented by adjusting the pH or chemical addition to suppress the formation of XO_3^-. Lower pH (below 5) promoted $Cl^{\bullet}$ to form ClO_3^- whereas pH > 5 resulted in the formation of $^{\bullet}OH$ due to the reaction between $Cl^{\bullet}$ and water (Faheem *et al.* 2020).

9. CONCLUSIONS AND FUTURE SCOPE

Here, a summarization on BC-supported catalysts and their functionalization products for their increased degradation efficiencies of PPCPs in SR-AOPs has been made. The presence of PFRs in the BC helps in the generation of radical species in SR-AOPs which result in better degradation of pollutants when BC is used as a catalyst. Pyrolysis temperature and feedstock material play a vital role in the efficiency of the BC to be used as a catalyst.

BC-supported catalyst can degrade PPCPs in SR-AOPs through radical ($^{\bullet}OH$, $SO_4^{\bullet -}$) and non-radical ($^1O_2$ and electron transfer) pathways which is influenced by the surface properties such as the SFGs on the surface of BC. For increasing the catalytic ability of the BC by the non-radical pathway, formation of oxidized carbon groups can pose a problem for activating PS and PMS in SR-AOP treatments as they can alter the active sites of the BC surface and deactivate the BC in the long run which can pose a challenge for the *in-situ* applications of the BC-supported catalyst. Suggestions to reduce the negative influence of oxidized groups on the BC surface by the non-radical pathway of degradation in SR-AOPs may require further studies.

Overall, BC-supported catalyst in SR-AOPs is a promising technology for removing PPCPs from water bodies. Feasibility studies on field implementations require a cost-benefit analysis and the long-term efficiency of BC-supported catalyst needs further evaluation.

ACKNOWLEDGEMENTS

The authors are thankful to the Director, CSIR-NEERI, Nagpur, India for encouraging and kind permission for publishing this article.

DATA AVAILABILITY STATEMENT

All relevant data are included in the paper or its Supplementary Information.

CONFLICT OF INTEREST

The authors declare there is no conflict.

REFERENCES

Ahmad, M., Rajapaksha, A. U., Lim, J. E., Zhang, M., Bolan, N., Mohan, D., Vithanage, M., Lee, S. S. & Ok, Y. S. 2014 Biochar as a sorbent for contaminant management in soil and water: a review. *Chemosphere* **99**, 19–33.

Anjali, R. & Shanthakumar, S. 2019 Insights on the current status of occurrence and removal of antibiotics in wastewater by advanced oxidation processes. *Journal of Environmental Management* **246**, 51–62.

Ashfaq, M., Li, Y., Rehman, M. S. U., Zubair, M., Mustafa, G., Nazar, M. F., Yu, C. P. & Sun, Q. 2019 Occurrence, spatial variation and risk assessment of pharmaceuticals and personal care products in urban wastewater, canal surface water, and their sediments: a case study of Lahore, Pakistan. *Science of the Total Environment* **688**, 653–663.

Avramiotis, E., Frontistis, Z., Manariotis, I. D., Vakros, J. & Mantzavinos, D. 2021 Oxidation of sulfamethoxazole by rice husk biochar-activated persulfate. *Catalysts* **11** (7), 850.

Babu, D. S., Srivastava, V., Nidheesh, P. V. V. & Kumar, M. S. 2019 Detoxification of water and wastewater by advanced oxidation processes. *Science of The Total Environment* **696**, 133961.

Bai, M., Niu, D., Xia, L., Yin, Y. & Wan, J. 2022 Efficient degradation of phenol with high salinity wastewater by catalytic persulfate activation using chitosan biochar. *Reaction Kinetics, Mechanisms and Catalysis* **135** (1), 425–439.

Balakrishna, K., Rath, A., Praveenkumarreddy, Y. & Siri, K. 2020 A review of the occurrence of pharmaceuticals and personal care products in Indian water bodies. *Ecotoxicology and Environmental Safety* **137**, 113–120.

Blair, B., Nikolaus, A., Hedman, C., Klaper, R. & Grundl, T. 2015 Evaluating the degradation, sorption, and negative mass balances of pharmaceuticals and personal care products during wastewater treatment. *Chemosphere* **134**, 395–401.

Brienza, M. & Katsoyiannis, I. 2017 Sulfate radical technologies as tertiary treatment for the removal of emerging contaminants from wastewater. *Sustainability* **9** (9), 1604.

Brown, S., Kennedy, L., Cullington, M., Mihle, A. & Lono-Batura, M. 2019 Relating pharmaceuticals and personal care products in biosolids to home exposure. *Urban Agriculture & Regional Food Systems* **4** (1), 1–14.

Chang, X., Meyer, M. T., Liu, X., Zhao, Q., Chen, H., Chen, J. A., Qiu, Z., Yang, L., Cao, J. & Shu, W. 2010 Determination of antibiotics in sewage from hospitals, nursery and slaughter house, wastewater treatment plant and source water in Chongqing region of Three Gorge Reservoir in China. *Environmental Pollution* **158** (5), 1444–1450.

Chen, Y. S., Zhang, H. B., Luo, Y. M., Hu, G. J., Zhao, Y. G. & Song, J. 2011 Investigation of 14 selected antibiotics in sediments of the typical cross sections of Tiaoxi River. *Huan Jing ke Xue=Huanjing Kexue* **32** (3), 667–672.

Chen, H., Li, X. & Zhu, S. 2012 Occurrence and distribution of selected pharmaceuticals and personal care products in aquatic environments: a comparative study of regions in China with different urbanization levels. *Environmental Science and Pollution Research* **19** (6), 2381–2389.

Chen, J., Yu, X., Li, C., Tang, X. & Sun, Y. 2020 Removal of tetracycline via the synergistic effect of biochar adsorption and enhanced activation of persulfate. *Chemical Engineering Journal* **382**, 122916.

Cheng, N., Wang, B., Wu, P., Lee, X., Xing, Y., Chen, M. & Gao, B. 2021 Adsorption of emerging contaminants from water and wastewater by modified biochar: a review. *Environmental Pollution* **273**, 116448.

Cui, X., Zhang, S.-S., Geng, Y., Zhen, J., Zhan, J., Cao, C. & Ni, S.-Q. 2021 Synergistic catalysis by Fe_3O_4-biochar/peroxymonosulfate system for the removal of bisphenol a. *Separation and Purification Technology* **276**, 119351.

Czech, B., Kończak, M., Rakowska, M. & Oleszczuk, P. 2021 Engineered biochars from organic wastes for the adsorption of diclofenac, naproxen and triclosan from water systems. *Journal of Cleaner Production* **288**, 125686.

Delgado, F., Gutierrez, V. S., Dennehy, M. & Alvarez, M. 2020 Stable and efficient metal-biochar supported catalyst: degradation of model pollutants through sulfate radical-based advanced oxidation processes. *Biochar* **2** (3), 319–328.

Dey, S., Bano, F. & Malik, A. 2019 *Pharmaceuticals and Personal Care Product (PPCP) Contamination – A Global Discharge Inventory.* Elsevier Inc., Delhi, India.

Ding, Y., Liu, Y., Liu, S., Li, Z., Tan, X., Huang, X., Zeng, G., Zhou, L. & Zheng, B. 2016 Biochar to improve soil fertility: a Review. *Agronomy for Sustainable Development* **36** (2), 36.

Duan, X., Yang, S., Wacławek, S., Fang, G., Xiao, R. & Dionysiou, D. D. 2020 Limitations and prospects of sulfate-radical based advanced oxidation processes. *Journal of Environmental Chemical Engineering* **8** (4), 103849.

Evgenidou, E. N., Konstantinou, I. K. & Lambropoulou, D. A. 2015 Occurrence and removal of transformation products of PPCPs and illicit drugs in wastewaters: a review. *Science of The Total Environment* **505**, 905–926.

Ewadh, H. M., Abdullah, S. R. S., Hasan, H. A., Anwar, N. & Ewadh, M. J. 2018 The effect of pharmaceuticals and personal care products (PPCPs) on human embryonic kidney (HEK 293) cells proliferation. *Journal of Pharmaceutical Sciences and Research* **10** (12), 3228–3232.

Faheem, M., Du, J., Kim, S. H., Hassan, M. A., Irshad, S. & Bao, J. 2020 Application of biochar in advanced oxidation processes: supportive, adsorptive, and catalytic role. *Environmental Science and Pollution Research* **27** (30), 37286–37312.

Flanagan, J., Griffith, W. P. & Skapski, A. C. 1984 The active principle of Caro's acid, HSO_5^-: X-ray crystal structure of $KHSO_5{\cdot}H_2O$. *Journal of the Chemical Society, Chemical Communications* (23), 1574–1575.

Gao, L., Shi, Y., Li, W., Liu, J. & Cai, Y. 2012 Occurrence, distribution and bioaccumulation of antibiotics in the Haihe River in China. *Journal of Environmental Monitoring* **14** (4), 1247–1254.

Gao, J., Han, D., Xu, Y., Liu, Y. & Shang, J. 2020 Persulfate activation by sulfide-modified nanoscale iron supported by biochar (S-nZVI/BC) for degradation of ciprofloxacin. *Separation and Purification Technology* **235**, 116202.

Ghaffar, A., Zhu, X. & Chen, B. 2018 Biochar composite membrane for high performance pollutant management: fabrication, structural characteristics and synergistic mechanisms. *Environmental Pollution* **233**, 1013–1023.

Gogoi, A., Mazumder, P., Tyagi, V. K., Tushara Chaminda, G. G., An, A. K. & Kumar, M. 2018 Occurrence and fate of emerging contaminants in water environment: a review. *Groundwater for Sustainable Development* **6**, 169–180.

Guerra-Rodríguez, S., Rodríguez, E., Singh, D. & Rodríguez-Chueca, J. 2018 Assessment of sulfate radical-based advanced oxidation processes for water and wastewater treatment: a review. *Water* **10** (12), 1828.

Gujar, S. K., Divyapriya, G., Gogate, P. R. & Nidheesh, P. V. 2022 Environmental applications of ultrasound activated persulfate/peroxymonosulfate oxidation process in combination with other activating agents. *Critical Reviews in Environmental Science and Technology* **53** (6), 1–23.

He, X., Zheng, N., Hu, R., Hu, Z. & Yu, J. C. 2021 Hydrothermal and pyrolytic conversion of biomasses into catalysts for advanced oxidation treatments. *Advanced Functional Materials* **31** (7), 2006505.

Hu, X., He, K. & Zhou, Q. 2012 Occurrence, accumulation, attenuation and priority of typical antibiotics in sediments based on long-term field and modeling studies. *Journal of Hazardous Materials* **225**, 91–98.

Huber, S., Remberger, M., Kaj, L., Schlabach, M., Jörundsdóttir, H. Ó., Vester, J., Arnórsson, M., Mortensen, I., Schwartson, R. & Dam, M. 2016 A first screening and risk assessment of pharmaceuticals and additives in personal care products in waste water, sludge, recipient water and sediment from Faroe Islands, Iceland and Greenland. *Science of the Total Environment* **562**, 13–25.

Hung, C.-M., Chen, C.-W., Huang, C.-P., Shiung Lam, S. & Dong, C.-D. 2022 Peroxymonosulfate activation by a metal-free biochar for sulfonamide antibiotic removal in water and associated bacterial community composition. *Bioresource Technology* **343**, 126082.

Ji, Y., Wang, L., Jiang, M., Lu, J., Ferronato, C. & Chovelon, J.-M. 2017 The role of nitrite in sulfate radical-based degradation of phenolic compounds: an unexpected nitration process relevant to groundwater remediation by in-situ chemical oxidation (ISCO). *Water Research* **123**, 249–257.

Jiang, L., Hu, X., Yin, D., Zhang, H. & Yu, Z. 2011 Occurrence, distribution and seasonal variation of antibiotics in the Huangpu River, Shanghai, China. *Chemosphere* **82** (6), 822–828.

Jin, H., Capareda, S., Chang, Z., Gao, J., Xu, Y. & Zhang, J. 2014 Biochar pyrolytically produced from municipal solid wastes for aqueous As(V) removal: adsorption property and its improvement with KOH activation. *Bioresource Technology* **169**, 622–629.

Jing, X. R., Wang, Y. Y., Liu, W. J., Wang, Y. K. & Jiang, H. 2014 Enhanced adsorption performance of tetracycline in aqueous solutions by methanol-modified biochar. *Chemical Engineering Journal* **248**, 168–174.

Kanakaraju, D., Glass, B. D. & Oelgemöller, M. 2018 Advanced oxidation process-mediated removal of pharmaceuticals from water: a review. *Journal of Environmental Management* **219**, 189–207.

Kazemi Shariat Panahi, H., Dehhaghi, M., Ok, Y. S., Nizami, A. S., Khoshnevisan, B., Mussatto, S. I., Aghbashlo, M., Tabatabaei, M. & Lam, S. S. 2020 A comprehensive review of engineered biochar: production, characteristics, and environmental applications. *Journal of Cleaner Production* **270**, 122462.

Kibet, J., Khachatryan, L. & Dellinger, B. 2012 Molecular products and radicals from pyrolysis of lignin. *Environmental Science & Technology* **46** (23), 12994–13001.

Kim, J.-W., Ishibashi, H., Yamauchi, R., Ichikawa, N., Takao, Y., Hirano, M., Koga, M. & Arizono, K. 2009 Acute toxicity of pharmaceutical and personal care products on freshwater crustacean (*Thamnocephalus platyurus*) and fish (*Oryzias latipes*). *The Journal of Toxicological Sciences* **34** (2), 227–232.

Klüpfel, L., Keiluweit, M., Kleber, M. & Sander, M. 2014 Redox properties of plant biomass-derived black carbon (biochar). *Environmental Science & Technology* **48** (10), 5601–5611.

Kohantorabi, M., Moussavi, G. & Giannakis, S. 2021 A review of the innovations in metal- and carbon-based catalysts explored for heterogeneous peroxymonosulfate (PMS) activation, with focus on radical vs. non-radical degradation pathways of organic contaminants. *Chemical Engineering Journal* **411**, 127957.

Krasucka, P., Pan, B., Sik Ok, Y., Mohan, D., Sarkar, B. & Oleszczuk, P. 2021 Engineered biochar – a sustainable solution for the removal of antibiotics from water. *Chemical Engineering Journal* **405**, 126926.

Krishnan, R. Y., Manikandan, S., Subbaiya, R., Biruntha, M., Govarthanan, M. & Karmegam, N. 2021 Removal of emerging micropollutants originating from pharmaceuticals and personal care products (PPCPs) in water and wastewater by advanced oxidation processes: a review. *Environmental Technology and Innovation* **23**, 101757.

Kumar, M., Ambika, S., Hassani, A. & Nidheesh, P. V. 2022 Waste to catalyst: role of agricultural waste in water and wastewater treatment. *Science of The Total Environment* **858**, 159762.

Lee, J., Kim, K.-H. & Kwon, E. E. 2017 Biochar as a catalyst. *Renewable and Sustainable Energy Reviews* **77**, 70–79.

Li, W., Shi, Y., Gao, L., Liu, J. & Cai, Y. 2012 Occurrence of antibiotics in water, sediments, aquatic plants, and animals from Baiyangdian Lake in North China. *Chemosphere* **89** (11), 1307–1315.

Li, Z., Liu, D., Huang, W., Wei, X. & Huang, W. 2020a Biochar supported CuO composites used as an efficient peroxymonosulfate activator for highly saline organic wastewater treatment. *Science of The Total Environment* **721**, 137764.

Li, Z., Sun, Y., Yang, Y., Han, Y., Wang, T., Chen, J. & Tsang, D. C. W. 2020b Biochar-supported nanoscale zero-valent iron as an efficient catalyst for organic degradation in groundwater. *Journal of Hazardous Materials* **383**, 121240.

Li, Z., Sun, Y., Yang, Y., Han, Y., Wang, T., Chen, J. & Tsang, D. C. W. 2020c Comparing biochar- and bentonite-supported Fe-based catalysts for selective degradation of antibiotics: mechanisms and pathway. *Environmental Research* **183**, 109156.

Li, J., Liu, Y., Ren, X., Dong, W., Chen, H., Cai, T., Zeng, W., Li, W. & Tang, L. 2021 Soybean residue-based biochar prepared by ball milling assisted alkali activation to activate peroxydisulfate for the degradation of tetracycline. *Journal of Colloid and Interface Science* **599**, 631–641.

Lin, K. & Gan, J. 2011 Sorption and degradation of wastewater-associated non-steroidal anti-inflammatory drugs and antibiotics in soils. *Chemosphere* **83** (3), 240–246.

Lin, A. Y. C., Lin, C. F., Tsai, Y. T., Lin, H. H. H., Chen, J., Wang, X. H. & Yu, T. H. 2010 Fate of selected pharmaceuticals and personal care products after secondary wastewater treatment processes in Taiwan. *Water Science and Technology* **62** (10), 2450–2458.

Liu, J. L. & Wong, M. H. 2013 Pharmaceuticals and personal care products (PPCPs): a review on environmental contamination in China. *Environment International* **59**, 208–224.

Liu, M., Yin, H. & Wu, Q. 2019 Occurrence and health risk assessment of pharmaceutical and personal care products (PPCPs) in tap water of Shanghai. *Ecotoxicology and Environmental Safety* **183**, 109497.

Liu, H., Liu, Y., Tang, L., Wang, J., Yu, J., Zhang, H., Yu, M., Zou, J. & Xie, Q. 2020 Egg shell biochar-based green catalysts for the removal of organic pollutants by activating persulfate. *Science of The Total Environment* **745**, 141095.

Luo, K., Pang, Y., Wang, D., Li, X., Wang, L., Lei, M., Huang, Q. & Yang, Q. 2021 A critical review on the application of biochar in environmental pollution remediation: role of persistent free radicals (PFRs). *Journal of Environmental Sciences* **108**, 201–216.

Lykoudi, A., Frontistis, Z., Vakros, J., Manariotis, I. D. & Mantzavinos, D. 2020 Degradation of sulfamethoxazole with persulfate using spent coffee grounds biochar as activator. *Journal of Environmental Management* **271**, 111022.

Minh, T. B., Leung, H. W., Loi, I. H., Chan, W. H., So, M. K., Mao, J. Q., Choi, D., Lam, J. C. W., Zheng, G., Martin, M., Lee, J. H. W., Lam, P. K. S. & Richardson, B. J. 2009 Antibiotics in the Hong Kong metropolitan area: ubiquitous distribution and fate in Victoria Harbour. *Marine Pollution Bulletin* **58** (7), 1052–1062.

Mohan, D., Sarswat, A., Ok, Y. S. & Pittman, C. U. 2014 Organic and inorganic contaminants removal from water with biochar, a renewable, low cost and sustainable adsorbent – a critical review. *Bioresource Technology* **160**, 191–202.

Mohan, H., Rajput, S. S., Jadhav, E. B., Sankhla, M. S., Sonone, S. S., Jadhav, S. & Kumar, R. 2021 Ecotoxicity, occurrence, and removal of pharmaceuticals and illicit drugs from aquatic systems. *Biointerface Research in Applied Chemistry* **11** (5), 12530–12546.

Muter, O., Pērkons, I. & Bartkevičs, V. 2019 Removal of pharmaceutical residues from wastewater by woodchip-derived biochar. *Desalination and Water Treatment* **159**, 110–120.

Naghdi, M., Taheran, M., Brar, S. K., Rouissi, T., Verma, M., Surampalli, R. Y. & Valero, J. R. 2017 A green method for production of nanobiochar by ball milling-optimization and characterization. *Journal of Cleaner Production* **164**, 1394–1405.

Nakarmi, K. J., Daneshvar, E., Eshaq, G., Puro, L., Maiti, A., Nidheesh, P. V., Wang, H. & Bhatnagar, A. 2022 Synthesis of biochar from iron-free and iron-containing microalgal biomass for the removal of pharmaceuticals from water. *Environmental Research* **214**, 114041.

Narumiya, M., Nakada, N., Yamashita, N. & Tanaka, H. 2013 Phase distribution and removal of pharmaceuticals and personal care products during anaerobic sludge digestion. *Journal of Hazardous Materials* **260**, 305–312.

Ngqwala, N. P. & Muchesa, P. 2020 Occurrence of pharmaceuticals in aquatic environments: a review and potential impacts in South Africa. *South African Journal of Science* **116** (8), 1–7.

Ni, N., Kong, D., Wu, W., He, J., Shan, Z., Li, J., Dou, Y., Zhang, Y., Song, Y. & Jiang, X. 2020 The role of biochar in reducing the bioavailability and migration of persistent organic pollutants in soil–plant systems: a review. *Bulletin of Environmental Contamination and Toxicology* **104** (2), 157–165.

Nidheesh, P. V. 2017 Graphene-based materials supported advanced oxidation processes for water and wastewater treatment: a review. *Environmental Science and Pollution Research* **24** (35), 27047–27069.

Nidheesh, P. V. & Rajan, R. 2016 RSC advances removal of rhodamine B from a water medium using hydroxyl and sulphate radicals generated by iron loaded activated carbon. *RSC Advances* **6** (4), 5330–5340.

Nidheesh, P. V., Gopinath, A., Ranjith, N., Praveen Akre, A., Sreedharan, V. & Suresh Kumar, M. 2021 Potential role of biochar in advanced oxidation processes: a sustainable approach. *Chemical Engineering Journal* **405**, 126582.

Nidheesh, P. V., Couras, C., Karim, A. V. & Nadais, H. 2022a A review of integrated advanced oxidation processes and biological processes for organic pollutant removal. *Chemical Engineering Communications* **209** (3), 390–432.

Nidheesh, P. V., Divyapriya, G., Ezzahra Titchou, F. & Hamdani, M. 2022b Treatment of textile wastewater by sulfate radical based advanced oxidation processes. *Separation and Purification Technology* **293**, 121115.

Ntaflou, M. & Vakros, J. 2020 Transesterification activity of modified biochars from spent malt rootlets using triacetin. *Journal of Cleaner Production* **259**, 120931.

Ntzoufra, P., Vakros, J., Frontistis, Z., Tsatsos, S., Kyriakou, G., Kennou, S., Manariotis, I. D. & Mantzavinos, D. 2021 Effect of sodium persulfate treatment on the physicochemical properties and catalytic activity of biochar prepared from spent malt rootlets. *Journal of Environmental Chemical Engineering* **9** (2), 105071.

Odinga, E. S., Waigi, M. G., Gudda, F. O., Wang, J., Yang, B., Hu, X., Li, S. & Gao, Y. 2020 Occurrence, formation, environmental fate and risks of environmentally persistent free radicals in biochars. *Environment International* **134**, 105172. https://doi.org/10.1016/j.envint.2019.105172.

Ouyang, D., Chen, Y., Yan, J., Qian, L., Han, L. & Chen, M. 2019 Activation mechanism of peroxymonosulfate by biochar for catalytic degradation of 1,4-dioxane: important role of biochar defect structures. *Chemical Engineering Journal* **370**, 614–624.

Patel, M., Kumar, R., Kishor, K., Mlsna, T., Pittman, C. U. & Mohan, D. 2019 Pharmaceuticals of emerging concern in aquatic systems: chemistry, occurrence, effects, and removal methods. *Chemical Reviews* **119** (6), 3510–3673.

Patel, N., Khan, Z. A., Shahane, S., Rai, D., Chauhan, D., Kant, C. & Chaudhary, V. K. 2020 Emerging pollutants in aquatic environment: source, effect, and challenges in biomonitoring and bioremediation – a review. *Pollution* **6** (1), 99–113.

Peng, X., Yu, Y., Tang, C., Tan, J., Huang, Q. & Wang, Z. 2008 Occurrence of steroid estrogens, endocrine-disrupting phenols, and acid pharmaceutical residues in urban riverine water of the Pearl River Delta, South China. *Science of the Total Environment* **397** (1–3), 158–166.

Peng, X., Tang, C., Yu, Y., Tan, J., Huang, Q., Wu, J., Chen, S. & Mai, B. 2009 Concentrations, transport, fate, and releases of polybrominated diphenyl ethers in sewage treatment plants in the Pearl River Delta, South China. *Environment International* **35** (2), 303–309.

Peng, X., Zhang, K., Tang, C., Huang, Q., Yu, Y. & Cui, J. 2011 Distribution pattern, behavior, and fate of antibacterials in urban aquatic environments in South China. *Journal of Environmental Monitoring* **13** (2), 446–454.

Rathi, B. S., Kumar, P. S. & Show, P.-L. 2021 A review on effective removal of emerging contaminants from aquatic systems: current trends and scope for further research. *Journal of Hazardous Materials* **409**, 124413.

Razmi, R., Ardjmand, M., Ramavandi, B. & Heydarinasab, A. 2019 Optimization of phenol removal from wastewater by activation of persulfate and ultrasonic waves in the presence of biochar catalyst modified by lanthanum chloride. *Water and Environment Journal* **33** (4), 499–507.

Reyes, N. J. D. G., Geronimo, F. K. F., Yano, K. A. V., Guerra, H. B. & Kim, L. 2021 Matrices: occurrence, pathways, and treatment processes. *Water* **13** (9), 1159.

Roberts, J., Kumar, A., Du, J., Hepplewhite, C., Ellis, D. J., Christy, A. G. & Beavis, S. G. 2016 Pharmaceuticals and personal care products (PPCPs) in Australia's largest inland sewage treatment plant, and its contribution to a major Australian river during high and low flow. *Science of the Total Environment* **541**, 1625–1637.

Rubeena, K. K., Hari Prasad Reddy, P., Laiju, A. R. & Nidheesh, P. V. 2018 Iron impregnated biochars as heterogeneous Fenton catalyst for the degradation of acid red 1 dye. *Journal of Environmental Management* **226**, 320–328.

Scaria, J. & Nidheesh, P. V. 2022a The role of non-radical pathway in heterogeneous activation of persulfate and H_2O_2 by superparamagnetic magnetite-reduced graphene oxide nanocomposite. *Separation and Purification Technology* **299**, 121764.

Scaria, J. & Nidheesh, P. V. 2022b Comparison of hydroxyl-radical-based advanced oxidation processes with sulfate radical-based advanced oxidation processes. *Current Opinion in Chemical Engineering* **36**, 100830.

Scaria, J. & Nidheesh, P. V. 2022c Magnetite–reduced graphene oxide nanocomposite as an efficient heterogeneous Fenton catalyst for the degradation of tetracycline antibiotics. *Environmental Science: Water Research & Technology* **8** (6), 1261–1276.

Scaria, J., Anupama, K. V. & Nidheesh, P. V. 2021a Tetracyclines in the environment: an overview on the occurrence, fate, toxicity, detection, removal methods, and sludge management. *Science of The Total Environment* **771**, 145291.

Scaria, J., Gopinath, A. & Nidheesh, P. V. 2021b A versatile strategy to eliminate emerging contaminants from the aqueous environment: heterogeneous Fenton process. *Journal of Cleaner Production* **278**, 124014.

Scaria, J., Gopinath, A., Ranjith, N., Ravindran, V., Ummar, S., Nidheesh, P. V. & Kumar, M. S. 2022 Carbonaceous materials as effective adsorbents and catalysts for the removal of emerging contaminants from water. *Journal of Cleaner Production* **350**, 131319.

Song, H., Li, Q., Ye, Y., Pan, F., Zhang, D. & Xia, D. 2021 Degradation of cephalexin by persulfate activated with magnetic loofah biochar: performance and mechanism. *Separation and Purification Technology* **272**, 118971.

Sun, S. P., Zeng, X., Li, C. & Lemley, A. T. 2014 Enhanced heterogeneous and homogeneous Fenton-like degradation of carbamazepine by nano-Fe_3O_4/H_2O_2 with nitrilotriacetic acid. *Chemical Engineering Journal* **244**, 44–49.

Sun, P., Zhang, K., Gong, J., Khan, A., Zhang, Y., Islama, M. S. & Zhang, Y. 2019 Sunflower stalk-derived biochar enhanced thermal activation of persulfate for high efficient oxidation of p-nitrophenol. *Environmental Science and Pollution Research* **26** (26), 27482–27493.

Tan, X. F., Liu, Y. G., Gu, Y. L., Xu, Y., Zeng, G. M., Hu, X. J., Liu, S. B., Wang, X., Liu, S. M. & Li, J. 2016 Biochar-based nano-composites for the decontamination of wastewater: a review. *Bioresource Technology* **212**, 318–333.

Taylor, D. & Senac, T. 2014 Human pharmaceutical products in the environment – the 'problem' in perspective. *Chemosphere* **115** (1), 95–99.

Tomczyk, A., Sokołowska, Z. & Boguta, P. 2020 Biochar physicochemical properties: pyrolysis temperature and feedstock kind effects. *Reviews in Environmental Science and Biotechnology* **19** (1), 191–215.

Tong, C., Zhuo, X. & Guo, Y. 2011 Occurrence and risk assessment of four typical fluoroquinolone antibiotics in raw and treated sewage and in receiving waters in Hangzhou, China. *Journal of Agricultural and Food Chemistry* **59** (13), 7303–7309.

Ushani, U., Lu, X., Wang, J., Zhang, Z., Dai, J., Tan, Y., Wang, S., Li, W., Niu, C., Cai, T., Wang, N. & Zhen, G. 2020 Sulfate radicals-based advanced oxidation technology in various environmental remediation: a state-of-the-art review. *Chemical Engineering Journal* **402**, 126232.

Valcárcel, Y., Alonso, S. G., Rodríguez-Gil, J. L., Gil, A. & Catalá, M. 2011 Detection of pharmaceutically active compounds in the rivers and tap water of the Madrid Region (Spain) and potential ecotoxicological risk. *Chemosphere* **84** (10), 1336–1348.

Wang, J. & Wang, S. 2016 Removal of pharmaceuticals and personal care products (PPCPs) from wastewater: a review. *Journal of Environmental Management* **182**, 620–640.

Wang, S. & Wang, J. 2019 Activation of peroxymonosulfate by sludge-derived biochar for the degradation of triclosan in water and wastewater. *Chemical Engineering Journal* **356**, 350–358.

Wang, B., Gao, B. & Fang, J. 2017 Recent advances in engineered biochar productions and applications. *Critical Reviews in Environmental Science and Technology* **47** (22), 2158–2207.

Wang, Y., Li, Y., Hu, A., Rashid, A., Ashfaq, M., Wang, Y., Luo, H., Yu, C. P. & Sun, Q. 2018 Monitoring, mass balance and fate of pharmaceuticals and personal care products in seven wastewater treatment plants in Xiamen City, China. *Journal of Hazardous Materials* **354**, 81–90.

Wang, B., Li, Y. & Wang, L. 2019a Metal-free activation of persulfates by corn stalk biochar for the degradation of antibiotic norfloxacin: activation factors and degradation mechanism. *Chemosphere* **237**, 124454.

Wang, H., Guo, W., Yin, R., Du, J., Wu, Q., Luo, H., Liu, B., Sseguya, F. & Ren, N. 2019b Biochar-induced Fe(III) reduction for persulfate activation in sulfamethoxazole degradation: insight into the electron transfer, radical oxidation and degradation pathways. *Chemical Engineering Journal* **362**, 561–569.

Wang, R. Z., Huang, D. L., Liu, Y. G., Zhang, C., Lai, C., Wang, X., Zeng, G. M., Gong, X. M., Duan, A., Zhang, Q. & Xu, P. 2019c Recent advances in biochar-based catalysts: properties, applications and mechanisms for pollution remediation. *Chemical Engineering Journal* **371**, 380–403.

Wang, J., Shen, M., Wang, H., Du, Y., Zhou, X., Liao, Z., Wang, H. & Chen, Z. 2020 Red mud modified sludge biochar for the activation of peroxymonosulfate: singlet oxygen dominated mechanism and toxicity prediction. *Science of The Total Environment* **740**, 140388.

Wang, Z., Tian, Q., Guo, J., Wu, R., Zhu, H. & Zhang, H. 2021 Co-pyrolysis of sewage sludge/cotton stalks with K_2CO_3 for biochar production: improved biochar porosity and reduced heavy metal leaching. *Waste Management* **135**, 199–207.

Wilkinson, J. L., Hooda, P. S., Barker, J., Barton, S. & Swinden, J. 2016 Ecotoxic pharmaceuticals, personal care products, and other emerging contaminants: a review of environmental, receptor-mediated, developmental, and epigenetic toxicity with discussion of proposed toxicity to humans. *Critical Reviews in Environmental Science and Technology* **46** (4), 336–381.

Wu, Y., Guo, J., Han, Y., Zhu, J., Zhou, L. & Lan, Y. 2018 Insights into the mechanism of persulfate activated by rice straw biochar for the degradation of aniline. *Chemosphere* **200**, 373–379.

Ye, S., Xiong, W., Liang, J., Yang, H., Wu, H., Zhou, C., Du, L., Guo, J., Wang, W., Xiang, L., Zeng, G. & Tan, X. 2021 Refined regulation and nitrogen doping of biochar derived from ramie fiber by deep eutectic solvents (DESs) for catalytic persulfate activation toward non-radical organics degradation and disinfection. *Journal of Colloid and Interface Science* **601**, 544–555.

Yu, Y., Wu, L. & Chang, A. C. 2013 Seasonal variation of endocrine disrupting compounds, pharmaceuticals and personal care products in wastewater treatment plants. *Science of the Total Environment* **442**, 310–316.

Yu, J., Tang, L., Pang, Y., Zeng, G., Wang, J., Deng, Y., Liu, Y., Feng, H., Chen, S. & Ren, X. 2019 Magnetic nitrogen-doped sludge-derived biochar catalysts for persulfate activation: internal electron transfer mechanism. *Chemical Engineering Journal* **364**, 146–159.

Yu, J., Tang, L., Pang, Y., Zeng, G., Feng, H., Zou, J., Wang, J., Feng, C., Zhu, X., Ouyang, X. & Tan, J. 2020 Hierarchical porous biochar from shrimp shell for persulfate activation: a two-electron transfer path and key impact factors. *Applied Catalysis B: Environmental* **260**, 118160.

Zhang, Y., Geißen, S. U. & Gal, C. 2008 Carbamazepine and diclofenac: removal in wastewater treatment plants and occurrence in water bodies. *Chemosphere* **73** (8), 1151–1161.

Zhang, D., Lin, L., Luo, Z., Yan, C. & Zhang, X. 2011 Occurrence of selected antibiotics in Jiulongjiang River in various seasons, south China. *Journal of Environmental Monitoring* **13** (7), 1953–1960.

Zhang, R., Zhang, G., Zheng, Q., Tang, J., Chen, Y., Xu, W., Zou, Y. & Chen, X. 2012 Occurrence and risks of antibiotics in the Laizhou Bay, China: impacts of river discharge. *Ecotoxicology and Environmental Safety* **80**, 208–215.

Zhang, L., Hu, J., Zhu, R., Zhou, Q. & Chen, J. 2013 Degradation of paracetamol by pure bacterial cultures and their microbial consortium. *Applied Microbiology and Biotechnology* **97** (8), 3687–3698.

Zhang, R., Li, Y., Wang, Z., Tong, Y. & Sun, P. 2020 Biochar-activated peroxydisulfate as an effective process to eliminate pharmaceutical and metabolite in hydrolyzed urine. *Water Research* **177**, 115809.

Zhang, Y., Xu, M., Liu, X., Wang, M., Zhao, J., Li, S. & Yin, M. 2021 Regulation of biochar mediated catalytic degradation of quinolone antibiotics: important role of environmentally persistent free radicals. *Bioresource Technology* **326**, 124780.

Zhang, P., Duan, W., Peng, H., Pan, B. & Xing, B. 2022 Functional biochar and its balanced design. *ACS Environmental Au* **2** (2), 115–127.

Zhao, J. L., Ying, G. G., Liu, Y. S., Chen, F., Yang, J. F. & Wang, L. 2010a Occurrence and risks of triclosan and triclocarban in the Pearl River system, South China: from source to the receiving environment. *Journal of Hazardous Materials* **179** (1–3), 215–222.

Zhao, J. L., Ying, G. G., Liu, Y. S., Chen, F., Yang, J. F., Wang, L., Yang, X. B., Stauber, J. L. & Warne, M. S. J. 2010b Occurrence and a screening-level risk assessment of human pharmaceuticals in the Pearl River system, South China. *Environmental Toxicology and Chemistry* **29** (6), 1377–1384.

Zhao, Y., Yuan, X., Li, X., Jiang, L. & Wang, H. 2021 Burgeoning prospects of biochar and its composite in persulfate-advanced oxidation process. *Journal of Hazardous Materials* **409**, 124893.

Zheng, Q., Zhang, R., Wang, Y., Pan, X., Tang, J. & Zhang, G. 2012 Occurrence and distribution of antibiotics in the Beibu Gulf, China: impacts of river discharge and aquaculture activities. *Marine Environmental Research* **78**, 26–33.

Zhong, Q., Lin, Q., Huang, R., Fu, H., Zhang, X., Luo, H. & Xiao, R. 2020 Oxidative degradation of tetracycline using persulfate activated by N and Cu codoped biochar. *Chemical Engineering Journal* **380**, 122608.

Zhou, X., Zeng, Z., Zeng, G., Lai, C., Xiao, R., Liu, S., Huang, D., Qin, L., Liu, X., Li, B., Yi, H., Fu, Y., Li, L., Zhang, M. & Wang, Z. 2020 Insight into the mechanism of persulfate activated by bone char: unraveling the role of functional structure of biochar. *Chemical Engineering Journal* **401**, 126127.

Zhou, X., Zhu, Y., Niu, Q., Zeng, G., Lai, C., Liu, S., Huang, D., Qin, L., Liu, X., Li, B., Yi, H., Fu, Y., Li, L., Zhang, M., Zhou, C. & Liu, J. 2021 New notion of biochar: a review on the mechanism of biochar applications in advanced oxidation processes. *Chemical Engineering Journal* **416**, 129027.

Zhu, J., Song, Y., Wang, L., Zhang, Z., Gao, J., Tsang, D. C. W., Ok, Y. S. & Hou, D. 2022 Green remediation of benzene contaminated groundwater using persulfate activated by biochar composite loaded with iron sulfide minerals. *Chemical Engineering Journal* **429**, 132292.

Zou, J., Yu, J., Tang, L., Ren, X., Pang, Y., Zhang, H., Xie, Q., Liu, Y., Liu, H. & Luo, T. 2020 Analysis of reaction pathways and catalytic sites on metal-free porous biochar for persulfate activation process. *Chemosphere* **261**, 127747.

First received 4 July 2022; accepted in revised form 24 February 2023. Available online 8 March 2023

doi: 10.2166/wst.2022.424

Thiabendazole degradation by photo-NaOCl/Fe and photo-Fenton like processes, using copper slag as an iron catalyst, in spiked synthetic and real secondary wastewater treatment plant effluents

Reyna García-Estrada, Sandra Arzate and Rosa-María Ramírez-Zamora *

Coordinación de Ingeniería Ambiental, Instituto de Ingeniería, Universidad Nacional Autónoma de México, Circuito Escolar s/n, Ciudad Universitaria, Coyoacán, Ciudad de México 04510, México

*Corresponding author. E-mail: rramirezz@iingen.unam.mx

SA, 0000-0002-1522-693X; R-MR-Z, 0000-0003-1129-4385

ABSTRACT

Thiabendazole degradation (TBZ_D) in diferent types of water matrices was assessed by applying two Advanced Oxidation Processes, both using simulated solar light (SSL), copper slag (CS) as an iron based catalyst, and separately H_2O_2 or NaOCl as oxidants. First, optimum conditions for TBZ_D were evaluated in distilled water, $TBZ_D = 90\%$ at 60 min for CS-H_2O_2-SSL, and 92% of TBZ_D in a twelfth of the time by the system CS-NaOCl-SSL; minimum TBZ depletion variations were observed between the first and the fifth reuse test: $88 \pm 2\%$ for CS-H_2O_2-SSL (60 min) and $90 \pm 1\%$ for CS-NaOCl-SSL (5 min). Those conditions were tested using a synthetic (SE) and a real secondary effluent (RE) from a wastewater treatment plant. The CS-H_2O_2-SSL system achieved TBZ_D of 88 and 77% after 90 min for SE and RE, with kinetic constants of 0.024 and 0.016 min^{-1}, respectively, whereas photo-NaOCl/Fe showed values of 0.365 and 0.385 min^{-1} for SE and RE, achieving a 94% TBZ_D removal in both types of water at 10 min. That might be related to the formation of $Cl^{\bullet}$ and $HO^{\bullet}$ during the photo-NaOCl/Fe process, highlighting that the CS-NaOCl-SSL is an attractive option that has great possibilities for scaling up by a better knowledge in real aqueous matrices.

Key words: fungicide, heterogeneous photo-Fenton, liquid chromatography, pollutant degradation, raceway pond reactor, water matrix

HIGHLIGHTS

- Copper slag was successfully used as catalyst in two photo assisted iron based AOPs.
- 90% TBZ degradation was achieved with the Fe based AOPs in a WWTP secondary effluent.
- Rate constant of photo/Fe/NaOCl system was three orders of magnitude larger than with H_2O_2.
- 94% TBZ removal was reached at 10 min in a WWTP secondary effluent by the photo-Fe/NaOCl.
- Nitrates and chlorides reduced the reaction rate of the photo-Fe/ H_2O_2 system by 30%.

GRAPHICAL ABSTRACT

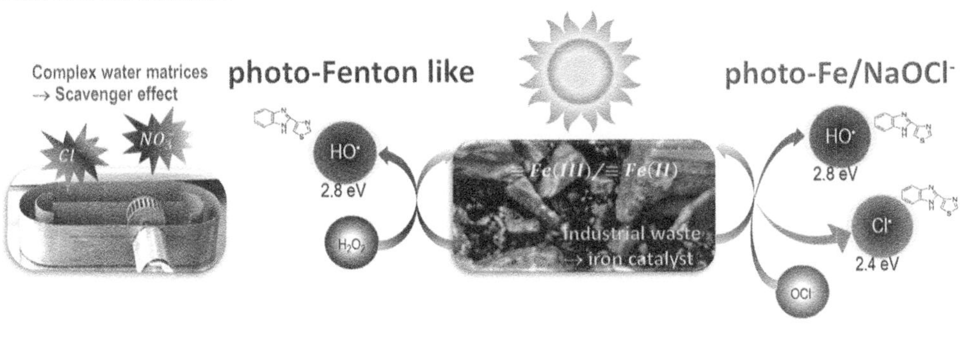

INTRODUCTION

The photo-Fenton process has been extensively studied for the degradation of recalcitrant organic contaminants in effluents of wastewater treatment plants (Ameta *et al.* 2018; Contreras-Bustos *et al.* 2019). In fact, several engineering-scale demonstration plants have been erected in order to exhibit the potentiality of the installation of this technology at full scale (Contreras-Bustos *et al.* 2019; Gualda-Alonso *et al.* 2022). One of the reasons why this technology is attractive is its capability of using solar light as a source of UV radiation, instead of lamps, and in this way, diminishing the energetic consumption (Cabrera-Reina *et al.* 2014; Hassani *et al.* 2021). Though, some drawbacks related to the pH dependency of the process, when it is performed in homogeneous systems, and the cost of the oxidant have pointed out the need to diversify operation strategies.

Thus, the implementation of a Fenton-like catalyst seems to be a good option to solve the pH drawback, due to the large pH operation range of these materials (Lima *et al.* 2017). Diverse iron-based compounds have been tested in photo-Fenton applications, like nanomaterials (nanoparticles and nano-zero-valent iron) (Carraro *et al.* 2018; Zhu *et al.* 2019; Ahmadi *et al.* 2021) and natural minerals (magnetite (Fe_3O_4), goethite ($\alpha FeOOH$), maghemite (γFe_2O_3) and hematite (αFe_2O_3) (Zhu *et al.* 2019).

Nevertheless, thinking on avoiding complex chemical synthesis and considering that some industrial wastes have similar characteristics to those of natural minerals, copper slags (CS) have been successfully utilized as a Fenton-like catalyst in the degradation process of various organic pollutants and pathogens (Solís-López *et al.* 2014; Arzate-Salgado *et al.* 2016). This is because they are composed mainly of iron oxides (Solís-López *et al.* 2014) that exhibit photo-activity under visible-light irradiation due to their mineralogical phases and band gap values (Arzate-Salgado *et al.* 2016). Also, the metallurgical slags are highly dense, and that allows an easy separation, in contrast to other nanoparticles that require an additional treatment step like coagulation to separate the nanoparticles from the treated effluent (Ghanbari *et al.* 2020). Hence, use of this material as a Fenton-like catalyst also includes the possibility to face two environmental problems at the same time: the water treatment and the industrial by-products disposal, since CS is produced in abundance, approximately 2 tons per ton of produced copper (Kim *et al.* 2013), and it represents an important environmental problem to attend to.

However, it is important to mention that this material, like the majority of heterogeneous catalysts, has been commonly evaluated in distilled water. Nevertheless, it has been reported that the presence of ions and organic matter in water matrices diminishes the effectivity of pollutant degradation (Tokumura *et al.* 2016; Lado Ribeiro *et al.* 2019); for this reason, the assessment of their effectivity in synthetic and real water matrices is an important step to understand the feasibility of the process proposed.

Regarding the oxidant, the use of chlorine instead of H_2O_2 has been proposed recently (Portilla-Sangabriel *et al.* 2021). The results show that a system composed of the magnetic fraction of a low-grade titanium ore (MFTO), NaOCl and simulated sunlight (SSL), MFTO-NaOCl-SSL, halves the reaction time necessary to degrade 97% of thiabendazole (TBZ) ($TBZ_0 =$ 37 μM) in distilled water. The successful degradation of thiabendazole by the MFTO-NaOCl-SSL is related to the production of strong radicals like chlorine radicals ($Cl^{\bullet}$ (2.4 eV)), hydroxyl radicals ($HO^{\bullet}$ (2.8 eV)), $Cl_2^{\bullet-}$ (2.0 eV) and $ClO^{\bullet}$ (1.4 eV) (Armstronga *et al.* 2013; Guo *et al.* 2017), whose formation is explained in the supplementary material (Section S.1).

In this context, this work aims to assess the degradation of thiabendazole using two Advanced Oxidation systems composed by CS-oxidant-SSL, the oxidants being H_2O_2 (CS-H_2O_2-SSL, photo-Fenton-like) and NaOCl (CS-NaoOCl-SSL, photo-NaOCl/Fe), in three aqueous matrices: distilled, synthetic, and real secondary effluent from a wastewater treatment plant (WWTP).

MATERIALS AND METHODS

Reagents

Sodium hypochlorite (5.25% w/w) was purchased from Golden Bell. Hydrogen peroxide (30% w/w), sulfuric acid, sodium hydroxide (ACS grade) and $MgSO_4{\cdot}7H_2O$ were acquired from J.T. Baker. N-Diethyl phenylenediamine was bought from Hach. Thiabendazole (>99%), beef extract (>90%), sodium thiosulphate (0.1 N), titanium (IV) oxysulfate solution (1.9–2.1%), phosphoric acid (85%), acetonitrile and water HPLC grade were supplied by Sigma–Aldrich. NaCl (95%), $CaSO_4{\cdot}2H_2O$ (98%), $MgSO_4$ (>98%), KCl (>99%), $CaCl_2{\cdot}H_2O$ (>99%), urea (>99%) and $NaHCO_3$ (95–100%) were provided by Meyer. Finally, peptone was acquired from Lab (Grade for Bacteriology).

Metallurgical copper slag

A metallurgical copper slag sample, issued from a northern Mexican smelter, was milled and sieved; the fraction with a particle size between 180 and 250 μm (60–80-mesh) was used in this study. CS characterization was previously reported by our group in the works of Solís-López *et al.* (2014); Arzate-Salgado *et al.* (2016) and García Estrada *et al.* (2020). Briefly, scanning electron microscopy (SEM) images were recorded to identify the morphology of CS (Philips XL20 SEM equipped with an Oxford 7593 solid state energy-dispersive X-ray (EDX) detector). The phase composition of the solid phase was determined by X-ray diffraction (XRD) on an Empyrean diffractometer with focused Cu-Kα radiation and a Ni filter in a 2θ interval from 4° to 70° in steps of 0.003° at 40 s per step. And XRD was used for the identification of the mineral phases. The chemical composition was determined by X-ray fluorescence (XRF) using an SRS 3X00 sequential X-ray spectrometer. The band gap of the CS was determined by applying the Tauc-plot method, considering direct interactions and using the UV-vis diffuse reflectance spectroscopy technique (UV-Vis-DRS) with a Cary 500 UV/vis/near infrared (NIR) spectrophotometer (Varian Inc.). The surface area of CS was determined by the adsorption-desorption of N_2 gas in a Bel-Sorp mini II apparatus (Belsorp Inc., Japan) using the BET (Brunauer-Emmett-Teller) isotherm methodology.

Water matrices

Three different aqueous matrices were used to run the experiments: distilled water, synthetic effluents and real secondary effluents from WWTPs. All of them were spiked with thiabendazole as model pollutant. TBZ is a fungicide usually used during postharvest storage to preserve fruits and vegetables. It has been used as a model pollutant because it is highly recalcitrant; in fact, concentrations of 0.11–70 μg/L have been found in wastewater treatment plant effluents (Bernabeu *et al.* 2011).

Distilled water used to prepared TBZ solutions was produced *in situ* and complies with Type II water characteristics, according to ASTM D1193-99e1 (ASTM 1999), with resistivity of >1 MΩ-cm, a conductivity of <1 μS/cm and <50 ppb of total organic carbon (TOC).

Synthetic effluent (SE) was prepared using the composition reported by Klamerth *et al.* (2010), which represents the effluent from aerobic sewage treatment according to OECD Guidelines (OECD 2001). Real effluent (RE) was collected in the downstream tank of activated sludge, a secondary treatment from the WWTP Cerro de la Estrella, located on the southeast side of Mexico City, Mexico. The main features of both synthetic and real effluents are shown in Table 1.

Considering that carbonates (HCO_3^-/CO_3^{2-}) present in the secondary effluents act as radical scavengers (Pignatello *et al.* 2006), a concentration of 15 mg/L, expressed as Total Inorganic Carbon (TIC), was set in both matrices; in SE, the concentration was adjusted by the addition of $NaHCO_3$, while in RE, a stoichiometric amount of sulfuric acid was added, under agitation, to strip the carbonates.

Analytical techniques

High-performance liquid chromatography (HPLC) was used to measure TBZ concentration in water, using a 1100 Agilent chromatography apparatus equipped with a photo-diode array detector (Agilent) and an HPLC column ACE5 C18-AR (150 mm × 4.6 mm). The mobile phase was a mixture of methanol, acetonitrile, and phosphoric acid (0.1%, v/v) at a

Table 1 | Characterization of water matrices

Parameter	Real effluent	Synthetic effluent
COD (mg/L)	22	51
BOD_5 (mg/L)	10	25
DOC (mg/L)	19	17
Cl^- (mg/L)	56	8
NO_3^- (mg/L)	265	53
SO_4^{2-} (mg/L)	3	3
pH (-)	7.34	7.94
Conductivity (μS/cm)	846	298
Turbidity (NTU)	0.8	1.7

10:8:82 ratio (v/v), pumped at a flow rate of 1.0 mL/min. The sample volume of injection was 30 μL and TBZ was detected at a wavelength (λ) of 270 nm. The limit of detection (LOD) was 100 μg/L (4.97×10^{-1} μM).

Oxidant concentrations were measured throughout each experiment; hydrogen peroxide concentration was determined with the DIN 38 402 H15 colorimetric method using a titanium (IV) oxysulfate solution; while the NaOCl content was quantified by measuring free chlorine concentration with an HI 96701 Free Chlorine device (ISM-HANNA Instruments), which adapts the USEPA method 330.5 and Standard Method 4500-Cl G.

Total leached iron concentration was measured using the Merck Kit (Spectroquant 1.14761.001) with an LOD of 0.3 mg/L, at the end of each experiment. The method implies the reduction of all the iron species in dissolution to Fe^{2+}, in a thioglycolate-buffered medium, to react with a triazine to form a red-violet complex.

Chemical oxygen demand (COD) was determined according to the standardized method 5220 D (AWWA/APHA/WEF 2012); the acid digestion was run in a COD digester, and the spectrophotometric measurement was made in a HACH LANGE DR/4000 U, UV-VIS Spectrophotometer. Biochemical oxygen demand (BOD_5) was measured using a Respirometric OxiTop measuring system, which follows the standardized method 5210 (AWWA/APHA/WEF 2012). Dissolved organic carbon (DOC) and inorganic carbon analysis were performed using the standardized method 5310B (AWWA/APHA/WEF 2012) with a TOC analyser (Shimadzu TOC-L CSN). Ion chromatography (Thermo Scientific Dionex) was used for measuring the concentration of anions. Conductivity and pH were measured with a Benchtop Meter (Thermo Scientific Orion Versa Star); and turbidity was quantified using a Merck Turbiquant.

Experimental setup

The oxidation experiments were performed using SSL provided from a Xenon lamp (400–800 nm), with 5–6% photon emissions between 290 and 400 nm. The profile of the emitted photons between 400 and 800 nm at a light intensity simulates the solar spectrum. The lamp is installed inside a solar simulator chamber (SUNTEST CPS + ATLAS) (Figure 1(a)), as previously described in Arzate-Salgado *et al.* (2016). A Raceway Pond Reactor (RPR) of 1 L capacity (Figure 1(b)) was operated at 5 cm liquid depth; its dimensions are length of 0.27 m and width of 0.10 m (Portilla-Sangabriel *et al.* 2021).

In all the experiments, the light intensity was set constant at 400 W/m^2, in order to simulate the global solar radiation in Mexico (hours of sunlight) (Almanza & López 1978).

Optimization of operating conditions

A central composite experimental design was used to determine the optimum values of CS and oxidant doses for the CS-H_2O_2-SSL and CS-NaOCl-SSL systems for a given initial concentration of thiabendazole (TBZ_0) in distilled water; this procedure is explained in detail in the supplementary material (Section S.2.1). The optimum values obtained were tested in five reuse cycles. CS was separated from the treated effluent by settling, because of its high density; then the catalyst was placed in tubes with water, stirred and centrifuged at 500 rpm for 10 min; the supernatant was decanted, and the sediment was re-suspended in water (Arzate-Salgado *et al.* 2016). Then the CS was dried at 105 °C before each cycle.

Related micropollutant (MP) removal mechanisms

Considering that the experimental systems are composed of three elements, CS, oxidant and SSL, several mechanisms that involve just one or two of them may occur during the reaction; thus, in order to clarify the contribution of each mechanism to

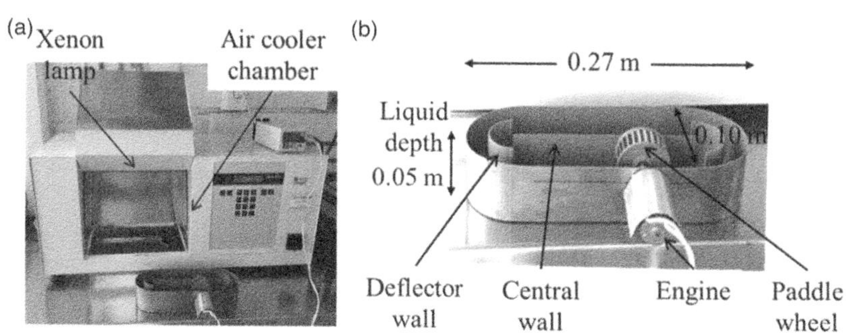

Figure 1 | Reaction system for the TBZ photocatalytic degradation tests: (a) solar simulator chamber and (b) Raceway Pond Reactor.

MP removal, control experiments were carried out, conserving the optimum values of factors determined with the statistical second-order model. Table 2 shows the mechanism analysed and the value of each factor.

TBZ degradation in complex water matrices

Experiments in synthetic and real WWTP secondary effluents were carried out. In both cases, TBZ initial concentration was adjusted to 7.36 mg/L in order to appraise better its concentration changes. CS and oxidant doses were set up by using the response surface graphs, looking for an oxidant/TBZ ratio that complied with 90% degradation and less oxidant consumption; in this way, 1.11 g CS and 50 mg H_2O_2/L or 98 mg NaOCl/L were used in the systems CS-H_2O_2-SSL and CS-NaOCl-SSL, respectively. Tests were run in triplicate.

In the sequence, reaction order was determined, and rate constants were calculated. Since TBZ degradation exhibited a pseudo-first-order reaction, the value was estimated using Equation (1); where k' is the pseudo-first-order rate constant, C_0 the initial concentration, C is the concentration at time t.

$$ln\frac{C}{C_0} = k't \tag{1}$$

RESULTS AND DISCUSSION

Copper slag characterization

A detailed description of the copper slag characterization is available in Solís-López *et al.* (2014), Arzate-Salgado *et al.* (2016) and García Estrada *et al.* (2020).

Briefly, the morphologies of CS samples obtained by SEM are shown in Figure 2. The sample is composed by small particles with angular shape; in the surface of the material large cavities are observed without an apparent pattern. The images show a non-porous material. This observation is in line with the superficial area 14.94 m^2/g (Solís-López *et al.* 2014), which allows us to expect low contaminant adsorption on the surface of the material (Carmalin & Eder 2018) and limitations on the production of $HO^{\bullet}$ by heterogeneous catalyst when the material is exposed to light (Larralde *et al.* 2019).

Nevertheless, since the CS surface has a high amount of iron in the surface, 55.82% as Fe_2Ox (Table 3), another mechanism like the heterogeneous Fenton can lead to radical formation (Appendix A.1, supplementary material).

The major mineralogical phase present in the CS is fayalite (Fe_2SiO_4) (Solís-López *et al.* 2014). For this reason, this material shows a band gap of 2.20 eV; this value is associated with the main crystalline phase and allows its activation under visible-light irradiation (García-Muñoz *et al.* 2016).

Optimization of operating conditions

The models presented in Equations (2) and (3) were obtained by the optimization procedure and statistical analysis described in A.2.2. They were solved, and experiments were carried out to verify the values of TBZ_D. Figure 3 shows experimental

Table 2 | Factor values of the control assays for TBZ degradation

	Factor			
Mechanism	**TBZ_0 (mg/L)**	**CS (g/L)**	**Oxidant (mg/L)**	**SSL (W/m^2)**
Adsorption	2.01	1.17	–	–
Photolysis	2.01	–	–	400
Heterogeneous photocatalysis	2.01	1.17	–	400
Chemical oxidation (H_2O_2)	2.01	–	29.7	–
Heterogeneous Fenton-type	2.01	1.17	29.7	–
Chemical oxidation photo assisted (H_2O_2-SSL)	2.01	–	29.7	400
Chemical oxidation (NaOCl)	1.23		70.1	–
Heterogeneous oxidation (Fe/NaOCl)	1.23	1.11	70.1	–
Chemical photo assisted oxidation (NaOCl-SSL)	1.23	–	70.1	400

(a) (b)

Figure 2 | Micrographs recorded using a LABE detector (low-angle electron backscattered detector, low angle backscattered electrons) by SEM. (a) Slag particles captured at a magnification of 600× and (b) 2000 × .

Table 3 | Chemical composition of CS determined by XRF (first published by García Estrada *et al.* (2020))

Compound	Wt. % ($n = 4$)
Fe_2O_3	59.26 ± 2.56
SiO_2	26.34 ± 1.54
Al_2O_3	4.2 ± 0.51
SO_3	1.33 ± 0.16
K_2O	1.15 ± 0.03
CaO	0.46 ± 0.02
TiO_2	0.31 ± 0.03
Na_2O	0.28 ± 0.04
MgO	0.30 ± 0.02
P_2O_5	0.03 ± 0.01
MnO	0.05 ± 0.01

kinetics of TBZ and oxidant concentrations in function of reaction time for the systems CS-H_2O_2-SSL (photo-Fenton like) (2) and CS-NaOCl-SSL (photo-Fe/NaOCl) (3), carried out under optimum conditions.

$$\begin{aligned} TBZ_{D-H_2O_2} = {} & 90.1762 + 0.9317\ H_2O_2 - 1.3424\ CS - 5.6541\ TBZ_0 \\ & - 0.03450\ H_2O_2^2 + 0.176875\ H_2O_2\ CS + 0.4360\ H_2O_2{*}TBZ_0 \\ & + 2.7723\ CS^2 - 2.7388\ CS{*}TBZ_0 - 0.9830\ TBZ_0^2 \end{aligned} \tag{2}$$

$$\begin{aligned} TBZ_{D-NaOCl} = {} & 58.0316 + 1.3283\ NaOCl - 3.4435\ CS - 9.4210\ TBZ_0 \\ & - 0.0078\ NaOCl^2 + 0.0826\ NaOCl{*}CS + 0.0108\ NaOCl{*}TBZ_0 \\ & - 1.8922\ CS^2 + 1.1725\ CS{*}TBZ_0 + 0.2545\ TBZ_0^2 \end{aligned} \tag{3}$$

It is evident that the TBZ degradation is faster when NaOCl is used instead of H_2O_2, showing that the first system allows an efficiency of 92% at 5 minutes, while the time in the second system should be increased 12 times to achieve 90% TBZ_D. Besides, the TBZ degradation was >99.9% at 7.5 min and 97% at 90 min for the systems CS-NaOCl-SSL and CS-H_2O_2-SSL, respectively. A similar tendency is followed by the oxidant concentration; free chlorine is not observed at

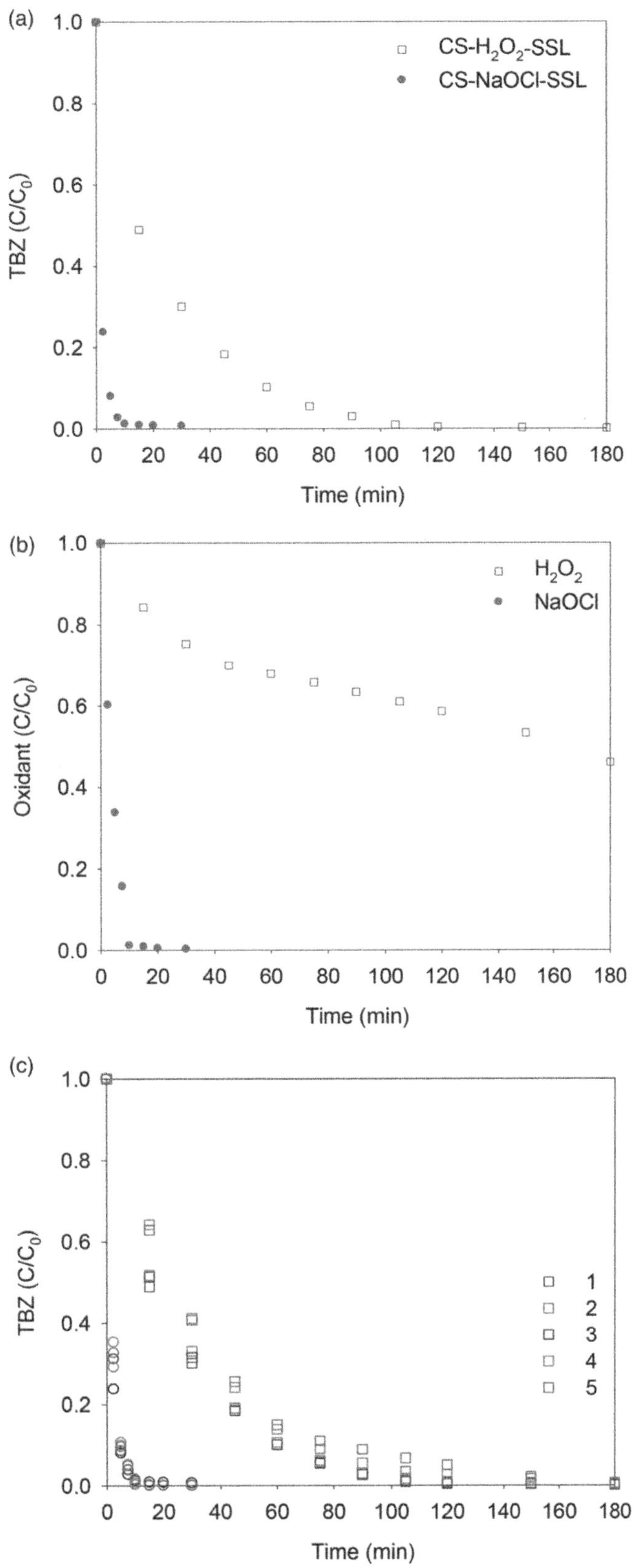

Figure 3 | TBZ degradation kinetics (a) and oxidant consumption (b) using the systems CS-H_2O_2-SSL (blue square) and CS-NaOCl-SSL (green circle), (c) reuse cycles for the TBZ degradation kinetics using the systems CS-H_2O_2-SSL and CS-NaOCl-SSL.

10 min of reaction, whereas H_2O_2 showed a slow conversion, and even at 180 min, 46% of the initial concentration remains in the reactor.

On one hand, the difference in the degradation is related to the reactivity of the radicals produced in each system. By the action of SSL, both oxidants may produce hydroxyl radicals (2.8 eV) (reactions A.4, A.5, explained in Appendix A.1, supplementary material), but chlorine radical (2.4 eV) is also produced when NaOCl is illuminated (Kong *et al.* 2016); whereby the CS-NaOCl-SSL system has this advantage over the CS-H_2O_2-SSL. On the other hand, photolytic decomposition of H_2O_2 occurs at 200–280 nm (Georgiou *et al.* 2002), while for the aqueous chlorine, decomposition is observed at 320 and 340 nm (Nowell & Hoigni 1992); thus, considering the distribution of the sunlight spectrum, the reaction using chlorine species is more favoured.

$$HOCl/OCl^- + \lambda \rightarrow HO^\bullet/O^{\bullet -} + Cl^\bullet \tag{4}$$

$$H_2O_2 + \lambda \rightarrow 2HO^\bullet \tag{5}$$

Additionally, the reaction between ferrous ion and NaOCl has been reported to be three orders of magnitude faster in comparison with H_2O_2 in homogeneous reactions (Behin *et al.* 2017). Even though there are no reports that measure the reaction of Fe^{2+} and NaOCl in heterogeneous systems, a faster reaction between them in comparison with Fenton-like reaction is expected, because the speed of the catalytic decomposition of H_2O_2 in the presence of iron species depends on the oxidation state of this metal (Neyens & Baeyens 2003; Sandoval-Gonález *et al.* 2021).

Thus, concerning TBZ degradation (Figure 4), Portilla Sangabriel *et al.* (2022) elucidated the transformation products (TPs) according to the oxidant used, H_2O_2 or NaOCl (Figure 4). When the catalyst-NaOCl-sunlight system was used only four TPs were detected, in two reaction routes: (1) the breaking of the thiazole ring, which generates TP161 and TP162. Then the HO˙ radical attack on the C1′ and the consequent C–C chain breakage form benzimidazole (TP118), also seen in TBZ degradation by photolysis (Murthy *et al.* 1996) and heterogeneous photocatalysis (Calza *et al.* 2003). But due to the high velocity of the reaction, TP 161 is not observed, unlike the heterogeneous photo-Fenton system; and (2) the formation of TP127 and TP128

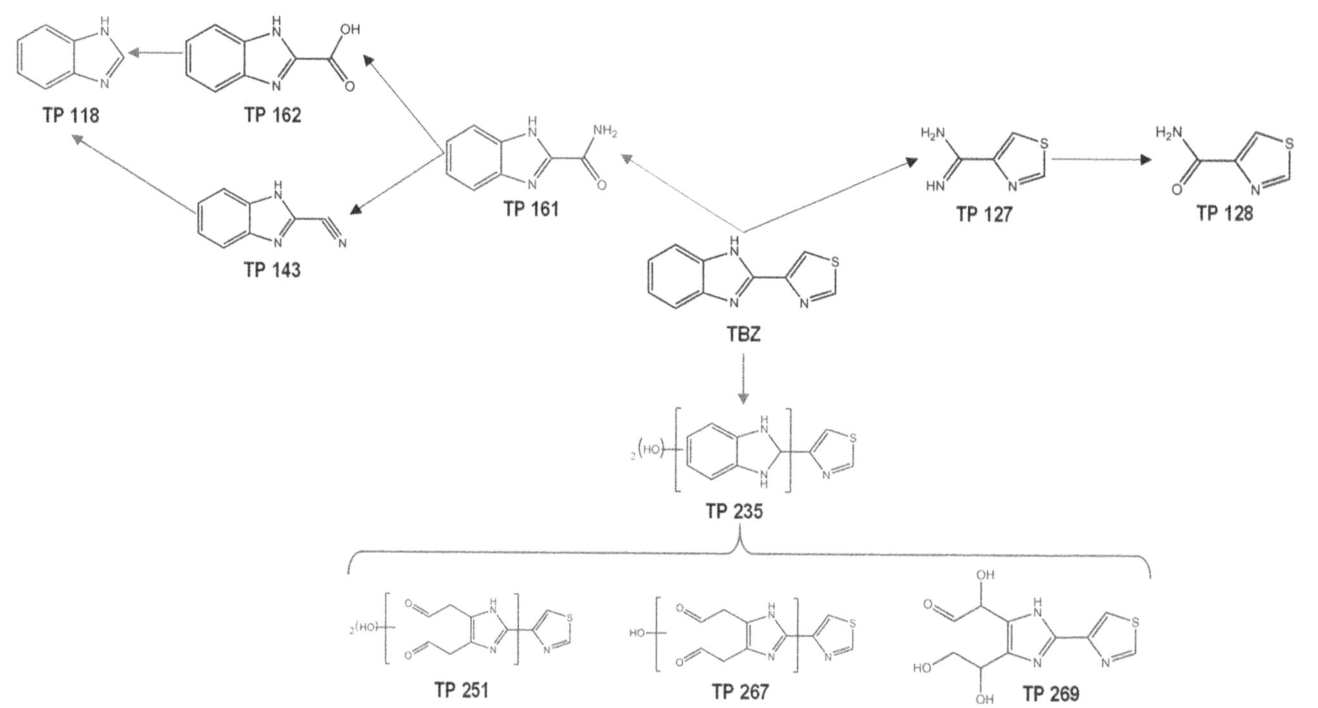

Figure 4 | Tentative transformation pathway of TBZ degradation: black intermediates occur when the system is composed of catalyst-NaOCl-light and the blue ones for MFTO-H_2O_2-light.

for the possible direct breakage of the TBZ molecule by the two N atoms in the imidazole ring. This reaction route is observed in both systems.

Then, a third reaction route is observed for the heterogeneous photo-Fenton. It starts by the formation of the TP235 by the addition of $HO^{\bullet}$ to the benzene ring by electrophilic substitution of two hydrogen atoms. Then, the benzene ring is opened, and new bonds with hydrogen and oxygen molecules are generated, forming TP251, TP267, and TP269 (Sirtori *et al.* 2014). Those TPs may lose their aliphatic chains and can be considered an intermediate step in the formation of TP127 and its subsequent oxidized derivative TP128, according to Calza *et al.* (2003). Nevertheless, a direct breakage of the TBZ molecule by the two N atoms in the imidazole ring could also generate these two TPs, because these sites are likely to be attacked by $HO^{\bullet}$ radicals due to their high electron density (Ji *et al.* 2013).

On another aspect and considering the importance of the reuse cycle of a catalyst, five repetitions using optimum conditions were performed to determine if there is any difference in MP degradation when CS is reused. Figure 3(c) presents the TBZ degradation kinetics using normalized MP concentrations for each reaction cycle. In both systems, dispersion of data is observed during the first minutes, but such difference is diminished as the reaction advances and loses speed.

Results of the CS-NaOCl-SSL system presented low dispersion in the points from 5 to 30 min, while for CS-H_2O_2-SSL two groups of data during the first 30 min (TBZ_D 64 $\pm$ 5%) can be observed, followed by irrelevant differences until the end of the reaction. Consequently, the degradation time required to achieve degradations up 80% (>45 min) is similar (standard deviation of 3%) regardless of the number of the number of times that CS is used, from one to five cycles, which is a satisfactory value according to other bench-scale studies (Bel Hadjltaief *et al.* 2014; Omri *et al.* 2020).

Related MP removal mechanisms

Figure 5 presents the kinetics of TBZ degradation using normalized TBZ concentrations corresponding to the removal mechanisms that may occur by one or the combination of two of the factors of the system. Photo-assisted oxidations (H_2O_2-SSL and NaOCl-SSL) present the best performances with respect to the rest of the mechanisms, achieving 92% of TBZ degradation at 75 min when the oxidant is H_2O_2 and 97% at 7.5 minutes when NaOCl is used instead, which fits with the kind of radicals produced by the photolysis of each oxidant. Complete degradations were achieved at 105 and 20 min using SSL with H_2O_2 or NaOCl, respectively; these reaction times are longer than data of CS-H_2O_2-SSL and CS-NaOCl-SSL.

Photolysis only contributed to 30% of TBZ degradation after 180 min due to the high stability of this contaminant (da Costa *et al.* 2019). A slight improvement of 4% by heterogeneous photo-catalysis was observed, reinforcing the idea that iron oxides are not the best semiconductors for this kind of process, since the recombination of the electron-hole pair is fast (Bin *et al.* 2018), and the porosity exhibits low values in the order of 14.94 m^2/g (Solís-López *et al.* 2014), which is responsible for the 5% of TBZ removal by adsorption.

Experiments carried out in darkness showed lower degradations, since only 11% of TBZ degradation was achieved by chemical oxidation with NaOCl and Fe/NaOCl at 30 min; while, at the same time, 3% of degradation was observed by oxidation with H_2O_2 and Fenton reaction, indicating again a higher reactivity of NaClO in contrast with H_2O_2. However, it is

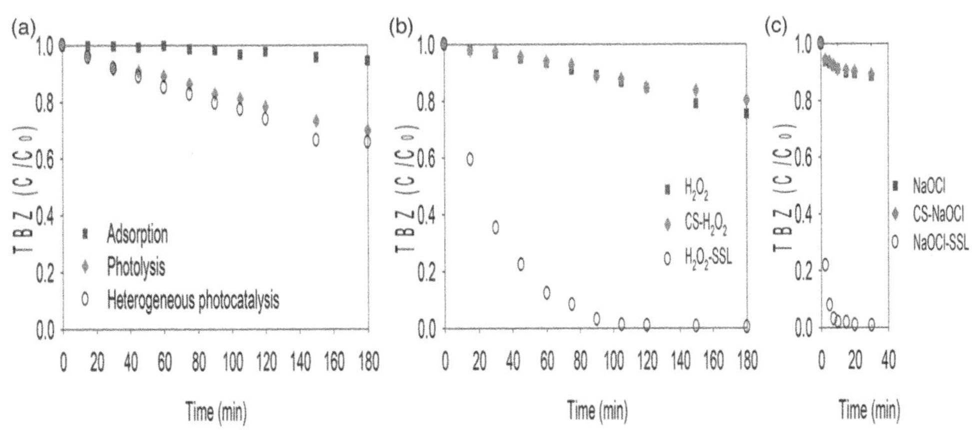

Figure 5 | TBZ degradation kinetics during control assays: (a) processes without oxidant, and processes related to (b) H_2O_2 and (c) NaOCl.

also worth mentioning the importance of SSL in the degradation mechanism to accelerate MP degradation (Zhang *et al.* 2019).

TBZ degradation in complex water matrices

Subsequently, systems composed of three elements were tested in a synthetic and a real WWTP secondary effluent. The composition between both effluents is different, as observed in Table 1; real effluent has higher concentrations of Cl^- and NO_3^- and lower contents of COD and BOD_5 in comparison whit the synthetic effluent, while COD/BOD_5 ratios are similar in both effluents (RE 2.2 and SE 2.0). The differences between the concentrations of some parameters are because the composition of the synthetic effluent corresponds to medium values of effluents issued from aerobic sewage treatment, according to OECD Guidelines (OCED 2001). Such effluents came from different WWTP installations, which do not necessarily follow the same operation strategy, but more importantly, influent compositions are not similar around the world; consequently, the sample from a real WWTP differs in some characteristics (Pescod 1992).

Despite this, the composition of such a synthetic effluent was frequently used to evaluate the performance of the photo-Fenton process in complex water matrices, having a constant composition throughout the experimentation (Klamerth *et al.* 2009). In this work, its use has been proposed in order to give an idea of how the scavenger effect of a more complex matrix, especially inorganic ions, impacts on TBZ degradation, bearing in mind that the degradation observed in distilled water may not be achieved in real applications (Loures *et al.* 2013). In fact, some studies have pointed out that ions naturally present in WWTP secondary effluents, like HCO_3^-, SO_4^{2-}, NO_3^- and Cl^-, could act as both radical scavengers and radical promoters (Gultekin & Ince 2004; Zhang *et al.* 2017).

The TBZ degradation kinetics rates using normalized MP concentrations are shown in Figure 6(a) and 6(c). For the system CS-H_2O_2-SSL, MP removal was slower in RE, with degradations of 61% at 60 min, in comparison with 75% attained in SE; then, by extending the reaction time until 90 min, the degradation values increased to 88 and 77% for SE and RE, respectively. The difference in the speed of MP removal in synthetic and real matrices was expected, due to the differences of the chemical complexity in each matrix. A similar observation was pointed out by de la Obra *et al.* (2017), who have mentioned that real effluents show a much more complex chemical composition than simulated effluents, and thus, the scavenger effect must be major.

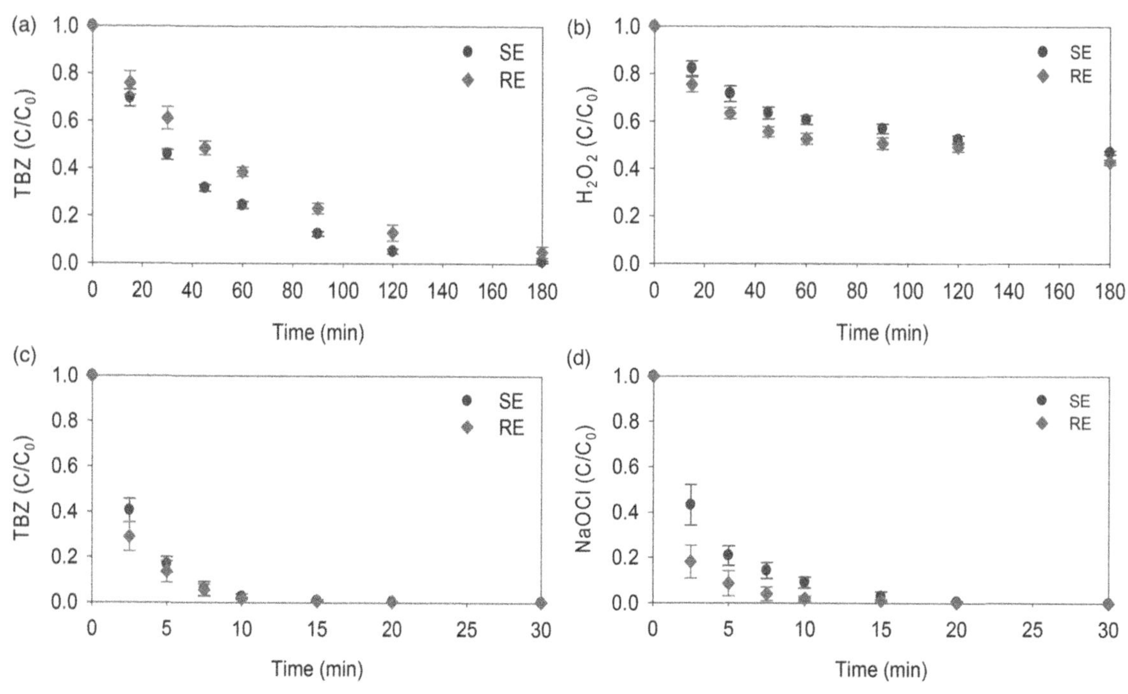

Figure 6 | Effect of the water matrix on the TBZ degradation kinetics, by the systems (a) CS-H_2O_2-SSL and (c) CS-NaOCl-SSL, and oxidant consumption (b) H_2O_2 and (d) NaOCl.

In contrast, the effect of water matrix was hardly evident during the first 7.5 min in the CS-NaOCl-SSL system; at that time, the TBZ degradation was only affected by 3%, the efficiencies being 83 and 86% on synthetic and real effluents, respectively; after 10 min of reaction time, TBZ_D was 94% in both matrices. As oxidant consumption was slightly faster in RE than SE (Figure 6(b) and 6(d)), the water matrix contributed more to oxidant consumption in RE than in SE using the CS-NaOCl-SSL system. This may indicate that the water matrix is not only scavenging radicals but also reacts with the oxidants (Ksibi 2006).

The possibility to take advantage of the sulfate radical formation to accelerate the reaction has been reported for oxidation processes based on sulfate radicals (Wu *et al.* 2015), and UV photolysis (Zhang *et al.* 2017). However, since this ion was in a similar concentration in both SE and RE, differences between the water matrices may be related to concentrations of the anions Cl^- and NO_3^-, which are seven and five times higher, respectively, in RE than SE. On this subject, reports pointed out that the presence of these ions represents a diminution of the degradation rate in AOPs like the UV/H_2O_2 process (Gultekin & Ince 2004) and photo-Fenton process using zero-valent iron (Gomathi Devi *et al.* 2012) or Fe-EDDS complex (Soriano-Molina *et al.* 2019). Nevertheless, there are no reports about the effect of ions present in the water matrix on the kinetics and degradation efficiency of MP using the solar NaOCl/Fe process.

Thus, calculating the kinetic constants for TBZ degradation (Table 4), it is evident that differences between water matrices are near to 30% in the CS-H_2O_2-SSL system, while the CS-NaOCl-SSL system has a difference for TBZ degradation of only 5%. That shows that the ionic concentration has a more important effect on the system whose oxidant is H_2O_2 in comparison to NaOCl. Moreover, results show that SE can be used as a tool to evaluate the performance of the process in real conditions under controlled conditions, but verification in RE is needed.

On the other hand, the values obtained for the CS-NaOCl-SSL system are one order of magnitude faster than CS-H_2O_2-SSL, and the same tendency is followed by the oxidant consumption, indicating that the use of NaOCl instead to H_2O_2 increases considerably the reaction rate.

In this regard, TBZ degradation in synthetic effluents have been reported using a homogeneous photo-Fenton process at both acidic (Rivas *et al.* 2015) and neutral pH (iron additions) (Carra *et al.* 2014). TBZ was fully degraded in 30 and 45 min, respectively; this time is significantly longer in comparison with the CS-NaOCl-SSL system.

The comparison between homogeneous photo-Fenton and the CS-NaOCl-SSL system opens a debate about which environmental impact is more important to avoid. On one hand, operation at both acidic pH implies the acidification and neutralization of the effluent before discharge into the environment (Papoutsakis *et al.* 2015); then, the addition of iron means that this catalyst turns into waste (Rizzo *et al.* 2019), while the extension of reaction times involves high energy consumption. But since the CS-NaOCl-SSL system complies with operation at neutral pH, short reaction time, and the feasibility to incorporate an industrial waste as an iron catalyst, such a process could be the key to diminishing the impacts of MP degradation by Fenton processes.

CONCLUSIONS

Complete TBZ degradation was achieved in three water matrices: distilled water, synthetic secondary effluent, and real secondary effluent of WWTP, by the use of two photo-assisted advanced oxidation processes: CS-H_2O_2-SSL and CS-NaOCl-SSL, separately. We present for the first time the effect of water matrix on the pollutant degradation and the kinetic constants of TBZ degradation using these systems in a comparative way.

Optimum operation conditions were determined in distilled water: $TBZ_0 = 7.36$ mg/L, 1.11 g CS /L, and 50 mg H_2O_2/L or 98 mg NaOCl/L. Results showed up to 90% TBZ degradation at 5 and 60 min when the oxidants were NaOCl and H_2O_2,

Table 4 | Results of the first-order kinetics for TBZ degradation and oxidant consumption

	TBZ_D				H_2O_2		NaOCl	
	CS-H_2O_2-SSL		CS-NaOCl-SSL		CS-H_2O_2-SSL		CS-NaOCl-SSL	
Water matrix	K_{TBZ_D} (1/min)	R^2	K_{TBZ_D} (1/min)	R^2	$K_{H_2O_2}$ (1/min)	R^2	K_{NaOCl} (1/min)	R^2
SE	0.024	0.99	0.365	1.00	0.008	0.96	0.237	0.97
RE	0.016	1.00	0.385	0.99	0.011	0.94	0.379	0.96

respectively. Reaction time differences are related to the type and quantity of radical species formed for each system. In the CS-NaOCl-SSL system, the formation of strong chlorine radicals like $Cl^{\bullet}$ (2.4 eV) and $Cl_2^{\bullet-}$ (2.0 eV) besides $HO^{\bullet}$ radical improved the rate of pollutant degradation, in comparison with the rate for the photo-Fenton-like reaction. Thus, in the CS-H_2O_2-SSL system, the presence of hydroxyl radical ($HO^{\bullet}$) is mainly responsible for the reaction performance, and its speed is strongly linked to the mass transfer of the heterogeneous system.

Slight differences in TBZ degradation efficiency were observed during five reuse cycles, showing that for the CS-H_2O_2-SSL system, a depletion of 80 $\pm$ 3% can be achieved in the same reaction time, 45 min, while at 5 min, an efficiency of 90 $\pm$ 1% was achieved using the CS-NaOCl-SSL system, regardless of the number of times that CS had been used (one to five). This means that CS is a chemically stable iron catalyst.

Finally, both processes were tested in SE and RE to observe the scavenger effect of the water matrix. Cl^- and NO_3^- concentrations were seven and five times higher, respectively, in RE than in SE. Results highlighted that the TBZ degradation in RE was slower than in SE for the heterogeneous photo-Fenton-like process (CS-H_2O_2-SSL), while the heterogeneous photo-NaOCl/Fe process was scarcely affected. In this way, if NaOCl was used instead, 94% TBZ degradation was reached at 10 min in both matrices, while for the CS-H_2O_2-SSL system TBZ degradation had values of 88 and 77% for SE and RE, respectively, at 90 min. It implies differences between the kinetic constants of 30 and 5% between RE and SE for the system oxidized by H_2O_2 and NaOCl, respectively.

Hence, as the CS-NaOCl-SSL system achieves complete degradation quickly and is applicable at neutral pH, is not greatly affected by ion concentrations and the iron catalyst is an industrial waste, this process looks like a feasible alternative for MP degradation, which deserves to be studied in depth in further work.

ACKNOWLEDGEMENTS

Reyna García Estrada acknowledges CONACYT for her Ph.D. scholarship. Sandra Arzate thanks DGAPA for her postdoctoral fellowship. The authors acknowledge the financial support granted by PAPIIT, UNAM (IT100921). Authors are also grateful to MSc. Leticia García Montes de Oca for the HPLC analysis.

DATA AVAILABILITY STATEMENT

All relevant data are included in the paper or its Supplementary Information.

CONFLICT OF INTEREST

The authors declare there is no conflict.

REFERENCES

Ahmadi, A., Zarei, M., Hassani, A., Ebratkhahan, M. & Olad, A. 2021 Facile synthesis of iron(II) doped carbonaceous aerogel as a three-dimensional cathode and its excellent performance in electro-Fenton degradation of ceftazidime from water solution. *Separation and Purification Technology* **278** (June), 119559. https://doi.org/10.1016/j.seppur.2021.119559.

Almanza, R. & López, S. 1978 Total solar radiation in Mexico using sunshine hours and meteorological data. *Solar Energy* **21** (5), 441–448. Available from: https://www.sciencedirect.com/science/article/pii/0038092X78901780?via%3Dihub.

Ameta, R., Chohadia, A. K., Jain, A. & Punjabi, P. B. 2018 Fenton and Photo-Fenton Processes. pp. 49–87 in Advanced Oxidation Processes for Waste Water Treatment, edited by Ameta, S. C. & Ameta, R. Academic Press.

Armstronga, D. A., Huie, R. E., Lymar, S., Koppenol, W. H., Merényi, G., Neta, P., Stanbury, D. M., Steenken, S. & Wardman, P. 2013 Standard electrode potentials involving radicals in aqueous solution: inorganic radicals. *BioInorganic Reaction Mechanisms* **9** (1–4), 59–61.

Arzate-Salgado, S.-Y., Morales-Pérez, A.-A., Solís-López, M. & Ramírez-Zamora, R.-M. 2016 Evaluation of metallurgical slag as a Fenton-type photocatalyst for the degradation of an emerging pollutant: diclofenac. *Catalysis Today* **266**, 126–135.

ASTM 1999 *ASTM D1193-99e1 Standard Specification for Reagent Water*. Available from: https://www.astm.org/DATABASE.CART/HISTORICAL/D1193-99E1.htm.

AWWA/APHA/WEF 2012 In: *Standard Methods for the Examination of Water and Wastewater* (Rice, E. W., Baird, R. B., Eaton, A. D. & Clesceri, L. S. C., eds). American Public Health Association, American Water Works Association, and Water Environment Federation, USA. Available from: https://www.techstreet.com/standards/standard-methods-for-the-examination-of-water-and-wastewater-22nd-edition-out-of-print?product_id=1817564.

Behin, J., Akbari, A., Mahmoudi, M. & Khajeh, M. 2017 Sodium hypochlorite as an alternative to hydrogen peroxide in Fenton process for industrial scale. *Water Research* **121**, 120–128. Available from: https://www.sciencedirect.com/science/article/abs/pii/S0043135417303743?via%3Dihub (accessed 20 December 2019).

Bel Hadjltaief, H., Da Costa, P., Beaunier, P., Gálvez, M. E. & Ben Zina, M. 2014 Fe-clay-plate as a heterogeneous catalyst in photo-Fenton oxidation of phenol as probe molecule for water treatment. *Applied Clay Science* **91–92**, 46–54.

Bernabeu, A., Vercher, R. F., Santos-Juanes, L., Simón, P. J., Lardín, C., Martínez, M. A., Vicente, J. A., González, R., Llosá, C., Arques, A. & Amat, A. M. 2011 Solar photocatalysis as a tertiary treatment to remove emerging pollutants from wastewater treatment plant effluents. *Catalysis Today* **161**, 235–240.

Bin, R., Mun, K., Wei, C., Pan, G. T., Yang, T. C. K. K., Ching, J., Lee, R. B., Lee, K. M., Lai, C. W., Pan, G. T., Yang, T. C. K. K. & Juan, J. C. 2018 The relationship between iron and Ilmenite for photocatalyst degradation. *Advanced Powder Technology* **29** (8), 1779–1786. https://doi.org/10.1016/j.apt.2018.04.013.

Cabrera Reina, A., Casas López, J. L., Maldonado Rubio, M. I., Santos-Juanes Jordá, L., García Sánchez, J. L. & Sánchez Pérez, J. A. 2014 Effects of environmental variables on the photo-Fenton plant design. *Chemical Engineering Journal* **237**, 469–477. http://dx.doi.org/10.1016/j.cej.2013.10.046 (accessed 24 July 2017).

Calza, P., Baudino, S., Aigotti, R., Baiocchi, C. & Pelizzetti, E. 2003 Ion trap tandem mass spectrometric identification of thiabendazole phototransformation products on titanium dioxide. *Journal of Chromatography A* **984** (1), 59–66.

Carmalin, S. A. & Eder, C. L. 2018 Removal of emerging contaminants from the environment by adsorption. *Ecotoxicology and Environmental Safety* **150** (December 2017), 1–17. https://doi.org/10.1016/j.ecoenv.2017.12.026.

Carra, I., Malato, S., Jiménez, M., Maldonado, M. I. & Sánchez Pérez, J. A. 2014 Microcontaminant removal by solar photo-Fenton at natural pH run with sequential and continuous iron additions. *Chemical Engineering Journal* **235**, 132–140. Available from: http://linkinghub.elsevier.com/retrieve/pii/S138589471301200X.

Carraro, P. M., Benzaquén, T. B., Alfano, O. M., Oliva, M. I. & Eimer, G. A. 2018 Advanced oxidation technologies to remove endocrine disruptors in water effluents based on zinc supported mesoporous catalysts. *Revista Mexicana de Ingeniería Quimica* **17** (3), 1135–1146.

Contreras-Bustos, R., Cárdenas Mijangos, J., Decoir-Espinoza, A., Rodríguez-García, A., Montoya-Herrera, L. & Jiménez-Becerril, J. 2019 Treatment of wastewater from the petrochemical industry by chemical Fenton process. *Revista Mexicana de Ingeniería Química* **19** (1), 523–532.

da Costa, E. P., Bottrel, S. E. C., Starling, M. C. V. M., Leão, M. M. D. & Amorim, C. C. 2019 Degradation of carbendazim in water via photo-Fenton in Raceway Pond Reactor: assessment of acute toxicity and transformation products. *Environmental Science and Pollution Research* **26** (5), 4324–4336. Available from: https://link.springer.com/article/10.1007/s11356-018-2130-z#citeas.

De la Obra, I., Ponce-Robles, L., Miralles-Cuevas, S., Oller, I., Malato, S. & Sánchez Pérez, J. A. 2017 Microcontaminant removal in secondary effluents by solar photo-Fenton at circumneutral pH in raceway pond reactors. *Catalysis Today* **287**, 10–14.

García-Estrada, R., Esteban García, B., Ramírez-Zamora, R. M. & Sánchez Pérez, J. A. 2020 Micropollutant degradation by the heterogeneous solar photo-Fenton process at circumneutral pH using copper slag. *Journal of Water Process Engineering* **38**, 101562. https://doi.org/10.1016/j.jwpe.2020.101562.

García-Muñoz, P., Pliego, G., Zazo, J. A., Bahamonde, A. & Casas, J. A. 2016 Ilmenite ($FeTiO_3$) as low cost catalyst for advanced oxidation processes. *Journal of Environmental Chemical Engineering* **4** (1), 542–548. https://doi.org/10.1016/j.jece.2015.11.037.

Georgiou, D., Melidis, P., Aivasidis, A. & Gimouhopoulos, K. 2002 Degradation of azo-reactive dyes by ultraviolet radiation in the presence of hydrogen peroxide. *Dyes and Pigments* **52** (2), 69–78.

Ghanbari, F., Zirrahi, F., Olfati, D., Gohari, F. & Hassani, A. 2020 TiO_2 nanoparticles removal by electrocoagulation using iron electrodes: catalytic activity of electrochemical sludge for the degradation of emerging pollutant. *Journal of Molecular Liquids* **310**, 113217. https://doi.org/10.1016/j.molliq.2020.113217.

Gomathi Devi, L., Girish Kumar, S., Mohan Reddy, K. & Munikrishnappa, C. 2012 Effect of various inorganic anions on the degradation of Congo Red, a di azo dye, by the photo-assisted Fenton process using zero-valent metallic iron as a catalyst. *Desalination and Water Treatment* **4** (1–3), 294–305. https://www.tandfonline.com/doi/pdf/10.5004/dwt.2009.478?needAccess = true.

Gualda-Alonso, E., Soriano-Molina, P., López, J. L. C., Sánchez, J. L. G., Plaza-Bolaños, P., Agüera, A. & Pérez, J. A. S. 2022 Large-scale raceway pond reactor for CEC removal from municipal WWTP effluents by solar photo-Fenton. *Applied Catalysis B: Environmental* 121908. Available from: https://linkinghub.elsevier.com/retrieve/pii/S0926337322008499 (accessed 5 September 2022).

Gultekin, I. & Ince, N. H. 2004 Degradation of reactive Azo Dyes by UV/H_2O_2: impact of radical scavengers. *Journal of Environmental Science and Health – Part A Toxic/Hazardous Substances and Environmental Engineering* **39** (4), 1069–1081.

Guo, K., Wu, Z., Shang, C., Yao, B., Hou, S., Yang, X., Song, W. & Fang, J. 2017 Radical chemistry and structural relationships of PPCP degradation by UV/Chlorine treatment in simulated drinking water. *Environmental Science and Technology* **51** (18), 10431–10439.

Hassani, A., Krishnan, S., Scaria, J., Eghbali, P. & Nidheesh, P. V. 2021 Z-scheme photocatalysts for visible-light-driven pollutants degradation: a review on recent advancements. *Current Opinion in Solid State and Materials Science* **25** (5), 100941. https://doi.org/10.1016/j.cossms.2021.100941.

Ji, Y., Zhou, L., Ferronato, C., Salvador, A., Yang, X. & Chovelon, J. M. 2013 Degradation of sunscreen agent 2-phenylbenzimidazole-5-sulfonic acid by TiO_2 photocatalysis: kinetics, photoproducts and comparison to structurally related compounds. *Applied Catalysis B: Environmental* **140–141**, 457–467. http://dx.doi.org/10.1016/j.apcatb.2013.04.046.

Kim, B. S., Jo, S. K., Shin, D., Lee, J. C. & Jeong, S. B. 2013 A physico-chemical separation process for upgrading iron from waste copper slag. *International Journal of Mineral Processing* **124**, 124–127. http://dx.doi.org/10.1016/j.minpro.2013.05.009.

Klamerth, N., Miranda, N., Malato, S., Agüera, A., Fernández-Alba, A. R., Maldonado, M. I. & Coronado, J. M. 2009 Degradation of emerging contaminants at low concentrations in MWTPs effluents with mild solar photo-Fenton and TiO_2. *Catalysis Today* **144** (1–2), 124–130. Available from: http://www.sciencedirect.com/science/article/pii/S0920586109000431 (accessed 14 April 2016).

Klamerth, N., Rizzo, L., Malato, S., Maldonado, M. I., Agüera, A. & Fernández-Alba, A. R. 2010 Degradation of fifteen emerging contaminants at microg L(-1) initial concentrations by mild solar photo-Fenton in MWTP effluents. *Water Research* **44** (2), 545–554. Available from: http://www.ncbi.nlm.nih.gov/pubmed/19853272 (accessed 8 November 2013).

Kong, X., Jiang, J., Ma, J., Yang, Y., Liu, W. & Liu, Y. 2016 Degradation of atrazine by UV/chlorine: efficiency, influencing factors, and products. *Water Research* **90**, 15–23.

Ksibi, M. 2006 Chemical oxidation with hydrogen peroxide for domestic wastewater treatment. *Chemical Engineering Journal* **119** (2–3), 161–165.

Lado Ribeiro, A. R., Moreira, N. F. F., Li Puma, G. & Silva, A. M. T. 2019 Impact of water matrix on the removal of micropollutants by advanced oxidation technologies. *Chemical Engineering Journal* **363** (October 2018), 155–173. Available from: https://www.sciencedirect.com/science/article/pii/S1385894719300968 (accessed 25 March 2019).

Larralde, A. L., Onna, D., Fuentes, K. M., Sileo, E. E., Hojamberdiev, M. & Aldabe Bilmes, S. 2019 Heterogeneous photo-Fenton process mediated by Sn-substituted goethites with altered OH-surface density. *Journal of Photochemistry and Photobiology A: Chemistry* **381** (May), 111856. https://doi.org/10.1016/j.jphotochem.2019.111856.

Lima, M. J., Silva, C. G., Silva, A. M. T., Lopes, J. C. B., Dias, M. M. & Faria, J. L. 2017 Homogeneous and heterogeneous photo-Fenton degradation of antibiotics using an innovative static mixer photoreactor. *Chemical Engineering Journal* **310**, 342–351. Available from: https://www.sciencedirect.com/science/article/pii/S1385894716304715#b0235 (accessed 1 December 2019).

Loures, C., Alcântara, M. & Filho, H. 2013 Advanced oxidative degradation processes: fundamentals and applications. *International Review of Chemical Engineering* **5** (March), 102–120. Available from: http://www.researchgate.net/publication/236735665_Advanced_Oxidative_Degradation_Processes_Fundamentals_and_Applications/file/e0b495192826d14e33.pdf.

Murthy, N. B. K., Moza, P. N., Hustert, K., Raghu, K. & Kettrup, A. 1996 Photolysis of thiabendazole in aqueous solution and in the presence of fulvic and humic acids. *Chemosphere* **33** (10), 1915–1920.

Neyens, E. & Baeyens, J. 2003 A review of classic Fenton's peroxidation as an advanced oxidation technique. *Journal of Hazardous Materials* **98** (1–3), 33–50.

Nowell, L. H. & Hoigni, J. U. R. G. 1992 Photolysis of aqueous chlorine at sunlight. *Production* **26** (5), 593–598. Available from: https://www.sciencedirect.com/science/article/pii/004313549290232S.

OECD 2001 *Simulation Test – Aerobic Sewage Treatment 303A*. Available from: http://biotecnologiebt.com/guide/OECD_227.pdf.

Omri, A., Hamza, W. & Benzina, M. 2020 Photo-Fenton oxidation and mineralization of methyl orange using Fe-sand as effective heterogeneous catalyst. *Journal of Photochemistry and Photobiology A: Chemistry* **393**, 112444.

Papoutsakis, S., Brites-Nóbrega, F. F., Pulgarin, C. & Malato, S. 2015 Benefits and limitations of using Fe(III)-EDDS for the treatment of highly contaminated water at near-neutral pH. *Journal of Photochemistry and Photobiology A: Chemistry* **303**, 1–7. Available from: http://www.sciencedirect.com/science/article/pii/S1010603015000398 (accessed 20 April 2017).

Pescod, M. B. 1992 Agricultural use of sewage sludge. In: *Wastewater Treatment and Use in Agriculture – FAO Irrigation and Drainage Paper 47*. Food and Agriculture Organization of the United Nations, 6.1. Available from: http://www.fao.org/3/t0551e/t0551e03.htm (accessed 22 May 2020).

Pignatello, J. J., Oliveros, E. & MacKay, A. 2006 Advanced oxidation processes for organic contaminant destruction based on the Fenton reaction and related chemistry. *Critical Reviews in Environmental Science and Technology* **36** (1), 1–84.

Portilla-Sangabriel, M., Arzate, S., Macías-Vargas, J. A. & Ramírez-Zamora, R. M. 2021 Assessment of the use of NaClO as an alternative to H_2O_2 in the oxidant-titanium ore-simulated solar light system for thiabendazole degradation. *Topics in Catalysis* **64**, 181–193. https://doi.org/10.1007/s11244-020-01388-8.

Portilla Sangabriel, M., Martínez Piernas, A. B., Agüera, A., Arzate, S., Sánchez-Pérez, J. A. & Ramírez-Zamora, R. M. 2022 Degradation of thiabendazole and its transformation products by two photo-assisted iron-based-processes in a raceway pond reactor. *Topics in Catalysis* **65**, 1113–1127. https://doi.org/10.1007/s11244-022-01638-x.

Rivas, G., Carra, I., García Sánchez, J. L., Casas López, J. L., Malato, S. & Sánchez Pérez, J. A. 2015 Modelling of the operation of raceway pond reactors for micropollutant removal by solar photo-Fenton as a function of photon absorption. *Applied Catalysis B: Environmental* **178**, 210–217.

Rizzo, L., Malato, S., Antakyali, D., Beretsou, V. G., Đolić, M. B., Gernjak, W., Heath, E., Ivancev-Tumbas, I., Karaolia, P., Lado Ribeiro, A. R., Mascolo, G., McArdell, C. S., Schaar, H., Silva, A. M. T. & Fatta-Kassinos, D. 2019 Consolidated vs new advanced treatment methods for the removal of contaminants of emerging concern from urban wastewater. *Science of The Total Environment* **655**, 986–1008. Available from: https://www.sciencedirect.com/science/article/pii/S0048969718346229 (accessed 25 June 2019).

Sandoval-Gonález, A., Álvarez-Gallegos, A., Hernández, J. A. & Silva-Martínez, S. 2021 Degradation of sucralose present in Splenda® sweetener by TiO_2 photocatalysis assisted with photo-Fenton. *Revista Mexicana de Ingeniería Química* **20** (1), 213–226.

Sirtori, C., Agüera, A., Carra, I. & Sanchéz Pérez, J. A. 2014 Identification and monitoring of thiabendazole transformation products in water during Fenton degradation by LC-QTOF-MS. *Analytical and Bioanalytical Chemistry* **406** (22), 5323–5337.

Solís-López, M., Durán-Moreno, A., Rigas, F., Morales, A. A., Navarrete, M. & Ramírez-Zamora, R. M. 2014 *Assessment of Copper Slag as A Sustainable Fenton-Type Photocatalyst for Water Disinfection*. Elsevier Inc. Available from: http://linkinghub.elsevier.com/retrieve/pii/B9780124116450000092

Soriano-Molina, P., Plaza-Bolaños, P., Lorenzo, A., Agüera, A., García Sánchez, J. L., Malato, S. & Sánchez Pérez, J. A. 2019 Assessment of solar raceway pond reactors for removal of contaminants of emerging concern by photo-Fenton at circumneutral pH from very different municipal wastewater effluents. *Chemical Engineering Journal* **366**, 141–149. Available from: https://www.sciencedirect.com/science/article/pii/S1385894719303043 (accessed 9 February 2019).

Tokumura, M., Sugawara, A., Raknuzzaman, M., Habibullah-Al-Mamun, M. & Masunaga, S. 2016 Comprehensive study on effects of water matrices on removal of pharmaceuticals by three different kinds of advanced oxidation processes. *Chemosphere* **159**, 317–325.

Wu, Y., Bianco, A., Brigante, M., Dong, W., De Sainte-Claire, P., Hanna, K. & Mailhot, G. 2015 Sulfate radical photogeneration using Fe-EDDS: influence of critical parameters and naturally occurring scavengers. *Environmental Science and Technology* **49** (24), 14343–14349.

Zhang, Y., Zhang, J., Xiao, Y., Chang, V. W. C. & Lim, T. T. 2017 Direct and indirect photodegradation pathways of cytostatic drugs under UV germicidal irradiation: process kinetics and influences of water matrix species and oxidant dosing. *Journal of Hazardous Materials* **324**, 481–488. http://dx.doi.org/10.1016/j.jhazmat.2016.11.016.

Zhang, M., Dong, H., Zhao, L., Wang, D. & Meng, D. 2019 A review on Fenton process for organic wastewater treatment based on optimization perspective. *Science of The Total Environment* **670**, 110–121. Available from: https://www.sciencedirect.com/science/article/pii/S0048969719311684 (accessed 21 March 2019).

Zhu, Y., Zhu, R., Xi, Y., Zhu, J., Zhu, G. & He, H. 2019 Strategies for enhancing the heterogeneous Fenton catalytic reactivity: a review. *Applied Catalysis B: Environmental* **255**, 117739. https://doi.org/10.1016/j.apcatb.2019.05.041.

First received 26 September 2022; accepted in revised form 9 December 2022. Available online 22 December 2022

© 2022 The Authors

doi: 10.2166/wst.2022.266

Visible light responsive Cu-N/TiO_2 nanoparticles for the photocatalytic degradation of bisphenol A

Sukanya Krishnan, Ansaf V. Karim and Amritanshu Shriwastav*
Environmental Science and Engineering Department, Indian Institute of Technology Bombay, Mumbai 400 076, India
*Corresponding author. E-mail: amritan@iitb.ac.in

AS, 0000-0002-1964-4207

ABSTRACT

Visible light active co-doped Cu-N/TiO_2 photocatalyst was synthesized by the sol-gel method. The synthesized catalysts were characterized by X-ray diffraction (XRD), field-emission transmission electron microscope (FE-TEM), X-ray photoelectron spectroscopy (XPS), Fourier transform infrared spectroscopy (FTIR), and UV–visible diffuse reflectance spectrophotometry (UV-vis DRS). The co-doping with Cu-N reduced the bandgap (~2.83 eV) and extended the optical absorption range of TiO_2 catalysts to the visible region. The incorporation of Cu and N on TiO_2 lattice results in sub-conduction and valence band formation, which enhanced the photoactivity and electron-hole generation rate. The visible light activity of Cu-N/TiO_2 was evaluated via photocatalytic degradation of bisphenol A (BPA) under blue LED illumination. The maximum BPA degradation of 42.7% was observed at 0.5 g L^{-1} catalyst dosage, initial pH of BPA solution = 8.2, and initial BPA concentration of 10 ppm. Further, a possible mechanism of photocatalytic degradation of BPA was also established.

Key words: bisphenol A, co-doping, Cu-N/TiO2, photocatalytic degradation, visible light activity

HIGHLIGHTS

- Co-doped Cu-N/TiO_2 nanoparticles were used for the photocatalytic degradation of bisphenol A.
- Enhanced visible light activity was achieved with co-doping of Cu and N on TiO_2.
- Comprehensive characterization of the catalyst and effect of major operational parameters on degradation efficiency were studied.
- At optimum conditions maximum bisphenol A degradation of 42.7% was observed.

GRAPHICAL ABSTRACT

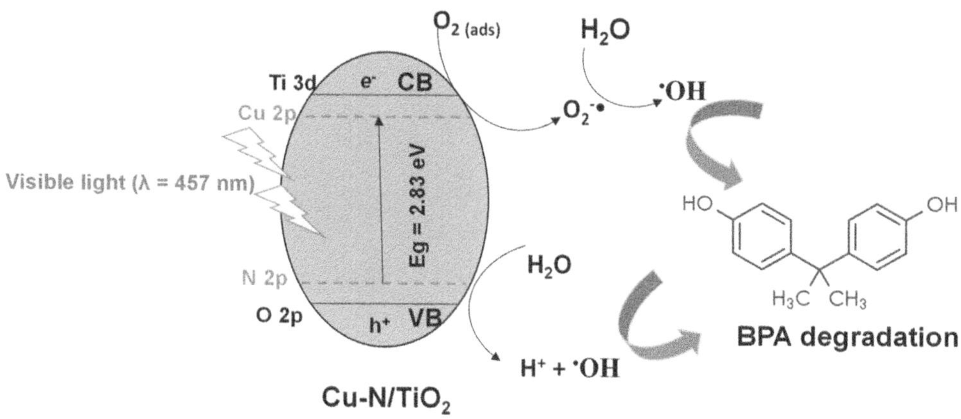

1. INTRODUCTION

Environmental pollution due to emerging contaminants (EC) in water is one of the globally pressing ecological problems (Ali *et al.* 2019; Kelly *et al.* 2019). Despite this fact, several organic compounds are discharged into water bodies without any prior treatment. Bisphenol A (BPA) is an EC widely present in wastewater and has been extensively used for the production of epoxy resins and polycarbonate plastics (Ali *et al.* 2020; Wang *et al.* 2021). They have been detected in various environmental matrices including surface water, landfill leachate, wastewater effluents, etc. (Zhang *et al.* 2018). Conventional treatment processes for water and wastewater cannot efficiently remove these pollutants and therefore advanced treatment processes are highly required (Hieu *et al.* 2018). The strong potential of advanced oxidation processes producing highly reactive and unselective radical species can be effectively used for the degradation of complex organic compounds (Murgolo *et al.* 2015; Divyapriya *et al.* 2018; Davuluri *et al.* 2019; Mohan *et al.* 2021). With the development of nanostructured materials with fascinating structural and catalytic properties, heterogeneous photocatalysis had been proved to be a good approach for the removal of organic and inorganic contaminants (Noorimotlagh *et al.* 2018; Ye *et al.* 2018). Owing to their low cost, higher stability, and environmental friendliness, TiO_2-based catalysts had been widely used in heterogeneous photocatalysis (Hieu *et al.* 2018). However, the relatively wider bandgap of the TiO_2 catalyst, which requires UV irradiation for their activation, limits their practical use (Parnicka *et al.* 2017).

Consequently, various attempts including doping, dye sensitization, and coupling semiconductors were developed to modify the visible light activity of TiO_2 for extending their absorption range into the visible light region (Zhang *et al.* 2016; Menon *et al.* 2019; Surenjan *et al.* 2019; Kovalevskiy *et al.* 2020). The visible light responses in doped materials arise from localized energy levels of dopants or the defects associated with oxygen vacancies (Banerjee *et al.* 2014). The formation of defects on the surface of the catalyst surface could also prevent charge recombination and improve photocatalytic performance (Ghazalian *et al.* 2017). The doping of TiO_2 with transition metals such as Fe, Cu, Co, etc. can enhance the photocatalytic properties of TiO_2 by promoting the separation of holes and electrons by trapping photogenerated charge carriers (Khaki *et al.* 2017). For instance, Cu doping on TiO_2 can replace some of the Ti^{4+} ions on the substitutional sites and reduce the bandgap of the catalyst due to the narrow bandgap energies of CuO and Cu_2O (Ganesh *et al.* 2014; Martínez Vargas *et al.* 2015). Similarly, non-metal doping with N, C, P, etc., can effectively narrow the bandgap of TiO_2 by raising their bandgap level or bandgap alteration by forming intermediate energy levels (Etacheri *et al.* 2015; Chang & Hu 2020). N doping has shown a promising effect on extending the activity of TiO_2 catalysts in visible light due to its comparable atomic radii of N and O, resulting in more substitutional doping (Banerjee *et al.* 2014). The 2p orbital of N can overlap with the 2p orbital of O, resulting in narrowing the bandgap (Asahi *et al.* 2001; Kongsong *et al.* 2018). However, the thermal instability and formation of charge recombination centers during metal doping and decrease in the surface area and absorbability of the catalysts during non-metal doping are major drawbacks (Zhang *et al.* 2008; Senthilnathan & Philip 2010).

In this regard, co-doping can improve the visible light activity due to the synergistic effect of both compounds tuning the electronic structure of the catalyst. Further, co-doping can help shift the absorption region to a lower energy level and create large amounts of single-electron-trapped oxygen vacancies (Liu *et al.* 2015; Eslami *et al.* 2016). Also, the co-doping improves the charge separation and hence can accelerate the transmission of photocurrent carriers (Cong *et al.* 2007). It is presumed that co-doping with N and Cu ions induced the formation of intermediate levels near the valence band of the catalyst resulting in narrowing the bandgap of the catalyst. Another vital factor that improves photocatalysis efficiency at the industrial level is using cost-effective light sources or solar irradiation for degradation studies. By enhancing the visible light activity of photocatalysts, low-power emitting LEDs with a longer life span, compatibility, and narrow light emission spectra can be used (Ye *et al.* 2018; Kelly *et al.* 2019).

In the present work, Cu-N/co-doped TiO_2 nanoparticles synthesized using a sol-gel method were used for the photocatalytic degradation of BPA under blue light irradiation. The degradation of BPA using Cu-N/TiO_2 nanoparticles under blue LED light irradiation has not been investigated earlier. A detailed characterization of the synthesized catalyst was done, followed by the BPA degradation studies. A mechanism for enhanced visible light photodegradation activity of Cu-N/TiO_2 catalyst was also proposed.

2. MATERIALS AND METHODS

2.1. Synthesis of co-doped Cu-N/TiO_2

The catalyst was prepared using a modified sol-gel process with titanium isopropoxide (97%, Sigma Aldrich, CAS no: 546-68-9) as the precursor for TiO_2, NH_4Cl (99%, Loba chemicals, CAS no. 12125-02-9), and $CuCl_2$ (98%, Sigma Aldrich, CAS no. 7447-39-4) as the N and Cu sources, respectively (Al-Muhtaseb & Khraisheh 2015; Cheng *et al.* 2016; Karim & Shriwastav 2020). In brief, the synthesis was started by mixing titanium isopropoxide and methanol (99.9%, Finar, CAS no. 67-56-1) in a volumetric ratio of 1:1. For the co-doped catalyst, the required amount of NH_4Cl and $CuCl_2$ were mixed with acidified deionized water (0.25 mL concentrated HNO_3 in 100 mL of deionized water) and were added to the solution containing titanium isopropoxide. The Ti:N:Cu during the synthesis of the nanoparticles was maintained as a 1:0.5:0.5 (molar ratio). The reaction mixture was stirred at 450 rpm, maintaining a solution temperature between 55–65 °C for 5 h. Then the reaction mixture was dried in a hot air oven, calcinated at 450 °C, and washed to obtain the doped nanoparticles. Similarly, pure TiO_2 particles were also synthesized as per our previous work (Karim & Shriwastav 2020) without the addition of Cu and N source chemicals. All the stock solutions for catalyst synthesis were prepared in deionized water.

2.2. Characterization of catalysts

The phase structure of the catalyst particles was identified using a X-ray diffractometer (X' Pert Pro, PANalytical) with Cu-Kα incident radiation beam ($\lambda = 1.54$ Ao) in the scan range of 20–80 °. The average crystallite size was determined using Scherer's equation (Li *et al.* 2016; Neogy *et al.* 2019). TEM images were recorded using a field emission gun electron microscope with a beam energy of 300 kV. FTIR spectrum was recorded in a Bruker Vertex 80 FTIR using the KBr pellet method. The elemental state and chemical composition of the elements present were detected by X-ray photoelectron spectroscopy (XPS) with AXIS Supra, Kratos Analytical, UK. The diffuse reflectance spectra were measured using a UV-Vis spectrophotometer (Carry 100, Agilent Technologies, USA) with barium sulphate as a standard to determine the bandgap of the particles. The pH drift method was employed to determine the point of zero charge (pHzc) of the catalysts (Khataee *et al.* 2017). In this method, pHzc of catalysts were determined by adjusting the pH of 0.1 M NaCl from 3 to 9 pH values by adding 0.1 M HCl or NaOH solutions. The catalysts measuring approximately 0.05 g were added in each of the pH adjusted (3, 4, 5, 6, 7, 8 and 9) 0.1 M NaCl solution in a centrifuge tube (50 mL) and were shaken in a shaking waterbath (OmniX 72) at 120 rpm for 24 hrs. The final pH was measured and the point at which pH(final) and pH(initial) curves intersect was taken as the pHzc of the catalysts.

2.3. Photocatalytic degradation experiments with bisphenol A

The BPA (98%, Loba, CAS no. 80-05-7) stock solution of 250 mg L^{-1} concentration was prepared with 10% methanol in distilled water and used for the photocatalytic experiments. In a typical photocatalytic experiment, catalyst concentration (0.5 g L^{-1}) was added to 250 mL of the aqueous solution of BPA with initial concentration of 10 mg L^{-1}, prepared from the stock solution of BPA (250 mg L^{-1}), at natural pH (~5.7). At the initial stage of photocatalysis, the dark adsorption was carried out for 30 minutes to reach adsorption equilibrium. Then the solution was exposed to light irradiation with blue LED light (λmax = 457 nm) under magnetic stirring at 450 rpm. The BPA concentration was analyzed after filtering the samples (~3 mL) collected in a time interval of 30 minutes through a 0.2-μm cellulose nitrate filter paper by using a UV-Vis spectrophotometer (Thermo-Scientific, Orion Aquamate 8000) at a maximum detection wavelength of 277 nm (Neogy *et al.* 2019). The effects of N, Cu, and co-doping Cu-N/TiO_2 as a baseline experiment were conducted. Further, detailed experiments were carried out by changing the initial pH of BPA solution, concentration of catalysts, and initial concentration of BPA solution. The reuse experiments were conducted at optimized condition, and after each run the catalyst particles were recovered by 0.45 μm cellulose nitrate filter paper followed by drying in a hot air oven. Then the recovered dried catalysts were used for the next cycle. The effect of loss of catalysts was accounted by adjusting the volume of the solution and distance of light from the water level, so that the catalyst concentration remained uniform in all the reuse cycles with uniform incident light.

Further, the scavenger experiments were also conducted with isopropyl alcohol (IPA) (99%, Loba Chemicals, CAS no. 67-63-0) for hydroxyl radicals, and formic acid (FA) (85%, Merck, CAS no. 64-18-6) for holes to examine the effect of reactive species on the degradation of BPA. All the photocatalytic experiments were conducted in duplicates.

3. RESULTS AND DISCUSSIONS

3.1. XRD

The XRD pattern of pure TiO_2, N-TiO_2, Cu-TiO_2 and co-doped Cu-N/TiO_2 are presented in Figure 1(a). The peaks observed for pure TiO_2 at 25.3°, 37.9°, 48.1°, 54.1°, 55.2°, 62.8°, 69.1°, 70.4° and 75.2° correspond to (101), (004), (200), (105), (211), (204), (116), (220) and (215) planes of anatase phase (Isari *et al.* 2020), which is in accordance with the JCPDS:21-1272 standard reference pattern. Also, the additional most prominent peak at 27.4° attributed to the rutile phase, which is mainly present in N, Cu, and co-doped TiO_2. Moreover, it was observed that anatase to rutile transformation occurred with

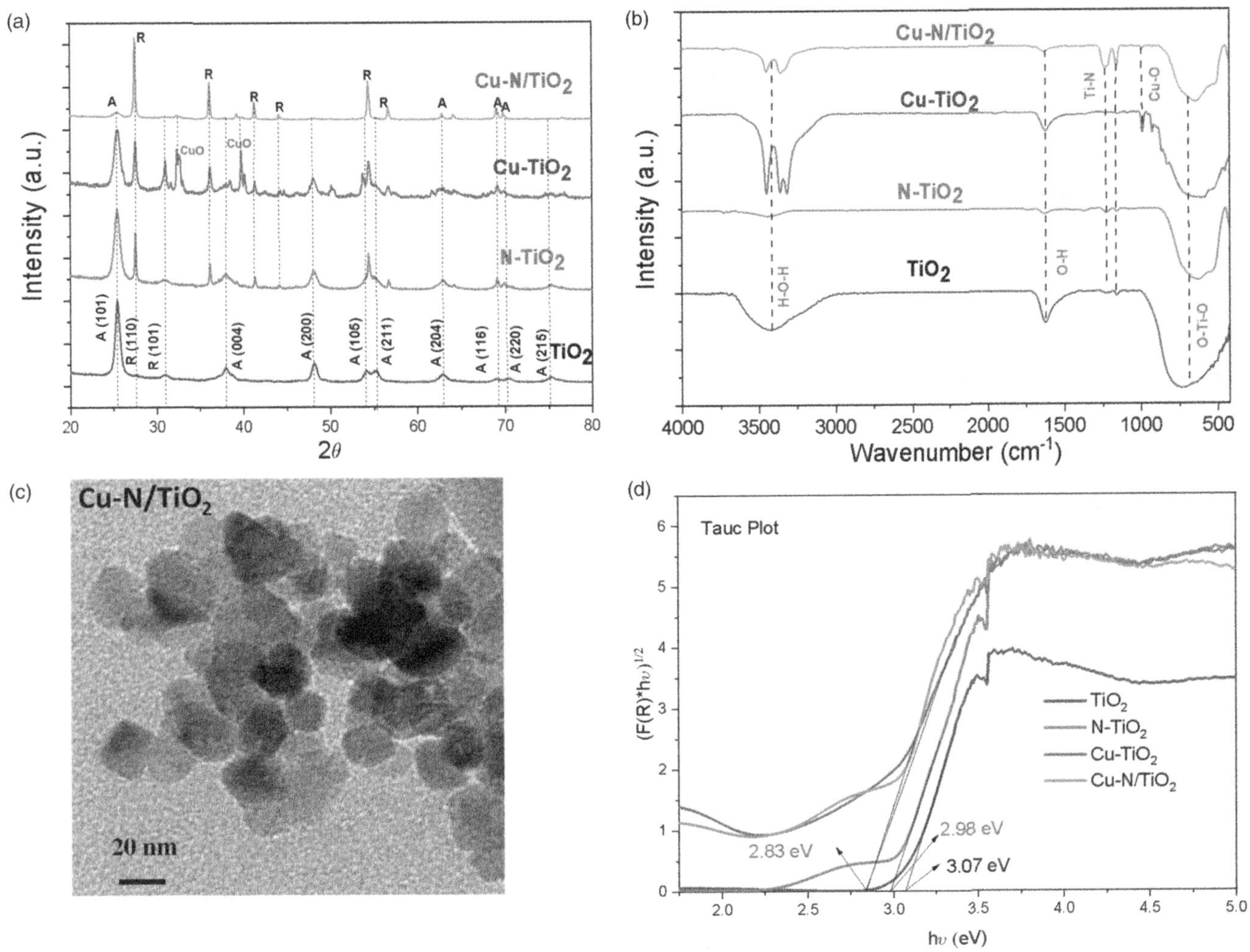

Figure 1 | (a) XRD spectra of TiO_2, N-TiO_2, Cu-TiO_2, and Co-doped Cu-N/TiO_2 nanoparticles; (b) FTIR spectrum of TiO_2, N-TiO_2, Cu-TiO_2, and co-doped Cu-N/TiO_2; (c) TEM image of co-doped Cu-N/TiO_2; and (d) Tauc plot for TiO_2, N-TiO_2, Cu-TiO_2, and co-doped Cu-N/TiO_2.

doping and co-doping with Cu and N (Thongpool *et al.* 2020). This is mainly because of some lattice distortion that happened with Cu doping, since Cu^{2+} has a relatively bigger size (0.087 nm) than Ti^{4+} (0.075 nm) and Cu^{2+} ions find their place at the Ti sites by replacement. In contrast, N doping causes oxygen deficiency in the TiO_2 lattice by replacing O_2^- with N owing to their identical atomic radii (Jaiswal *et al.* 2015). The average crystallite size (nm) was calculated for TiO_2 as 12 nm, for N-TiO_2 as 9 nm, for Cu-TiO_2 as 10 nm, and for co-doped Cu-N/TiO_2 as 11 nm. The study by Wang *et al.* (2014) reported that as the concentration of Cu dopant increases, it prevents crystal growth and decreases grain size. In addition, there is a small shift in the anatase (101) plane observed in Cu-TiO_2 compared with pure TiO_2 due to perturbation in the anatase crystal phase (Aguilar *et al.* 2013). Some characteristic peaks at $2\theta = 32.4°$ and 39.3° corresponding to CuO (tenorite) species were observed for Cu and co-doped Cu-N/TiO_2 crystalline phases (Meghana *et al.* 2015; Dasineh Khiavi *et al.* 2019; Yu *et al.* 2019; Reda *et al.* 2020).

3.2. FTIR

The FTIR spectra of the catalysts TiO_2, N-TiO_2, Cu-TiO_2 and co-doped Cu-N/TiO_2 are presented in Figure 1(b). The signature bands for TiO_2 at 400–800 cm^{-1}, 1,627 cm^{-1} and 3,200–3,600 cm^{-1} were present in all the doped and co-doped TiO_2. The broad band at 400–800 cm^{-1} was attributed to O-Ti-O vibrations and the other bands at 1,627 cm^{-1} and 3,200–3,600 cm^{-1} correspond to O-H-O vibrations by adsorbed water from moisture and -OH stretching and bending vibrations (Neogy *et al.* 2019; Thongpool *et al.* 2020; Asgari *et al.* 2021). The broadening of prominent bands of hydroxyl groups after Cu and N incorporation onto the TiO_2 can enhance their catalytic performance by generating highly reactive hydroxyl radicals during photocatalysis (Isari *et al.* 2020). Apart from the characteristic peaks for TiO_2, the Cu- and N-doped TiO_2 showed additional bands at 1,000 cm^{-1}, 1,232, and 1,160 cm^{-1} corresponding to Cu-O vibrations and Ti-N vibrations, respectively (Reda *et al.* 2020). Further, there was a shift in the O-Ti-O peak for Cu, N, and co-doped TiO_2 confirming the successful incorporation of Cu and N into the titania framework (Isari *et al.* 2020).

3.3. TEM

TEM images of the co-doped Cu-N/TiO_2 nanoparticles are presented in Figure 1(c). The morphology of Cu-N/TiO_2 nanoparticles is primarily comprised of nearly spherical shapes clustered together. The average particle size was calculated by measuring 100 particles using ImageJ software and found to be 14 ± 3 nm.

3.4. UV-Vis DRS

The Kubelka–Munk function and Tauc plot was used to determine the bandgap energy assuming indirect electronic transition (Isari *et al.* 2020; Pedroza-Herrera *et al.* 2020) as shown in Figure 1(d). The bandgap value of pure TiO_2 was around 3.07 eV which is close to the expected value of the anatase phase (3.2 eV) (Jaiswal *et al.* 2015). It was observed that a redshift in the reflection edge was caused by the doping of N and Cu into the TiO_2 lattice (see Fig. S1). After N and Cu doping, the bandgap of TiO_2 was reduced to 2.98 eV and 2.83 eV, respectively. The reduction in the bandgap for Cu-N/TiO_2 (~ 2.83 eV) indicated the formation of an intermediate band or delocalized states inside the TiO_2 bandgap, which is evident from the only one linear zone in the Tauc plot (Aguilar *et al.* 2013; Pedroza-Herrera *et al.* 2020). The reduced bandgap helps in higher visible light absorption and photocatalytic performance than pure TiO_2 nanoparticles for Cu-N/TiO_2 (Dashora *et al.* 2014; Thongpool *et al.* 2020). The co-doping with N and Cu results in various lattice defects and distortions, which improved the visible light absorption and reduced rate of charge recombination (Wang *et al.* 2014; Tahmasebizad *et al.* 2020). As per previous reports, nitrogen as a non-metal dopant creates a new energy state by hybridization of N 2p with O 2p above the valance band of titania, at the same time copper as a metallic dopant introduces a new energy state by Cu 3d hybridizing with Ti 3d near conduction band of TiO_2, enhancing the visible light absorption and thereby reducing the bandgap by the synergistic effect of Cu and N in TiO_2 (Wang *et al.* 2014; Isari *et al.* 2020).

3.5. XPS

The full spectra of XPS for TiO_2 and co-doped Cu-N/TiO_2 are shown in Figure 2(a). The XPS spectra of co-doped Cu-N/TiO_2 nanoparticles showed the additional peaks for Cu and N apart from the characteristic peaks for TiO_2 such as Ti 2p and O 1 s, which confirmed the presence of Cu and N in the catalyst. Figure 2(b) displays Cu 2p spectra of co-doped Cu-N/TiO_2 with Cu $2p_{3/2}$ peak at 931.4 eV and Cu $2p_{1/2}$ peak at 951.7 eV, and a small shake-up satellite peak at around 940.4 eV confirmed the presence of loaded copper in the form of Cu^{2+} in co-doped Cu-N/TiO_2 (Aguilar *et al.* 2013; Khalid *et al.* 2013; Zhao *et al.* 2020). The satellite peak for Cu 2p was observed due to the shake-up transition by a ligand–metal 3d charge transfer (Wang *et al.* 2014; Jaiswal *et al.* 2015). The deconvolution of Cu 2p and N 1 s was not performed due less intense peaks of Cu and N due to their small concentrations (Pedroza-Herrera *et al.* 2020). Figure 2(c) shows the weak peak located at 397.5 eV corresponding to the N 1 s state (Wang *et al.* 2014). This can be attributed to the Ti-N bond in Cu-N/TiO_2 nanoparticles (Zeng *et al.* 2014; Neogy *et al.* 2019). Therefore, it can be concluded that the N and Cu species have been successfully doped onto the TiO_2 catalysts. The deconvoluted peak for O 1 s is shown in Figure 2(d). The two peaks were attributed to lattice oxygen at 531.5 eV and adsorbed OH groups at 529.7 eV (Wang *et al.* 2014; Pongwan *et al.* 2016). The deconvoluted peaks for Ti 2p at 455.8 eV for Ti^{4+} $2p_{3/2}$ and 461.8 eV for Ti^{4+} $2p_{1/2}$ are shown in Figure 2(e). The broad peak at 455.8 eV for Ti^{4+} $2p_{3/2}$ is further deconvoluted to Ti^{3+} $2p_{1/2}$ with 457.6 eV binding energy, attributed to the presence of Ti^{4+} and Ti^{3+} species, respectively. The presence of the Ti^{3+} $2p_{1/2}$ peak signifies the presence of oxygen vacancies in the TiO_2 lattice (Dashora *et al.* 2014). These oxygen vacancies will be filled by N during co-doping in Cu-N/TiO_2 (Neogy *et al.* 2019).

3.6. Photocatalytic activity of catalysts

Figure 3(a) shows the degradation results of BPA under blue visible LED light irradiation for different catalyst materials. There was no noticeable degradation of BPA in the absence of catalyst under visible blue light LED irradiation suggesting that direct photolysis of BPA was insignificant in comparison with catalysts. The presence of the dopant ions improved the photocatalytic performance of pure TiO_2. However, the catalyst activity is dependent on the dopant type under the same experimental conditions. In the presence of N-TiO_2, an improved photocatalytic degradation of 32% was observed

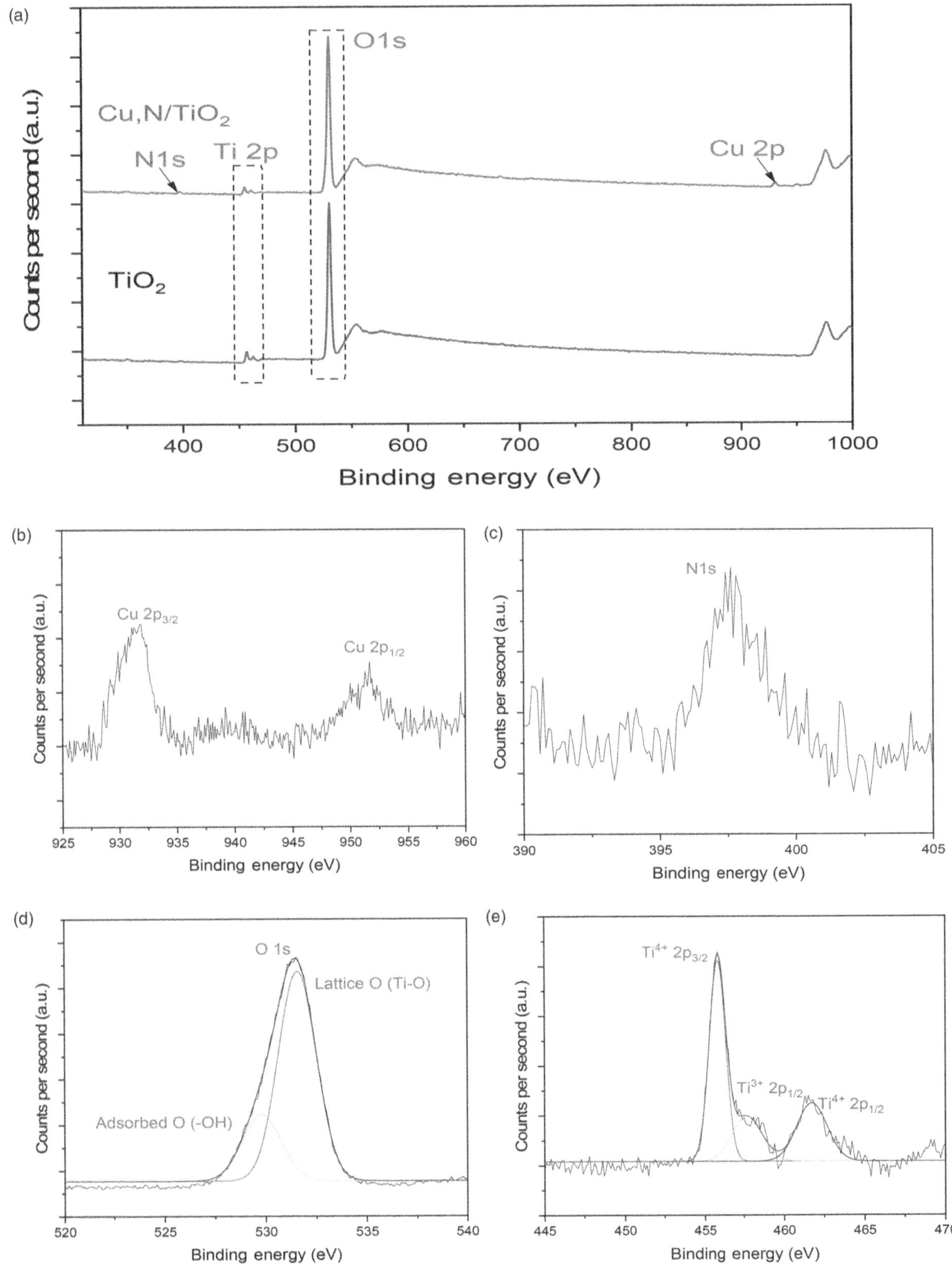

Figure 2 | (a) XPS full spectrum of TiO_2 and Cu-N/TiO_2; (b) Cu 2p spectra of Cu-N/TiO_2; (c) N 1 s spectra of Cu-N/TiO_2; (d) deconvoluted O 1 s peak of Cu-N/TiO_2; (e) deconvoluted Ti 2p peak of Cu-N/TiO_2.

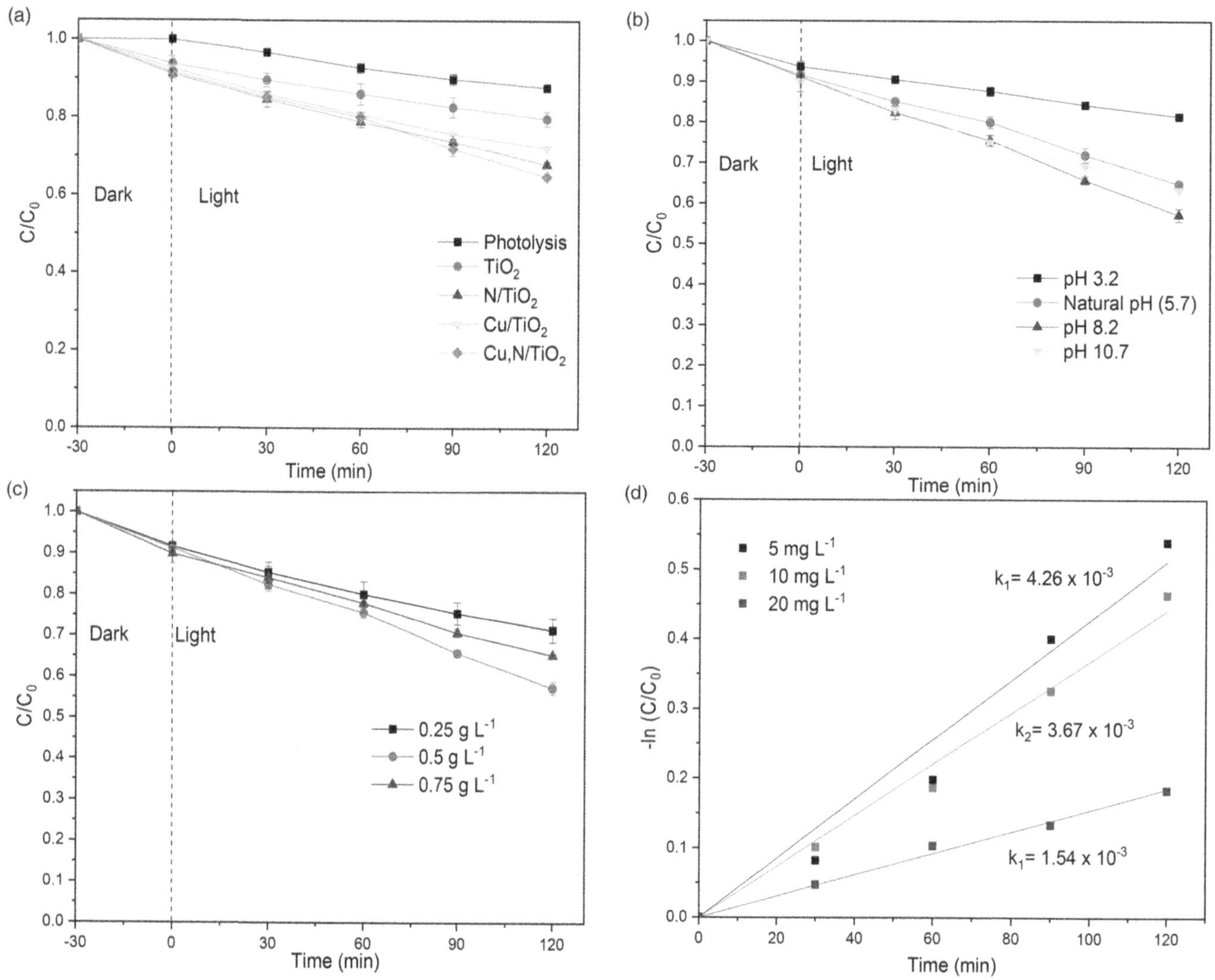

Figure 3 | (a) Baseline experiments with TiO_2, N-TiO_2, Cu-TiO_2, and co-doped Cu-N/TiO_2 nanoparticles (pH = natural pH (5.7), catalyst dosage 0.5 g L^{-1}, initial BPA concentration 10 mg L^{-1}). (b) Effect of initial pH of BPA solution with Cu-N/TiO_2 catalyst (catalyst dosage 0.5 g L^{-1}, initial BPA concentration 10 mg L^{-1}). (c) Effect of catalyst dosage (Initial pH = 8.2, catalyst dosage 0.5 g L^{-1}, initial BPA concentration 10 mg L^{-1}). (d) Effect of initial BPA concentration (catalyst dosage 0.5 g L^{-1}, initial pH = 8.2).

owing to their higher visible light activity due to the formation of a new energy level in the bandgap of TiO_2 (Li *et al.* 2015; Kalantari *et al.* 2016). The enhanced photocatalytic degradation of 27.8% observed for Cu-TiO_2 can be attributed to the homogeneous dispersion of dopant resulting in the interaction of Ti–O–Cu promoting visible light activity and electron trapping (Martínez Vargas *et al.* 2015; Chen *et al.* 2020). However, a synergistic effect on the photocatalytic degradation of BPA was observed for the Cu-N/TiO_2 for BPA degradation resulting in 35.2% degradation after 2 h of visible light irradiation. The higher degradation observed for the co-doped catalyst may be a result of a reduced bandgap of 2.83 eV induced by the formation of intermediate levels near to VB of the catalyst (Huang *et al.* 2017; Zhao *et al.* 2020). Further, the co-doping can also improve the charge separation and hence can accelerate the transmission of photocurrent carriers (Cong *et al.* 2007; Thongpool *et al.* 2020). For the subsequent studies, the co-doped catalyst was chosen and the degradation studies were conducted.

3.6.1. Effect of pH

The initial solution pH is an important parameter as it could influence the catalyst charge and the interaction between BPA molecules and Cu-N/TiO_2 particles (Oh *et al.* 2018). The pHzc of Cu-N/TiO_2 was estimated to be 6.22 by pH drift method (see Fig. S2). Moreover, the pK_a value of BPA is ~9.6 and the compound is positively charged at pH <9.6 and negatively charged at pH >9.6 (Bechambi *et al.* 2015; Hunge *et al.* 2021). Figure 3(b) displays lower photocatalytic degradation at extreme acidic conditions (pH 3.3) and increased with the initial pH solution. It was found that the maximum photocatalytic degradation of 42.7% was observed at pH 8.2 and then decreased at pH 10.3. At extreme acidic conditions (pH 3.3), both the catalyst and BPA are protonated, whereas, at the alkaline state (pH 10.3), the BPA molecules are deprotonated and become negatively charged (Daskalaki *et al.* 2011; Bechambi *et al.* 2016). The observed trend in BPA degradation correlates to the electrostatic repulsive action of the Cu-N/TiO_2 surface and BPA molecules (Hunge *et al.* 2021). Similarly, at natural pH (5.7), both the catalyst and BPA molecules are positively charged and this resulted in lesser degradation. However, at pH 8.2, where the BPA molecules have a positive charge, maximum adsorption of BPA occurred on the Cu-N/TiO_2 surface resulting in higher degradation.

3.6.2. Effect of catalyst dosage

The effect of Cu-N/TiO_2 nanoparticles dosage on the degradation of BPA was investigated at a solution pH of 8.2 and an initial BPA concentration of 10 mg L^{-1} and the results are shown in Figure 3(c). Experiments were performed by varying the catalyst dosage (0.25 g L^{-1} to 0.75 g L^{-1}), and the maximum BPA removal efficiency of 42.7% was found to be at 0.5 g L^{-1} catalyst dosage. The presence of an appropriate catalyst in the solution will be beneficial for electron–hole pairs and enhance BPA degradation (Bechambi *et al.* 2015). At 0.25 g L^{-1} catalyst dose, a lower photocatalytic activity was observed due to insufficient utilization of photon energy at a lower catalyst dosage. As the catalyst loading increases, more active sites were available for BPA adsorption and photocatalytic reactions (Hunge *et al.* 2021). However, with an increase in the catalyst dosage beyond an optimum level, a decline in photocatalytic activity was observed due to scattering of light or turbidity of solution, causing a decrease in the transparency of the solution (Huang *et al.* 2017; Prabavathi *et al.* 2021). Further, it can also cause the agglomeration of particles resulting in less irradiation of the catalyst surface due to the hindrance of light in the dense slurry (Neogy *et al.* 2019).

3.6.3. Effect of initial concentration

The effect of initial BPA concentration on degradation efficiency was studied at 5 mg L^{-1}, 10 mg L^{-1}, and 20 mg L^{-1} at initial solution pH of 8.2 and catalyst dosage of 0.5 g L^{-1} under blue LED light irradiation. After 120 min of irradiation, the degradation reached 51.2% for 5 mg L^{-1} BPA whereas the degradation efficiencies were 43.2% and 21.3% at 10 mg L^{-1} and 20 mg L^{-1} respectively. The linear correlation observed for the logarithmic value of BPA concentration and irradiation time, as shown in Figure 3(d), confirms that the pseudo-first-order-kinetic model fits well with the experimental data. The obtained rate constants are 4.3×10^{-3}, 3.7×10^{-3}, and 1.5×10^{-3} min^{-1} at BPA concentrations of 5 mg L^{-1}, 10 mg L^{-1}, and 20 mg L^{-1} respectively. At constant operational parameters, the amount of radicals generated remains constant, consequently, the ratio of reactive radicals/BPA molecules is reduced at higher concentrations, resulting in lower photodegradation (Tambat *et al.* 2016; Aoudjit *et al.* 2018).

3.6.4. Reuse study

The reuse potential of the synthesized Cu-N/TiO_2 catalysts was investigated without any regeneration of catalysts between the successive cycles. There was ~ 17% reduction of catalysts after each cycle during filtration and recovery process. It was observed that there was significant reduction in the removal after each cycle (see Fig S3). This could be due to the adsorption of BPA intermediates into the active sites of the catalysts surface, thereby reducing the degradation efficiency. This limitation can be overcome by using suitable regeneration techniques after each cycles such as H_2O_2 washing, treatment with UV irradiation, recalcination, washing with NaOH and NH_4OH etc. to recover the photocatalytic activity of the catalysts (Miranda-García *et al.* 2014; Gao *et al.* 2022).

3.7. Possible mechanism of photocatalytic degradation of BPA

The possible mechanism of photocatalysis degradation of BPA with co-doped Cu-N/TiO_2 is shown in Figure 4. Under visible light illumination, the excitation of electrons occurs from the valence band to the conduction band of Cu-N/TiO_2, generating electron and hole pairs. The doping with Cu and N on TiO_2 results in the formation of a sub-conduction band (Cu 2p) and valence band (N 2p), as shown in Figure 4, which enhances the visible light activity and electron-hole pair generation (Liu *et al.* 2013; Isari *et al.* 2020). Also, the excess concentration of Cu dopant in the TiO_2 lattice may reduce Cu^{2+} with holes and inhibit the electron–hole pair separation (Wang *et al.* 2014). These photogenerated

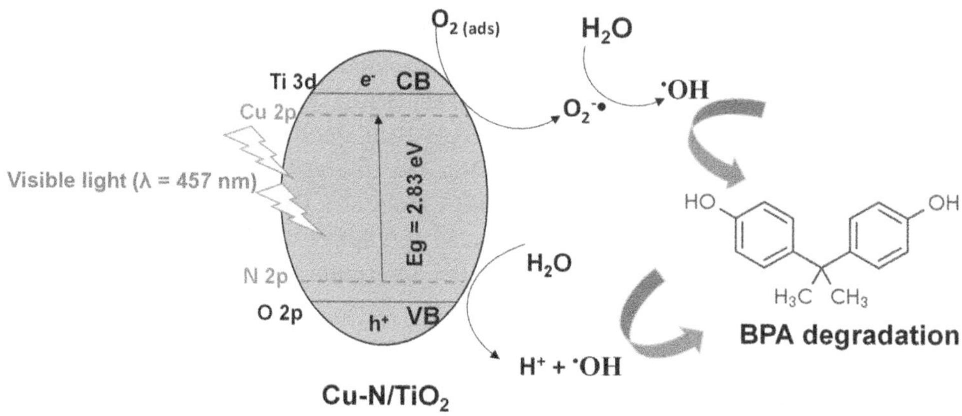

Figure 4 | Possible mechanism of degradation of bisphenol A (BPA) with co-doped Cu-N/TiO_2.

electrons generated by visible light irradiation first react with adsorbed O_2 may also lead to the formation of $O_2^{-\bullet}$ (superoxide) radicals. Simultaneously, the photogenerated holes will directly react with H_2O to form hydroxyl radicals (Zhao *et al.* 2020). In addition, the photogenerated holes can directly oxidize BPA molecules to form intermediates products of degradation (Yu *et al.* 2019). Further, radical scavenger experiments were conducted using isopropyl alcohol (IPA) as $^{\bullet}OH$ radical scavenger and FA as hole scavengers at optimized condition (initial concentration of BPA = 10 ppm, catalyst dosage = 0.5 g L^{-1}, initial pH = 8.2, concentration of scavengers = 5 mM). The results (see Figure 5) indicate that BPA degradation decreases in both the cases indicating that hydroxyl radical ($^{\bullet}OH$) and photogenerated holes (h^+) are the major reactive species in the degradation process of BPA. In addition, more decline in the BPA degradation occurred in the presence of IPA due to the quenching of hydroxyl radicals, which infers that hydroxyl radicals are the major species responsible for the degradation. Therefore, the presence of these reactive oxidizing species can effectively degrade the BPA in the photocatalytic system.

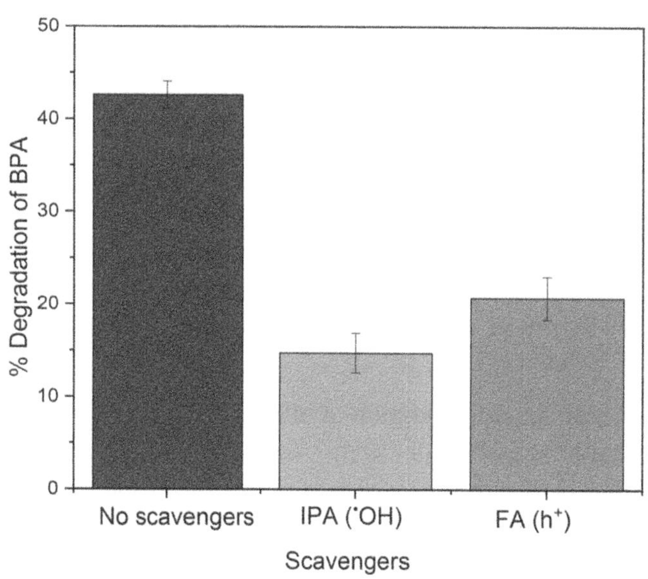

Figure 5 | Scavenger experiments at optimized condition (initial concentration of BPA = 10 ppm, catalyst dosage = 0.5 g L^{-1}, initial pH = 8.2, concentration of scavengers = 5 mM).

4. CONCLUSION

The co-doped Cu-N/TiO_2 nanoparticles were successfully synthesized by the sol-gel technique. The incorporation of Cu and N resulted in anatase to rutile transformation of TiO_2 lattice as evident from the XRD results. The Cu-N/TiO_2 nanoparticles showed an average crystalline size of 11 nm and optical band gap of 2.83 eV with high visible light responses. Doping with Cu and N ions results in the formation of defect sites and oxygen vacancies in the TiO_2 lattice, which is confirmed by the XPS spectroscopy. It was found that Cu ions substitute Ti ions, while N atoms substitute O atoms and occupy the interstitial sites of TiO_2. Maximum degradation of 42.7% was observed with an initial BPA concentration of 10 ppm at 0.5 g L^{-1} catalyst dosage and initial pH of BPA solution of 8.2. Also, the kinetic study showed that the degradation of BPA follows a pseudo-first order pattern. Further, a possible mechanism of BPA degradation was also proposed with Cu-N/TiO_2 catalysts.

ACKNOWLEDGEMENTS

The authors would like to thank the Sophisticated Analytical Instrument Facility (SAIF) and Metallurgical Engineering and Materials Science Department, IIT Bombay, India, for access to their analytical facility.

FUNDING

This study was supported by the funding from Department of Science and Technology (DST), Government of India (DST/TMD/EWO/WTI/2K19/EWFH/2019/183).

DATA AVAILABILITY STATEMENT

All relevant data are included in the paper or its Supplementary Information.

CONFLICT OF INTEREST

The authors declare there is no conflict.

REFERENCES

Aguilar, T., Navas, J., Alcántara, R., Fernández-Lorenzo, C., Gallardo, J. J., Blanco, G. & Martín-Calleja, J. 2013 A route for the synthesis of Cu-doped TiO_2 nanoparticles with a very low band gap. *Chemical Physics Letters* **571**, 49–53. http://dx.doi.org/10.1016/j.cplett.2013.04.007

Ali, S., Li, Z., Chen, S., Zada, A., Khan, I., Khan, I., Ali, W., Shaheen, S., Qu, Y. & Jing, L. 2019 Synthesis of activated carbon-supported TiO_2-based nano-photocatalysts with well recycling for efficiently degrading high-concentration pollutants. *Catalysis Today* **335**, 557–564.

Ali, W., Ullah, H., Zada, A., Muhammad, W., Ali, S., Shaheen, S., Alamgir, M. K., Ansar, M. Z., Khan, Z. U., Bilal, H. & Yap, P. S. 2020 Synthesis of TiO_2 modified self-assembled honeycomb ZnO/SnO_2 nanocomposites for exceptional photocatalytic degradation of 2,4-dichlorophenol and bisphenol A. *Science of the Total Environment* **746**, 141291.

Al-Muhtaseb, A. H. & Khraisheh, M. 2015 Photocatalytic removal of phenol from refinery wastewater: catalytic activity of Cu-doped titanium dioxide. *Journal of Water Process Engineering* **8**, 82–90.

Aoudjit, L., Martins, P. M., Madjene, F., Petrovykh, D. Y. & Lanceros-Mendez, S. 2018 Photocatalytic reusable membranes for the effective degradation of tartrazine with a solar photoreactor. *Journal of Hazardous Materials* **344**, 408–416.

Asahi, R., Morikawa, T., Ohwaki, T., Aoki, K. & Taga, Y. 2001 Visible-light photocatalysis in nitrogen-doped titanium oxides. *Science* **293** (5528), 269–271.

Asgari, E., Sheikhmohammadi, A., Nourmoradi, H., Nazari, S. & Aghanaghad, M. 2021 Degradation of ciprofloxacin by photocatalytic ozonation process under irradiation with UVA: comparative study, performance and mechanism. *Process Safety and Environmental Protection* **147**, 356–366.

Banerjee, S., Pillai, S. C., Falaras, P., O'shea, K. E., Byrne, J. A. & Dionysiou, D. D. 2014 New insights into the mechanism of visible light photocatalysis. *Journal of Physical Chemistry Letters* **5** (15), 2543–2554.

Bechambi, O., Sayadi, S. & Najjar, W. 2015 Photocatalytic degradation of bisphenol A in the presence of C-doped ZnO: effect of operational parameters and photodegradation mechanism. *Journal of Industrial and Engineering Chemistry* **32**, 201–210.

Bechambi, O., Jlaiel, L., Najjar, W. & Sayadi, S. 2016 Photocatalytic degradation of bisphenol A in the presence of Ce-ZnO: evolution of kinetics, toxicity and photodegradation mechanism. *Materials Chemistry and Physics* **173**, 95–105.

Chang, C. W. & Hu, C. 2020 Graphene oxide-derived carbon-doped $SrTiO_3$ for highly efficient photocatalytic degradation of organic pollutants under visible light irradiation. *Chemical Engineering Journal* **383**, 123116.

Chen, M., Wang, H., Chen, X., Wang, F., Qin, X., Zhang, C. & He, H. 2020 High-performance of Cu-TiO_2 for photocatalytic oxidation of formaldehyde under visible light and the mechanism study. *Chemical Engineering Journal* **390**, 124481.

Cheng, X., Yu, X., Xing, Z. & Yang, L. 2016 Synthesis and characterization of N-doped TiO_2 and its enhanced visible-light photocatalytic activity. *Arabian Journal of Chemistry* **29** (3), 291.

Cong, Y., Zhang, J., Chen, F., Anpo, M. & He, D. 2007 Preparation, photocatalytic activity, and mechanism of nano-TiO_2 Co-doped with nitrogen and iron (III). *Journal of Physical Chemistry C* **111** (28), 10618–10623.

Dashora, A., Patel, N., Kothari, D. C., Ahuja, B. L. & Miotello, A. 2014 Formation of an intermediate band in the energy gap of TiO_2 by Cu-N-codoping: first principles study and experimental evidence. *Solar Energy Materials and Solar Cells* **125**, 120–126. http://dx.doi.org/10.1016/j.solmat.2014.02.032.

Dasineh Khiavi, N., Katal, R., Kholghi Eshkalak, S., Masudy-Panah, S., Ramakrishna, S. & Jiangyong, H. 2019 Visible light driven heterojunction photocatalyst of CuO–Cu_2O thin films for photocatalytic degradation of organic pollutants. *Nanomaterials* **9** (7), 1011.

Daskalaki, V. M., Frontistis, Z., Mantzavinos, D. & Katsaounis, A. 2011 Solar light-induced degradation of bisphenol-A with TiO_2 immobilized on Ti. *Catalysis Today* **161** (1), 110–114.

Davuluri, S., P.V., N. & Kumar, M. S. 2019 Detoxification of water and wastewater by advanced oxidation processes. *Science of The Total Environment* **696**, 133961.

Divyapriya, G., Nambi, I. & Senthilnathan, J. 2018 Ferrocene functionalized graphene based electrode for the electro – Fenton oxidation of ciprofloxacin. *Chemosphere* **209**, 113–123.

Eslami, A., Amini, M. M., Yazdanbakhsh, A. R., Mohseni-Bandpei, A., Safari, A. A. & Asadi, A. 2016 N,S co-doped TiO_2 nanoparticles and nanosheets in simulated solar light for photocatalytic degradation of non-steroidal anti-inflammatory drugs in water: a comparative study. *Journal of Chemical Technology and Biotechnology* **91** (10), 2693–2704.

Etacheri, V., Di Valentin, C., Schneider, J., Bahnemann, D. & Pillai, S. C. 2015 Visible-light activation of TiO_2 photocatalysts: advances in theory and experiments. *Journal of Photochemistry & Photobiology, C: Photochemistry Reviews* **25**, 11–29.

Ganesh, I., Kumar, P. P., Annapoorna, I., Sumliner, J. M., Ramakrishna, M., Hebalkar, N. Y., Padmanabham, G. & Sundararajan, G. 2014 Preparation and characterization of Cu-doped TiO_2 materials for electrochemical, photoelectrochemical, and photocatalytic applications. *Applied Surface Science* **293**, 229–247.

Gao, P., Li, Z., Feng, L., Liu, Y., Du, Z. & Zhang, L. 2022 Construction of novel MWCNTs/$Bi_4O_5I_2$ nanosheets with enhanced adsorption and photocatalytic performance for the degradation of tetracycline: efficiency, mechanism and regeneration. *Chemical Engineering Journal* **429**, 132398. https://doi.org/10.1016/j.cej.2021.132398.

Ghazalian, E., Ghasemi, N. & Amani-Ghadim, A. R. 2017 Effect of gadollunium doping on visible light photocatalytic performance of Ag_3PO_4: evaluation of activity in degradation of an anthraquinone dye and mechanism study. *Journal of Molecular Catalysis A: Chemical* **426**, 257–270.

Hieu, C., Fu, C. & Juang, R. 2018 Degradation of methylene blue and methyl orange by palladium- doped TiO_2 photocatalysis for water reuse : efficiency and degradation pathways. *Journal of Cleaner Production* **202**, 413–427.

Huang, Z., Wu, P., Gong, B., Zhang, X., Liao, Z., Chiang, P. C., Hu, X. & Cui, L. 2017 Immobilization of visible light-sensitive (N, Cu) co-doped TiO_2 onto rectorite for photocatalytic degradation of p-chlorophenol in aqueous solution. *Applied Clay Science* **142**, 128–135.

Hunge, Y. M., Yadav, A. A., Khan, S., Takagi, K., Suzuki, N., Teshima, K., Terashima, C. & Fujishima, A. 2021 Photocatalytic degradation of bisphenol A using titanium dioxide@ nanodiamond composites under UV light illumination. *Journal of Colloid and Interface Science* **582**, 1058–1066.

Isari, A. A., Hayati, F., Kakavandi, B., Rostami, M., Motevassel, M. & Dehghanifard, E. 2020 N, Cu co-doped TiO_2@functionalized SWCNT photocatalyst coupled with ultrasound and visible-light: an effective sono-photocatalysis process for pharmaceutical wastewaters treatment. *Chemical Engineering Journal* **392**, 123685.

Jaiswal, R., Bharambe, J., Patel, N., Dashora, A., Kothari, D. C. & Miotello, A. 2015 Copper and nitrogen co-doped TiO_2 photocatalyst with enhanced optical absorption and catalytic activity. *Applied Catalysis B: Environmental* **168–169**, 333–341.

Kalantari, K., Kalbasi, M., Sohrabi, M. & Royaee, S. J. 2016 Synthesis and characterization of N-doped TiO_2 nanoparticles and their application in photocatalytic oxidation of dibenzothiophene under visible light. *Ceramics International* **42** (13), 14834–14842.

Karim, A. V. & Shriwastav, A. 2020 Degradation of ciprofloxacin using photo, sono, and sonophotocatalytic oxidation with visible light and low-frequency ultrasound: degradation kinetics and pathways. *Chemical Engineering Journal* **392**, 124853.

Kelly, J., Morrison, G., Skillen, N., Manesiotis, P. & Robertson, P. K. J. 2019 An investigation of the role of pH in the rapid photocatalytic degradation of MCPA and its primary intermediate by low-power UV LED irradiation. *Chemical Engineering Journal* **359**, 112–118.

Khaki, M. R. D., Shafeeyan, M. S., Raman, A. A. A. & Daud, W. M. A. W. 2017 Application of doped photocatalysts for organic pollutant degradation – A review. *Journal of Environmental Management* **198**, 78–94.

Khalid, N. R., Ahmed, E., Hong, Z., Ahmad, M., Zhang, Y. & Khalid, S. 2013 Cu-doped TiO_2 nanoparticles/graphene composites for efficient visible-light photocatalysis. *Ceramics International* **39** (6), 7107–7113.

Khataee, A., Kayan, B., Gholami, P., Kalderis, D. & Akay, S. 2017 Sonocatalytic degradation of an anthraquinone dye using TiO_2-biochar nanocomposite. *Ultrasonics Sonochemistry* **39**, 120–128.

Kongsong, P., Taleb, A., Masae, M., Jeenarong, A., Hansud, P. & Khumruean, S. 2018 Effect of nitrogen doping on the photocatalytic activity and hydrophobic property of rutile TiO_2 nanorods array. *Surface and Interface Analysis* **50** (12–13), 1271–1277.

Kovalevskiy, N., Selishchev, D., Svintsitskiy, D., Selishcheva, S., Berezin, A. & Kozlov, D. 2020 Synergistic effect of polychromatic radiation on visible light activity of N-doped TiO_2 photocatalyst. *Catalysis Communications* **134**, 105841.

Li, H., Hao, Y., Lu, H., Liang, L., Wang, Y., Qiu, J., Shi, X., Wang, Y. & Yao, J. 2015 A systematic study on visible-light N-doped TiO_2 photocatalyst obtained from ethylenediamine by sol-gel method. *Applied Surface Science* **344**, 112–118.

Li, L., Feng, Y., Liu, Y., Wei, B., Guo, J., Jiao, W., Zhang, Z. & Zhang, Q. 2016 Titanium dioxide nanoparticles modified by salicylic acid and arginine: structure, surface properties and photocatalytic decomposition of p-nitrophenol. *Applied Surface Science* **363**, 627–635.

Liu, Y., Liang, W., Zhang, W., Zhang, J. & Han, P. 2013 First principle study of CuN, Cu and N-doped anatase TiO_2. *Solid State Communications* **164**, 27–31.

Liu, X., Chen, Y. Y., Cao, C., Xu, J., Qian, Q., Luo, Y., Xue, H., Xiao, L., Chen, Y. Y. & Chen, Q. 2015 Electrospun nitrogen and carbon co-doped porous TiO_2 nanofibers with high visible light photocatalytic activity. *New Journal of Chemistry* **39** (9), 6944–6950.

Martínez Vargas, D. X., Rivera De la Rosa, J., Lucio-Ortiz, C. J., Hernández-Ramirez, A., Flores-Escamilla, G. A. & Garcia, C. D. 2015 Photocatalytic degradation of trichloroethylene in a continuous annular reactor using Cu-doped TiO_2 catalysts by sol-gel synthesis. *Applied Catalysis B: Environmental* **179**, 249–261.

Meghana, S., Kabra, P., Chakraborty, S. & Padmavathy, N. 2015 Understanding the pathway of antibacterial activity of copper oxide nanoparticles. *RSC Advances* **5** (16), 12293–12299.

Menon, N. G., Tatiparti, S. S. V. & Mukherji, S. 2019 Synthesis, characterization and photocatalytic activity evaluation of TiO_2–ZnO nanocomposites: elucidating effect of varying Ti:Zn molar ratio. *Colloids and Surfaces A: Physicochemical and Engineering Aspects* **565**, 47–58.

Miranda-García, N., Suárez, S., Maldonado, M. I., Malato, S. & Sánchez, B. 2014 Regeneration approaches for TiO_2 immobilized photocatalyst used in the elimination of emerging contaminants in water. *Catalysis Today* **230**, 27–34. http://dx.doi.org/10.1016/j.cattod.2013.12.048

Mohan, H., Ramalingam, V., Karthi, N., Malathidevi, S., Shin, T., Venkatachalam, J. & Seralathan, K. K. 2021 Enhanced visible light-driven photocatalytic activity of reduced graphene oxide/cadmium sulfide composite: methylparaben degradation mechanism and toxicity. *Chemosphere* **264**, 128481.

Murgolo, S., Petronella, F., Ciannarella, R., Comparelli, R., Agostiano, A., Curri, M. L. & Mascolo, G. 2015 UV and solar-based photocatalytic degradation of organic pollutants by nano-sized TiO_2 grown on carbon nanotubes. *Catalysis Today* **240** (PA), 114–124.

Neogy, A., Rani, S., Garg, A., Singhania, T., Sangal, V. K., Yadav, S. R., Singh, A., Sharma, S. & Garg, N. 2019 Photocatalytic degradation of bisphenol-A using N, Co codoped TiO_2 catalyst under solar light. *Scientific Reports* **9** (1), 1–13.

Noorimotlagh, Z., Kazeminezhad, I., Jaafarzadeh, N., Ahmadi, M., Ramezani, Z. & Silva Martinez, S. 2018 The visible-light photodegradation of nonylphenol in the presence of carbon-doped TiO_2 with rutile/anatase ratio coated on GAC: effect of parameters and degradation mechanism. *Journal of Hazardous Materials* **350**, 108–120.

Oh, W. D., Lok, L. W., Veksha, A., Giannis, A. & Lim, T. T. 2018 Enhanced photocatalytic degradation of bisphenol A with Ag-decorated S-doped g-C_3N_4 under solar irradiation: performance and mechanistic studies. *Chemical Engineering Journal* **333**, 739–749.

Parnicka, P., Mazierski, P., Grzyb, T., Wei, Z., Kowalska, E., Ohtani, B., Lisowski, W., Klimczuk, T. & Nadolna, J. 2017 Preparation and photocatalytic activity of Nd-modified TiO_2 photocatalysts: insight into the excitation mechanism under visible light. *Journal of Catalysis* **353**, 211–222.

Pedroza-Herrera, G., Medina-Ramírez, I. E., Lozano-Álvarez, J. A. & Rodil, S. E. 2020 Evaluation of the photocatalytic activity of copper doped TiO_2 nanoparticles for the purification and/or disinfection of industrial effluents. *Catalysis Today* **341**, 37–48.

Pongwan, P., Wetchakun, K., Phanichphant, S. & Wetchakun, N. 2016 Enhancement of visible-light photocatalytic activity of Cu-doped TiO_2 nanoparticles. *Research on Chemical Intermediates* **42** (4), 2815–2830.

Prabavathi, S. L., Saravanakumar, K., Park, C. M. & Muthuraj, V. 2021 Photocatalytic degradation of levofloxacin by a novel Sm_6WO_{12}/g-C_3N_4 heterojunction: performance, mechanism and degradation pathways. *Separation and Purification Technology* **257**, 117985.

Reda, S. M., Khairy, M. & Mousa, M. A. 2020 Photocatalytic activity of nitrogen and copper doped TiO_2 nanoparticles prepared by microwave-assisted sol-gel process. *Arabian Journal of Chemistry* **13** (1), 86–95.

Senthilnathan, J. & Philip, L. 2010 Photocatalytic degradation of lindane under UV and visible light using N-doped TiO_2. *Chemical Engineering Journal* **161** (1–2), 83–92.

Surenjan, A., Pradeep, T. & Philip, L. 2019 Application and performance evaluation of a cost-effective vis- LED based fluidized bed reactor for the treatment of emerging contaminants. *Chemosphere* **228**, 629–639.

Tahmasebizad, N., Hamedani, M. T., Shaban Ghazani, M. & Pazhuhanfar, Y. 2020 Photocatalytic activity and antibacterial behavior of TiO_2 coatings co-doped with copper and nitrogen via sol-gel method. *Journal of Sol-Gel Science and Technology* **93** (3), 570–578.

Tambat, S., Umale, S. & Sontakke, S. 2016 Photocatalytic degradation of Milling Yellow dye using sol-gel synthesized CeO_2. *Materials Research Bulletin* **76**, 466–472.

Thongpool, V., Phunpueok, A., Jaiyen, S. & Sornkwan, T. 2020 Synthesis and photocatalytic activity of copper and nitrogen co-doped titanium dioxide nanoparticles. *Results in Physics* **16**, 102948.

Wang, S., Yang, X. J., Jiang, Q. & Lian, J. S. 2014 Enhanced optical absorption and photocatalytic activity of Cu/N-codoped TiO_2 nanocrystals. *Materials Science in Semiconductor Processing* **24** (1), 247–253.

Wang, G., Dai, J., Luo, Q. & Deng, N. 2021 Photocatalytic degradation of bisphenol A by TiO_2@aspartic acid-β-cyclodextrin@reduced graphene oxide. *Separation and Purification Technology* **254**, 117574.

Ye, Y., Feng, Y., Bruning, H., Yntema, D. & Rijnaarts, H. H. M. 2018 Photocatalytic degradation of metoprolol by TiO_2 nanotube arrays and UV-LED: effects of catalyst properties, operational parameters, commonly present water constituents, and photo-induced reactive species. *Applied Catalysis B: Environmental* **220**, 171–181.

Yu, X., Zhang, J., Zhang, J., Niu, J., Zhao, J., Wei, Y. & Yao, B. 2019 Photocatalytic degradation of cipro fl oxacin using Zn-doped Cu_2O particles: analysis of degradation pathways and intermediates. *Chemical Engineering Journal* **374**, 316–327.

Zeng, L., Song, W., Li, M., Jie, X., Zeng, D. & Xie, C. 2014 Comparative study on the visible light driven photocatalytic activity between substitutional nitrogen doped and interstitial nitrogen doped TiO_2. *Applied Catalysis A: General* **488**, 239–247.

Zhang, G., Ding, X., Hu, Y., Huang, B., Zhang, X., Qin, X., Zhou, J. & Xie, J. 2008 Photocatalytic degradation of 4BS dye by N,S-codoped TiO_2 pillared montmorillonitne photocatalysts under visible-light irradiation. *Journal of Physical Chemistry C* **112** (46), 17994–17997.

Zhang, X., Peng, T., Song, S. & Chem, J. M. 2016 Recent advances in dye-sensitized semiconductor systems for photocatalytic hydrogen production. *Journal of Materials Chemistry A* **4** (7), 2365–2402.

Zhang, D., Wang, F., Cao, S. & Duan, X. 2018 Investigation on enhanced photocatalytic degradation of bisphenol A with bismuth oxyiodide catalyst using response surface methodology. *RSC Advances* **8** (11), 5967–5975.

Zhao, Z., Omer, A. A., Qin, Z., Osman, S., Xia, L. & Singh, R. P. 2020 Cu/N-codoped TiO_2 prepared by the sol-gel method for phenanthrene removal under visible light irradiation. *Environmental Science and Pollution Research* **27** (15), 17530–17540.

First received 15 June 2022; accepted in revised form 12 August 2022. Available online 23 August 2022

doi: 10.2166/wst.2022.255

Electrochemical oxidation process on palm oil mill effluent waste activated sludge: optimization by response surface methodology

Branda Jian Tong Yap[a], Gan Chin Heng [b,*] and Choon Aun Ng[a]

[a] Department of Environmental Engineering, Faculty of Engineering and Green Technology, Universiti Tunku Abdul Rahman, Jalan Universiti, Bandar Barat, Kampar 31900, Perak, Malaysia

[b] Department of Civil Engineering, Lee Kong Chian Faculty of Engineering and Science, Universiti Tunku Abdul Rahman, Jalan Sungai Long, Bandar Sungai Long, Cheras, Kajang 43000, Selangor, Malaysia

*Corresponding author. E-mail: ganch@utar.edu.my

GCH, 0000-0003-1007-3353

ABSTRACT

Biological-based treatment as the conventional treatment for palm oil mill effluent (POME) in open-ponding system face a well known rate-limiting step which is hydrolysis. In this study, electrochemical oxidation (EO) by a ruthenium oxide-coated titanium (Ti/RuO_2) electrode was introduced as a pre-treatment for POME waste activated sludge (WAS). Surface morphology and elemental analysis were investigated using field emission scanning electron microscopy and energy dispersive X-ray spectroscopy, respectively. Response surface methodology type central composite design was used in this study to understand the relationship between the independent and dependent variables. Analysis of variance (ANOVA) was used to validate the model of the studied variables. The correlation coefficients (R^2) indicated a close agreement between the experimental results and the predicted values, with high R^2 values of 0.9044–0.9773. Multiple response optimization suggested that the range of current density (17–27 mA/cm^2) and electrolysis time (55–75 min) at a fixed concentration of sodium chloride (10 g/L), resulted in mixed liquor volatile suspended solids (MLVSS) removal >20%, capillary suction timer (CST) reduction >43%, extracellular polymeric substances (EPS) increment <19% and soluble chemical oxygen demand (sCOD) increment >25%. EO appears to be an efficient pre-treatment as well as practical way to improve the POME WAS disintegration and dewaterability.

Key words: electrochemical oxidation, palm oil mill effluent, response surface methodology, Ti/RuO_2 electrodes, waste activated sludge

HIGHLIGHTS

- Electrochemical oxidation (EO) for waste activated sludge (WAS) from palm oil mill effluent (POME) was investigated.
- Current density and electrolysis time showed significant interactive effects.
- Optimum operating conditions based on CCD-RSM were ranges at current density 17–27 mA/cm^2 and electrolysis time 55–75 min, at a fixed 10 g/L of NaCl.
- Suitability of EO using Ti/RuO_2 prior to anaerobic digestion was confirmed.

GRAPHICAL ABSTRACT

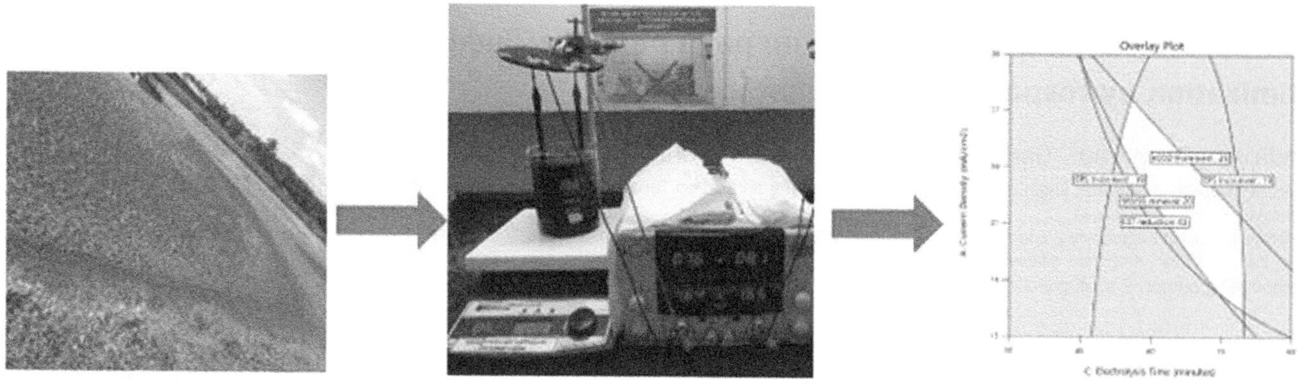

Palm Oil Mill Effluent Waste Activated Sludge

Electrochemical Oxidation Process

Process Optimization by RSM

1. INTRODUCTION

The palm oil industry has always been a very popular and crucial contributor to the economy of Malaysia. Over the years, the number of palm oil mills has increased gradually, consequently generating more and more by-products and waste from the palm oil mill effluent (POME). By-products have had significant impacts upon the environment as they generate excessive biological oxygen demand (BOD) and chemical oxygen demand (COD). Furthermore, POME contains excessive concentrations of organic nitrogen, phosphorus and different supplement substances. These have caused the need for treatment for POME to ensure that it stays within the threshold limit value (TLV) based on the standards set by the local environmental organisations, for instance Department of Environment (DOE), Malaysia before discharge into water bodies (Kamyab *et al.* 2018).

Biological treatment is the most common treatment technology being practiced to treat POME. However, waste activated sludge (WAS), an unwanted residue that contains high organic and solid contents with poor dewaterability will be generated after the biological treatment. Landfilling and incineration are both common ways to handle WAS, however these methods have their downside by either jeopardizing the environment or threating the living population. Anaerobic digestion (AD) has always been the conventional method for treating the sewage treatment plant (STP) WAS, whereby it could reduce its volume, to ensure that it is safe to be treated as bio-solids. In addition, AD has the capability to release biogas that can be used as a source of renewable energy. Nevertheless, due to the presence of hard-to-manage substances within the POME WAS, AD could only do so much for instance, the elimination of the biodegradable part (Nurhayati *et al.* 2020). Hydrolysis in AD has always been a drawback that limits the treatment's efficiency due to the presence of extracellular polymeric substances (EPS) like lipids, polysaccharides, fatty acids and nuclei acids that have high water binding ability. For that reason, pre-treatment is required to achieve a better performance on the overall treatment system for POME WAS.

Mechanical, thermal, chemical or biological treatments are a few common types of pre-treatment that serve the same purpose, which is to increase the solubility or reduce the size of organic compounds, so that the sludge can be degraded easily and further improve its dewatering properties (Heng *et al.* 2021). Among the pre-treatment methods, the advance oxidation process (AOP) is easily implemented and considered as a comprehensive technology in terms of its competence on degrading contaminants and its ability to conquer the shortcoming of conventional methods (Iskurt *et al.* 2021; Jiang *et al.* 2021). AOP is able to synergize with the AD process through the generation of radicals (•OH) by disrupting and destroying the cell walls that lead to the process of sludge cell disintegration (Pérez-Rodríguez *et al.* 2019). It helps in converting high molecular weight organic matter into low molecular weight matter, which then reduces the biodegradation difficulty and hence bypasses the rate-limiting step. Electrochemical oxidation (EO) is termed a clean and efficient AOP, as it has the ability to remove recalcitrant and persistent pollutants from wastewater and convert them into simpler forms (Nurhayati *et al.* 2020). The outcome of EO proves that it could be an environmentally friendly and efficient pre-treatment system for POME WAS. EO involves the utilization of energy input and electrolyte to interact with contaminants in order to degrade them successfully. The mechanisms of EO start with oxidation of water molecules to form •OH radicals and induce a reaction with metal oxides that generates a high oxidizing power electrode. Following that, redox reactions of RuO_3/RuO_2 oxidize the pollutants (P); decomposition of the oxidized electrode by the oxygen evolution by-reaction is shown in the following equations (Equations (1)–(4)) (Heng *et al.* 2021):

$$RuO_2 + H_2O \rightarrow RuO_2(\bullet OH) + H^+ + e^- \tag{1}$$

$$RuO_2(\bullet OH) \rightarrow RuO_3 + H^+ + e^- \tag{2}$$

$$RuO_3 + P \rightarrow RuO_2 + PO \tag{3}$$

$$2RuO_3 \rightarrow 2RuO_2 + O_2 \tag{4}$$

The nature of the electrodes used plays a crucial role as it could influence the performance of the EO. Bare metal as electrodes is treatable but corrosion could occur. These problems were able to be resolved by introducing a coating of transition metal oxides (TMO) and mixed metal oxides (MMO) (Ken & Sinha 2021). MMO helps to reduce the percentage of corrosion current through the coating which then increases the shelf life of the electrodes. MMO coated on metal plates are also termed as dimensionally stable anodes (DSA), they are known to have an extensive service life over a vast range of current density (Johnson & Kumar 2020). Lead dioxide (PbO_2) is able to excel in the treatment, but due to its lead dissolution properties it endangers the treatment system. Tin(IV) oxide (SnO_2) possesses a shorter life span when compared with others and iridium (IV) oxide (IrO_2) lacks O_2 over-potential that causes it to be less efficient. As for ruthenium (IV) oxide (RuO_2), it has been listed as one of the most versatile and finest oxidation catalyst due to its high stability and ohmic conductivity (Ajab *et al.* 2020). For excellent chlorine evolution properties (~2.0 V), mechanical and chemical resistance in high pH conditions is needed (Johnson & Kumar 2020). Comparing MMO, RuO_2 showed a promising performance. It covers a vast range of treatments for wastewater from pharmaceutical to oil mill and even dye removal with the assistance of chlorine generated during the treatment process (Jiang *et al.* 2021). With its stability, it tends to have a longer life span when compared with other DSA. Pechini's method is an effective process used in the fabrication of DSA as the chelated metal cation and glycol undergoes polyesterification, forming covalent and coordinative bonds (Kang *et al.* 2020).

The aim of this study was to identify the optimum operating conditions for EO leading to a pre-treated sludge that is well disintegrated and dewatered for the subsequent AD treatment. The efficiencies of the pre-treatment were tested based on four dependent variables i.e., mixed liquor volatile suspended solids (MLVSS) removal, capillary suction timer (CST) reduction, EPS increment and soluble COD (sCOD) and increment on three independent variables which were current density, electrolyte concentration and electrolysis time. Current density is known to be one of the most important operating variable in EO, as higher current density would allow more electrons to pass by generating more oxidizing species (•OH radicals) to biodegrade the organic pollutants (Moteshaker *et al.* 2020). One study showed that EO promoted no biodegradability in terms of sCOD at 30–40 mA/cm^2 (Pasalari *et al.* 2021). Electrolysis time determines the reaction rate and removal efficiency of the pre-treatment (Heidari *et al.* 2020). Stagnant sCOD solubilization would gradually start to occur at 90 minutes due to the accumulation of soluble organic matter, which builds up the resistance of the system thus reducing the current (Ye *et al.* 2016). Lastly, the type of electrolyte and its concentration affect the conductivity of the EO process. For instance, sodium chloride (NaCl) at the range of 1.8–20 g/L, is able to generate oxidant species that have the quality of showing a higher reaction rate with organic compounds (Fajardo *et al.* 2017). The chloride ions from NaCl would minimize the side effects of anions such as bicarbonate (HCO_3^-) and sulfate (SO_4^{2-}), which would cause resistance to the overall treatment efficiencies. Electrolyte concentration should be limited, as excessive active chlorine species would lead to the annihilation of organic matter (Zhu *et al.* 2019). The conventional methods of determining the relationship between dependent and independent variable are rather sophisticated and time consuming. Hence, the response surface methodology (RSM), a statistical and mathematical method, was implemented due to its reliability and accuracy. RSM is popular in experiment design, model buildings, the relationship of variables evaluation and condition optimization. The most well-known technique in RSM is central composite design (CCD) for identifying the optimum operating conditions for pre-treatment by making traits need fewer experiment runs (Darvishmotevalli *et al.* 2019).

2. EXPERIMENTAL

2.1. Materials

The POME WAS were collected from the local palm oil mill located at Ayer Kuning, Perak, Malaysia. The collected WAS were thickened and refrigerated at 4 °C prior to any analysis and pre-treatment. The characteristics of raw POME WAS are listed in Table 1. Different stages of the experiments required different types of chemicals, together with its purposes, which are concluded in Table 2 below. The electrodes used were titanium plates with the dimensions of 150 mm × 30 mm and a thickness of 2 mm.

2.2. Synthesizing of dimensionally stable anodes (DSA)

Pechini's method was chosen for the fabrication of titanium (Ti) electrodes coated with ruthenium oxide (RuO_2). RuO_2 was chosen due to its distinction of chlorine evolution over-potential and weakening of oxygen evolution, therefore generating free radicals for organic compound oxidation (Johnson & Kumar 2020). Three stages are involved in the fabrication process. Firstly, pre-treatment of electrodes using sandpaper and isopropyl alcohol for washing purposes then boiling with 20% chloride acid and 10% oxalic acid in order to remove the titanium's oxide hydroxide coating and improve its roughness at the same time. The second stage is the preparation of precursor solution, which was EG:CA:$RuCl_3$ with the molar ratio of 10:3:1. The final stage is the coating process, 20 layers of coating was done and completed with the final calcination at 400–500 °C for 1 hour. The morphology of the coated electrodes was analyzed using field emission scanning electron microscopy (FESEM) and energy dispersive X-ray (EDX) spectroscopy.

Table 1 | Characteristics of raw palm oil mill effluent (POME) waste activated sludge (WAS)

Characteristic	Unit	Mean value
pH	–	7.92
Total Solids	mg/L	100,240
Total Volatile Solids	mg/L	84,990
Mixed Liquor Suspended Solids	mg/L	43,000
Mixed Liquor Volatile Suspended Solids	mg/L	29,800
Soluble Chemical Oxygen Demand	mg/L	855.9
Extracellular Polymeric Substances	mg/L	580.79
Capillary Suction Time	s	95.2

Table 2 | Materials and theirs purpose for EO pre-treatment

Material	Purpose
Palm Oil Mill Effluent (POME) Waste Activated Sludge (WAS)	Subject to be treated
Isopropyl Alcohol (C_3H_8O)	Synthesizing of dimensionally stable anodes (DSA)
20% of Chloride Acid ($HClO_3$)	
10% of Oxalic Acids ($C_2H_2O_4$)	
Ruthenium (III) Chloride ($RuCl_3$)	
Citric Acid (CA)	
Ethylene Glycol (EG)	
Sodium Chloride (NaCl)	Electrolyte
Tris Hydrochloride	EPS test: protein and polysaccharides
Triton X-100	
Bicinchoninic Acid (BCA) Protein Assay Kit	
Sulfuric Acid	

2.3. Experimental procedures

Electrochemical oxidation was carried out in a 500-mL beaker with an electrode distance of 5 cm throughout the experiment in a fixed parallel manner with an effective electrode area of ~55 cm^2, to achieve high solubility of sludge (Ye *et al.* 2016). A constant speed of 300 rpm was applied throughout the experiment to ensure that the process was maintained at a homogeneous manner and to prevent any settlement of the sludge. The electrolyte used in this study was NaCl, where it has a trait in generating oxidant species that had the quality of showing higher reaction rates with organic compounds (Fajardo *et al.* 2017).

As the abovementioned, RSM type CCD was applied in order to identify the optimized operating conditions of the pre-treatment as a means to enhance sludge disintegration and dewaterability. CCD is known for its capability to analyze the relationship between independent factors and dependent responses. A five level factorial ($-\alpha$, -1, 0, $+1$, $+\alpha$) was applied in this analysis. Three independent variables were set: current density (A), electrolyte concentration (B) and electrolysis time (C). MLVSS removal, CST increment, EPS increment and sCOD increment were the responses of the pre-treatment as abovementioned. The range of each independent variables was inserted into Design Expert Software (version 11.1.2.0) to generate a set of 20 experimental runs that contained combinations of factorial runs, axial runs and center points. The number of runs was calculated using the equation (Equation (5)),

$$N = 2^n + 2n + n_c \qquad (5)$$

whereby N = number of runs, n = number of factors and n_c = number for center points. Calculations of axial points can be obtained through a series of equations whereby these steps were easily calculated through the software. The complete ranges and level of independent variables are shown in Table 3. Analysis of variance (ANOVA) was used by the software to calculate the results (Isam *et al.* 2019).

2.4. Analytical methods

Standard method 2540E was applied for the measurement of MLVSS and its reduction was calculated. Standard method 5220D was used for the measurement of sCOD level after pre-treatment (APHA 2005). Furthermore, a CST apparatus (model: 304M Triton) was used to measure the dewaterability of the pre-treated sludge in seconds. Protein and polysaccharide are both common indicators for the quantification of EPS. The EPS increment was calculated by the increase in the summation of both protein and polysaccharide. The BCA method and the sulfuric-UV method were used to identify the protein and polysaccharide, respectively. The detection methods for EPS and sCOD were through a UV-vis spectrophotometer (model: JASCO V-730) and were calculated with the equations generated from the calibration curve.

3. RESULTS AND DISCUSSION

3.1. Morphology of Ti/RuO_2 electrodes

Figure 1 shows the morphology of the ruthenium oxide-coated titanium electrodes (Ti/RuO_2) with a magnification of ×500 and ×5,000, respectively using FESEM. It can be seen in Figure 1(a), that there was a rather even coated and distributed surface, which represents that EO could be carried out smoothly due to the high levels of active sites. In Figure 1(b), there were some minor 'cracked mud' surfaces detected. The contrast of the rate of expansion between the metal oxide film and the titanium substrate was the cause of the crack (Bezerra *et al.* 2020). This is a common phenomenon when coating with DSA as it involves different temperatures of heating. This may be possibly caused by the high rate of the heating process (Gonzaga *et al.* 2020). From the FESEM morphology, a multilayer coating of RuO_2 and a minimum occurrence of crack, portrayed the

Table 3 | Independent variables and ranges for RSM type CCD

			Levels and ranges				
Independent variable	**Unit**	**Code**	**Minimum**	**−1**	**0**	**+1**	**Maximum**
Current Density	mA/cm^2	A	1.14	10.0	23.0	36.0	44.86
Electrolyte Concentration	g/L	B	1.23	6.0	13.0	20.0	24.77
Electrolysis Time	minutes	C	9.50	30.0	60.0	90.0	110.45

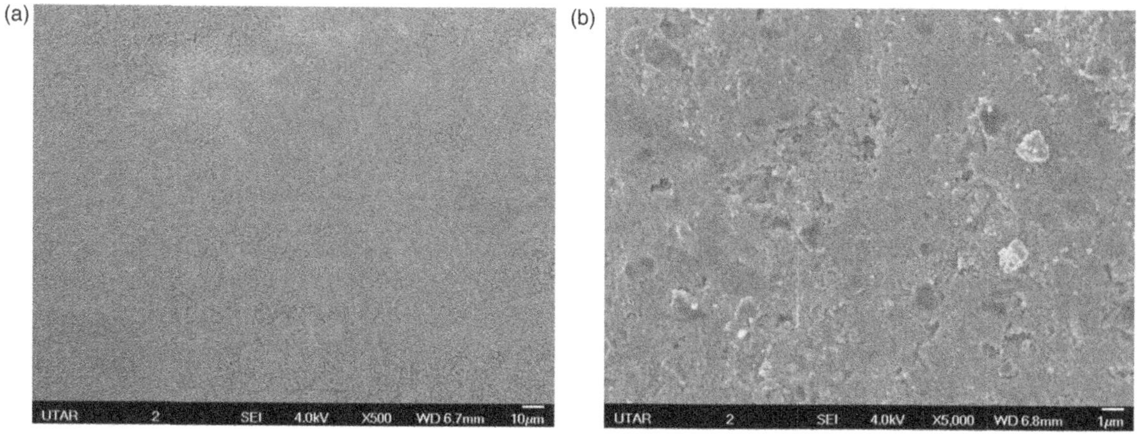

Figure 1 | FESEM image of Ti/RuO_2 at two magnifications (a) ×500 and (b) ×5,000.

stability of DSA as the electrolyte would be compromised when attacking the catalyst layer (Ajab *et al.* 2020). Moreover, from the outcome of EDX has shown an escalating percentage content of ruthenium which proves the coating's chemical composition (Bezerra *et al.* 2020). Another proof of even distribution and homogeneous coating can be seen from the energy dispersive spectroscopy (EDS) elemental mapping, as shown in Figure 2(a). In Figure 2(b), the elemental analysis revealed that the titanium electrode contained Ru (51.49% wt.), O (47.81% wt.) and Ti (0.70% wt.), indicating its high purity and suitability as a working electrode. The EDX result was consistent with the FESEM images, which clearly indicated that RuO_2 composite had been successfully deposited on the surface of the titanium plate.

3.2. RSM type CCD optimization

In this study, RSM type CCD was applied for optimization that generated 20 runs (14 non-center points and six center points), the experimental design data together with its responses are listed in Table 4.

The connection with the predictor variable and response variable were analyzed through the ANOVA. Four (4) regression polynomial equations (Equations (6)–(9)) were generated based on the four dependent variables with the indications of current density (A), electrolyte concentration (B) and electrolysis time (C) as the three independent variables. Based on the regression equations, there was an immediate positive effect between both the independent factors and the performance of dependent variables whereby the increase in the independent variable was able to improve the performance and vice versa (Moteshaker *et al.* 2020). Through the comparison of the coefficients among the regression equations in linear term, it helps to identify the relative impact of the variables (Hu *et al.* 2021). In this study, the coefficient of current density (A) had achieved the highest value among the predictor variables in terms of MLVSS removal, CST reduction and sCOD

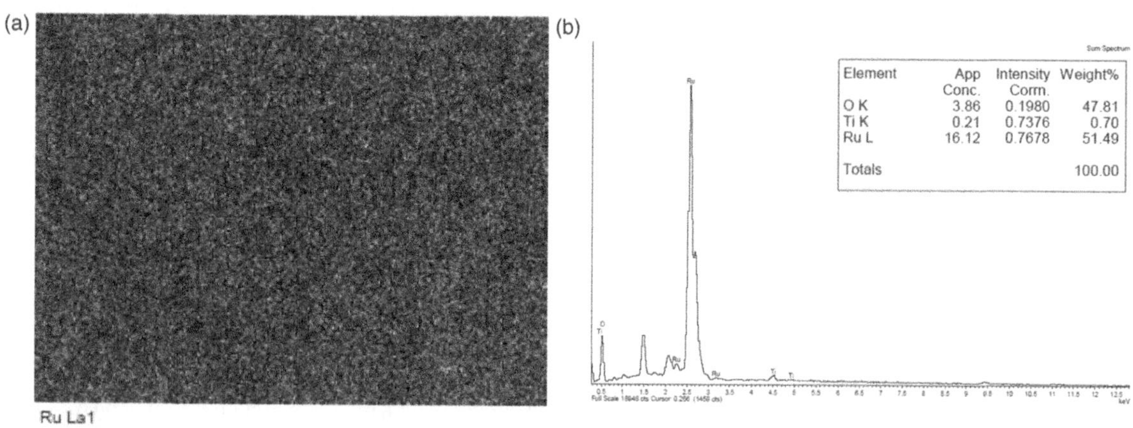

Figure 2 | (a) Elemental mapping with EDS analysis of coating and (b) elemental spectrum.

Table 4 | Experimental design and responses for the EO pre-treatment of POME WAS

	A: Current density (mA/cm^2)	B: Electrolyte concentration (g/L)	C: Electrolysis time (minutes)	MLVSS removal (%)	CST reduction (%)	EPS increment (%)	sCOD increment (%)
1	23	13	60	20.8	46.0	25.4	21.2
2	23	13	60	14.4	43.8	13.7	37.2
3	36	6	90	70.5	85.9	46.7	−10.2
4	10	6	90	21.5	27.5	38.2	45.7
5	23	24.77	60	24.6	40.0	30.4	39.4
6	23	13	9.5	5.6	21.3	93.4	46.5
7	1.14	13	60	7.0	8.5	10.4	18.1
8	23	13	60	19.8	40.6	12.4	24.1
9	23	13	60	20.7	45.7	18.4	35.0
10	36	6	30	20.5	22.6	64.3	25.8
11	10	6	30	19.0	11.5	43.8	64.7
12	23	13	110.5	57.0	78.3	61.2	5.2
13	23	1.23	60	30.2	38.8	7.1	25.1
14	44.86	13	60	63.1	84.9	25.9	−30.8
15	36	20	30	16.1	23.7	80.1	36.5
16	23	13	60	17.4	54.7	21.8	33.4
17	23	13	60	22.0	47.2	11.8	25.7
18	36	20	90	66.1	84.9	50.3	−18.9
19	10	20	90	7.4	19.9	48.0	15.6
20	10	20	30	19.4	12.6	65.9	33.7

increment of 14.66, 20.07 and 15.28, respectively. Hence, the equations showed that current density outperformed the electrolysis time and electrolyte concentration in terms of significance level. The coefficient of the variable signified its significance level in the further optimization study:

$$\text{MLVSS Removal} = 19.25 + 14.66A - 2.33B + 12.96C + 0.6162AB + 13.70AC - 1.82BC + 5.20A^2 + 2.49B^2 + 3.88C^2 \tag{6}$$

$$\text{CST Reduction} = 46.63 + 20.07A - 0.3201B + 17.83C + 0.8313AB + 12.65AC - 1.37BC - 1.81A^2 - 4.39B^2 - 0.7042C^2 \tag{7}$$

$$\text{EPS Increment} = 16.75 + 5.24A + 6.62B - 9.14C - 1.57AB - 2.99AC - 3.06BC + 3.56A^2 + 3.77B^2 + 24.45C^2 \tag{8}$$

$$\text{sCOD Increment} = 29.19 - 15.28A - 2.56B - 14.49C + 7.88AB - 6.78AC - 2.31BC - 11.05A^2 + 2.60B^2 + 0.3295C^2 \tag{9}$$

The ANOVA outputs were tabulated in Table 5. Coefficient of determination (R^2), level of significant (*p*-value) and the lack of fit test are the most representative indicators in proving the model's significance (Isam *et al.* 2019). R^2 and Adjusted R^2 (R^2_{adj}) were given more attention whereby R^2 is the extent of the level of response variability reduced through the independent variable of the model. R^2 reaching 1 is the most desirable value, to have a balanced match between experimental and predicted data. However, a minimum of 0.8 is sufficient (Darvishmotevalli *et al.* 2019). All of the dependent variables were able to achieve the minimum suggested value of R^2 i.e. $R^2_{MLVSS} = 0.9773$, $R^2_{CST} = 0.9610$, $R^2_{EPS} = 0.9230$ and

Table 5 | RSM type CCD ANOVA output table for four dependent variables

Source	Sum of squares	Degree of freedom	Mean square	F-value	*p*-value	Remarks
MLVSS Removal						
Model	7,228.77	9	825.42	47.74	<0.0001	Significant
Residual	172.90	10	17.29	–	–	–
Lack of Fit	134.00	5	26.80	3.44	0.1004	Not significant
$\mathbf{R^2 = 0.9773\ R^2_{adj} = 0.9568\ R^2_{pred} = 0.85347\ AP = 22.8731}$						
CST Reduction						
Model	1,451.71	9	27.39	27.39	<0.0001	Significant
Residual	464.57	10	46.46	–	–	–
Lack of Fit	353.93	5	70.79	3.20	0.1138	Not significant
$\mathbf{R^2 = 0.9610\ R^2_{adj} = 0.9259\ R^2_{pred} = 0.7614\ AP = 17.2987}$						
EPS Increment						
Model	10,953.85	9	1,217.08	13.32	0.0002	Significant
Residual	913.92	10	91.39	–	–	–
Lack of Fit	760.54	5	152.11	4.94	0.0521	Not significant
$\mathbf{R^2 = 0.9230\ R^2_{adj} = 0.8537\ R^2_{pred} = 0.4979\ AP = 12.5927}$						
sCOD Increment						
Model	9,032.58	9	1,003.62	10.51	0.0005	Significant
Residual	954.92	10	95.49	–	–	–
Lack of Fit	737.71	5	147.54	3.40	0.1029	Not significant
$\mathbf{R^2 = 0.9044\ R^2_{adj} = 0.8183\ R^2_{pred} = 0.4003\ AP = 11.9074}$						

$R^2_{sCOD} = 0.9044$. On top of that, the closeness between R^2 and R^2_{adj}, i.e. $R^2_{adj\ MLVSS} = 0.9568$, $R^2_{adj\ CST} = 0.9259$, $R^2_{adj\ EPS} = 0.8537$ and $R^2_{adj\ sCOD} = 0.8183$ indicated that there were no chances of insignificance within the model (Ken & Sinha 2021).

Moreover, a *p*-value <0.05 should be assured as it serves as evidence that the model is significant (Darvishmotevalli *et al.* 2019). The RSM type CCD ANOVA output tables for the four variables (linear, two-way interactions and quadratic) are attached in Tables S1–S4 as supplementary material. From the table, all variables were able to achieve a relatively low *p*-value where the *p*-values for MLVSS and CST were <0.0001, whereas the *p*-values for EPS and sCOD were 0.0002 and 0.0005, respectively. Furthermore, a lack of fit of test should be 'not significant' as it proves that the model is able to be fitted. All four dependent variables were able to achieve, however, a probability lack of fit (PLOF) of EPS that showed a lower percentage among the other three variables. This portrays that there was a lower chance that EPS will not be fitted due to noise. Lastly, the measurement of the ratio between signal to noise is termed as the adequate precision (AP) (Moteshaker *et al.* 2020). It shows the comparison between the ranges of values predicted at design points with the mean prediction error. All of the AP exceeded the minimum value of 4, means that the maneuvering of design space could take place through the models (Isam *et al.* 2019).

3.3. Three-dimensional surface plot and process optimization by RSM type CCD

The three-dimensional surface plots of each response variables are shown in Figure 3 with current density (mA/cm^2) and electrolysis time (minutes) as the axis due to their higher significance level, as discussed in the previous section.

3.3.1. Effect of current density

From the surface plots (Figure 3(a)–3(c)), MLVSS removal, CST reduction and EPS increment were shown to be increasing as the current density increased. This can be explained due to the ion species that were formed that possessed the robust affinity to treat the organic matter within the WAS. Higher current promotes more electrons flowing and thus enhancies the rate of oxidation. High resistance within the reaction is not the only factor to be affected. To further explain this, secondary oxidizing

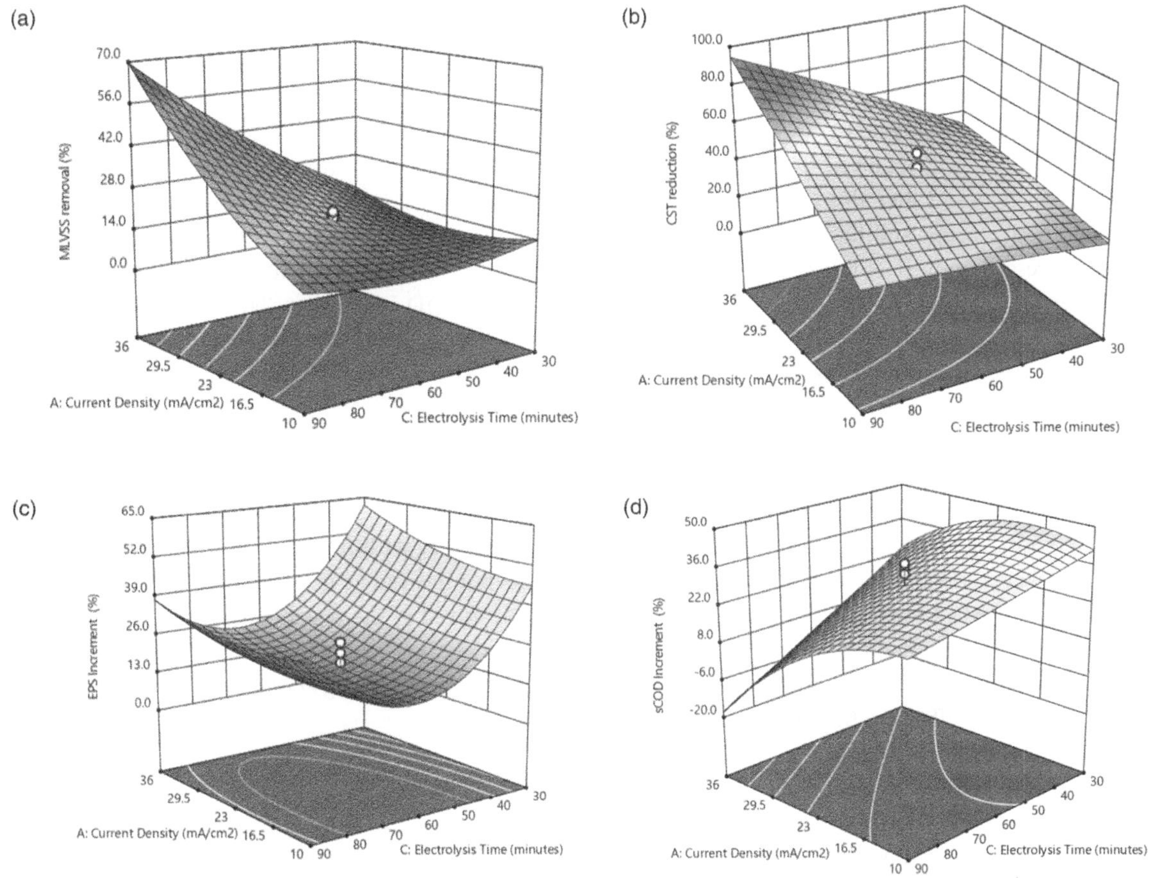

Figure 3 | Three-dimensional response surface plots of (a) MLVSS removal; (b) CST reduction; (c) EPS increment and (d) sCOD increment between current density and electrolysis time at 13 g/L of NaCl concentration.

agents will be created due to the oxidation of organic matter. These secondary oxidizing agents tend to be weaker and are unable to perform as well as •OH radicals, which will cause mineralization to take place and influence the post-treatment (Pérez-Rodríguez *et al.* 2019). As mentioned, pre-treatment efficiency is dependent on the amount of oxidizing agents being created, the increase in current density only affects direct oxidation as strong oxidizing agents are formed at the surface of the electrodes. Although, the higher the current density the better the efficiency, particularly in removing the WAS solid content. However, the amount of voltage being applied needs to be taken into consideration as it would be less economical or possibly side reactions might occur that would deteriorate the overall performance, especially on the sCOD solubilization beyond ~23 mA/cm^2, as shown in Figure 3(d). This is in close agreement with Pasalari and his co-researchers who found that no biodegradability would be promoted in terms of sCOD in the EO process when the current density reached 30–40 mA/cm^2 (Pasalari *et al.* 2021). It is important to understand that the restriction of current density is necessary as voltage increases, heat would be dissipated leading to other side effects (Hu *et al.* 2021). Not only that, the operating conditions of the EO pre-treatment should be applied to help in solubilizing the organic matter in order to enhance biodegradability and overcome the hydrolysis problem in AD. High current density would lead to the mineralization of organic molecules instead of solubilizing them (Barrios *et al.* 2021).

3.3.2. Effect of electrolysis time

Electrolysis time is considered as one of the few vital parameters that would affect the performance of EO. From the surface plots, MLVSS removal and CST reduction were directly proportional towards the electrolysis time, as shown in Figure 3(a) and 3(b), due to the increased level of oxidizing agents (•OH radicals). The electrolysis time had more effect with higher current density, in terms of MLVSS removal. Both of the independent variables had inseparable relationship whereby a balance needed to be achieved when the voltage of the system was higher, and a lower electrolysis time is needed to achieve a certain level of solubilization (Rodriguez *et al.* 2018). However, there would be decay in organic matter if the duration is too long. The sCOD increment was enhanced by electrolysis time at the beginning of the EO and dropped further after the peak at 30 min, as shown in Figure 3(d). According to Pasalari *et al.* (2021), sCOD concentration would increase sharply at the beginning of the pre-treatment, which showed that intracellular materials were released into the aqueous phase, thus enhancing the biodegradability of the WAS. This was due to the electromigration of ions at the beginning of the EO. Inorganic matter is released into the liquid phase through solubilization. The accumulation of ions or organic matters takes place at both electrodes and hence causes a blockage that increases the current resistance and restricts the process's ability to take place correctly (Heidari *et al.* 2020). Due to the increase in current resistance, the current drops gradually throughout the entire process, which affected the performance of the sCOD increment. The decrease in concentration of sCOD most likely occurred at extremely high current density at 44.86 mA/cm^2 or due to the high electrolysis time at 90 min from the results. This could be explained by the rapid sCOD mineralization of dissolved organic substances compared with solubilization.

3.3.3. Effect of electrolyte (NaCl) concentration

Electrolyte (NaCl) concentration was fixed at the middle level set due to its least significant level. NaCl helps in preventing the formation of oxidizing layers by other anions and causes ohmic resistance within the cell and thus prevents the decline in COD removal. However, depletion in sCOD due to mineralization is not desired as it would further affect the biogas generation in the subsequent AD. From the results, a slight drop in sCOD occurred as the NaCl concentration increased, as well as the EPS concentrations. This could be due to the presence of negative charges at the bacteria surfaces that generate a high electrostatic repulsive force that leads to the occurrence of sludge disintegration (Heng *et al.* 2021). MLVSS removal and CST reduction rate showed no effect on the increase in NaCl concentration.

3.4. Validation of the model

The applicability of the mathematical model can be assessed by diagnostic plots, to show the predicted versus observed values of percentage of each dependent variables (Figure 4). The predicted and experimental values are close to a straight line. This trend indicated that the predicted values using the calibrated models were highly correlated with the actual values of MLVSS removal, CST reduction, EPS increment and sCOD increment. It also indicated a close agreement between experimental results and predicted values, with a high R^2 of 0.9044–0.9773. Therefore, the models can be used to predict the studied responses, which confirmed the good predictive quality of the models.

3.5. Optimization

The optimization process was done by inserting the minimum or maximum desired values for the response variables. In order to have a better WAS disintegration and dewaterability, minimum MLVSS removal, CST reduction and sCOD increment but maximum EPS increment were set. All three factors were taken into considerations with different weightage, the criteria were set at MLVSS removal $>20\%$, CST reduction $>43\%$, EPS increment $<19\%$ and sCOD increment $>25\%$. An overlay plot shown in Figure 5 was generated, which was based on a specific criterion portraying a visual selection of ideal conditions. The shaded region showed that the criteria could be met when the range for the optimum operating conditions was followed, i.e. current density (17–27 mA/cm^2) and electrolysis time (55–75 minutes) at a fixed concentration of NaCl (10 g/L). The observed results were similar to those reported in the literature (Barrios *et al.* 2021; Pasalari *et al.* 2021), which recommended electrolysis working at a current density lower than 30 mA/cm^2, to prevent the acceleration of organic matter mineralization from happening. Thus, it is advisable to limit the current density to avoid adverse effects such as heat generation and higher power consumption.

Pre-treatments were performed aiming to enhance the digestibility of WAS, which in turn improved the performance of AD. The best way to indicate AD performance was through biogas yield and different studies have been carried out showing that pre-treatment was able to improve biogas yield. Huang *et al.* (2021) showed that an sCOD increment from 29 to 350 mg/L successfully increased the methane yield (from 105 to 134 mL CH_4/g VSS) using an electrochemical pre-treatment (Ti/RuO_2) on municipal WAS. Another study proved that biogas generation from 109 N-L CH_4/kg.VS in non-pretreated WAS increased to 311.9 $\pm$ 6 N-L CH_4/kg.VS using electrochemical pre-treatment at a current density of 28.6 mA/cm^2 (Barrios *et al.* 2021).

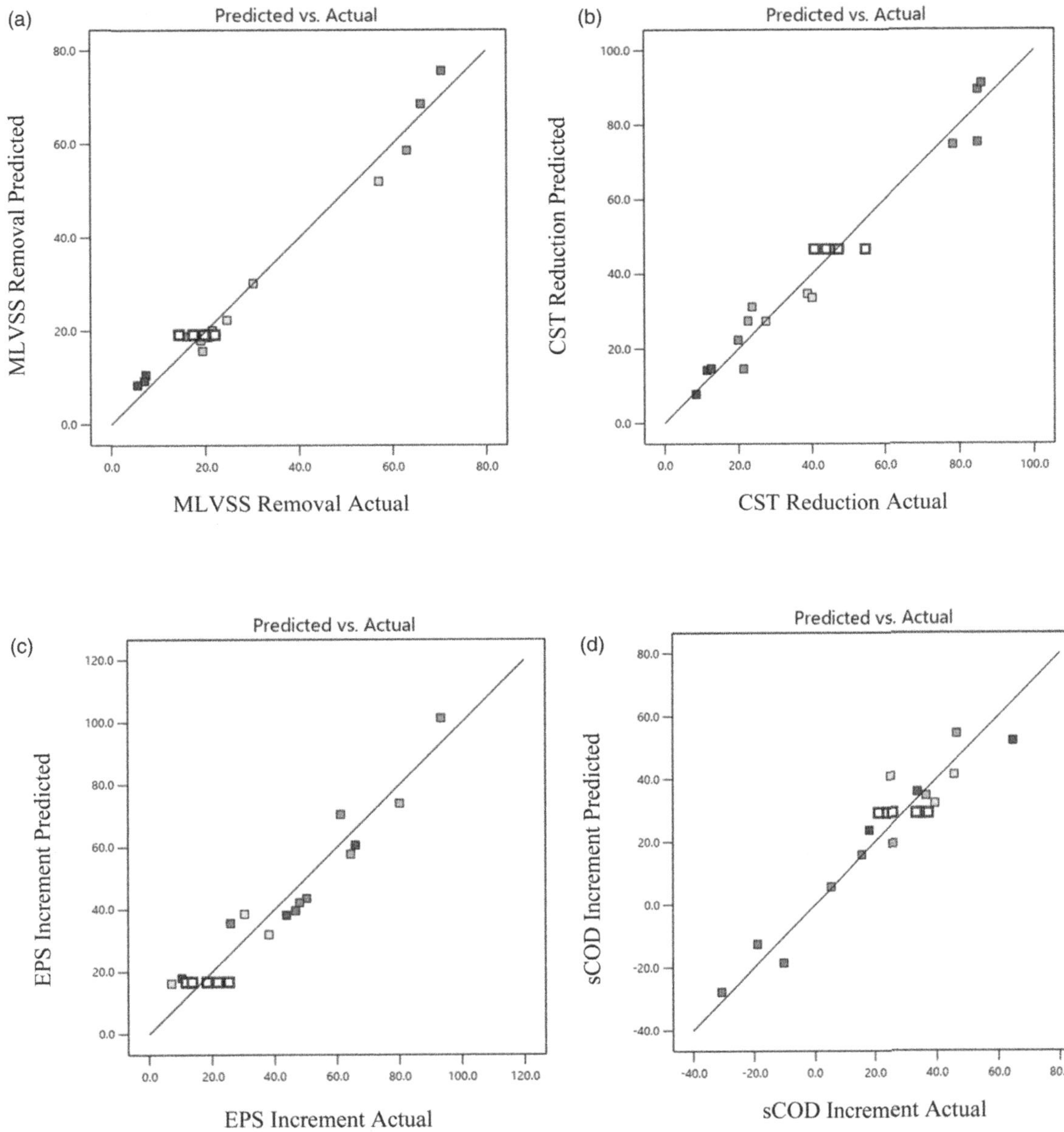

Figure 4 | Correlation between predicted and actual values for (a) MLVSS removal, (b) CST reduction, (c) EPS increment and (d) sCOD increment.

4. CONCLUSIONS

The generation of POME WAS has gradually increased over many years due to the development of the palm oil industry. It is important that the generated sludge is treated before being released into the environment. The EO process using Ti/RuO_2 was used in this study. FESEM surface images and EDS elemental analysis of the Ti/RuO_2 electrodes indicated that the preparation following Pechini's method can produce a uniform and homogeneous coating throughout the titanium plate. An RSM type CCD was successfully applied to describe the EO process using a second-order model. The most significant effects on WAS disintegration and dewaterability were found to be due to current density and electrolysis time, while the influence of NaCl concentration was relatively insignificant. The model exhibited that the suggested model agreed well with experimental results with high R^2 values (0.9044–0.9773) for all the studied dependent variables. The optimum operating conditions were in the range of current density 17–27 mA/cm^2 and electrolysis time 55–75 minutes, at a fixed

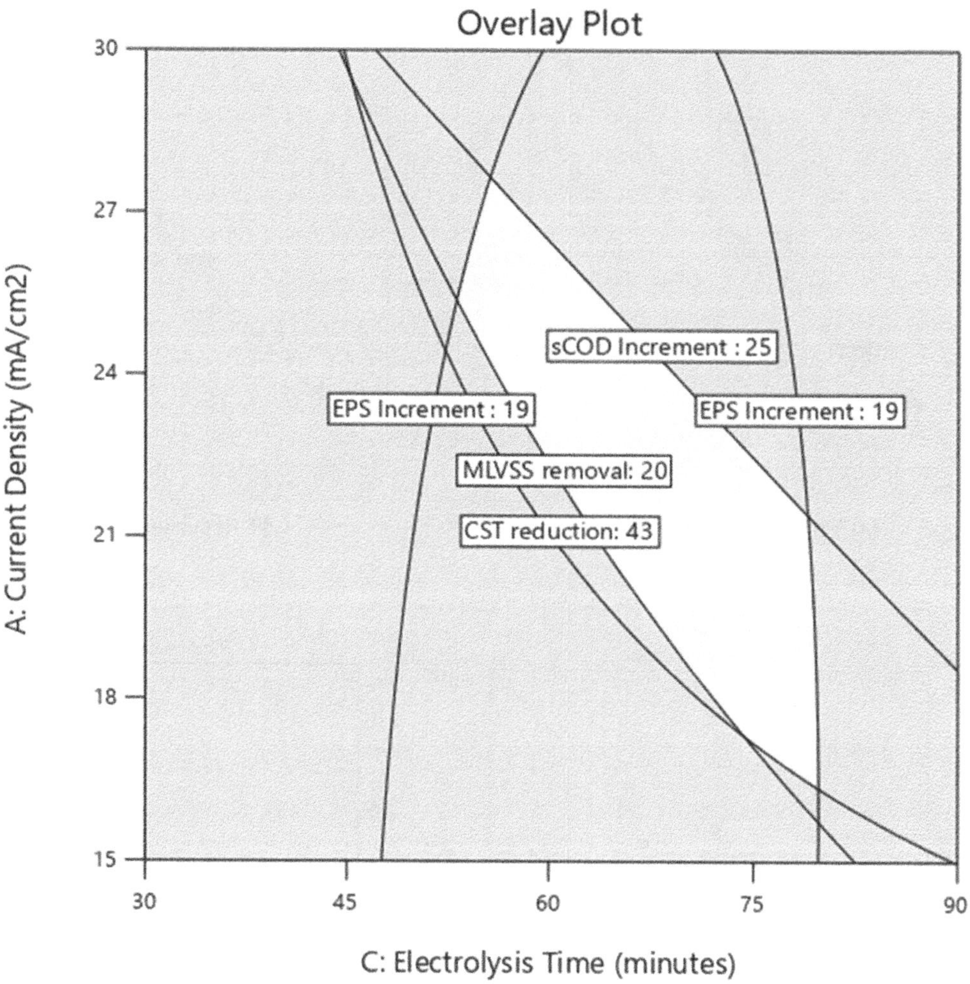

Figure 5 | Overlay plot for optimal region at electrolyte concentration 10 g/L.

concentration of NaCl (10 g/L), resulting in MLVSS removal >20%, CST reduction >43%, EPS increment <19% and sCOD increment >25%. Consequently, these results proved that Ti/RuO_2 could play an important role in the pre-treatment of POME WAS in sludge disintegration and adequate dewaterability, for the subsequent AD to further enhance the solid removal and biogas generation.

ACKNOWLEDGEMENTS

The authors are thankful to Universiti Tunku Abdul Rahman (UTAR), Malaysia for providing facilities and support for this research. This work was funded by the UTAR Research Fund under project number IPSR/RMC/UTARRF/2019-C1/G01.

DATA AVAILABILITY STATEMENT

All relevant data are included in the paper or its Supplementary Information.

CONFLICT OF INTEREST

The authors declare there is no conflict.

REFERENCES

Ajab, H., Isa, M. H. & Yaqub, A. 2020 Electrochemical oxidation using Ti/RuO_2 anode for COD and PAHs removal from aqueous solution. *Sustain. Mater. Technol.* **26**, e00225. doi:10.1016/j.susmat.2020.e00225.

APHA 2005 *Standard Methods for the Examination of Water and Wastewater*, 21st edn. American Public Health Association/American Water Works Association/Water Environment Federation, Washington DC, USA.

Barrios, J. A., Cano, A., Rivera, F. F., Cisneros, M. E. & Duran, U. 2021 Efficiency of integrated electrooxidation and anaerobic digestion of waste activated sludge. *Biotechnol. Biofuels* **14** (1), 81. doi:10.1186/s13068-021-01929-7.

Bezerra, C. W. A., Santos, G. O. S., Pupo, M. M. S., Gomes, M. A., Silva, R. S., Eguiluz, K. I. B. & Salazar-Banda, G. R. 2020 Novel eco-friendly method to prepare Ti/RuO_2–IrO_2 anodes by using polyvinyl alcohol as the solvent. *J. Electroanal. Chem.* **859**, 113822. doi:10.1016/j.jelechem.2020.113822.

Darvishmotevalli, M., Zarei, A., Moradnia, M., Noorisepehr, M. & Mohammadi, H. 2019 Optimization of saline wastewater treatment using electrochemical oxidation process: prediction by RSM method. *MethodsX* **6**, 1101–1113. doi:10.1016/j.mex.2019.03.015.

Fajardo, A. S., Seca, H. F., Martins, R. C., Corceiro, V. N., Freitas, I. F., Quinta-Ferreira, M. E. & Quinta-Ferreira, R. M. 2017 Electrochemical oxidation of phenolic wastewaters using a batch-stirred reactor with NaCl electrolyte and Ti/RuO_2 anodes. *J. Electroanal. Chem.* **785**, 180–189. doi:10.1016/j.jelechem.2016.12.033.

Gonzaga, I. M. D., Dória, A. R., Vasconcelos, V. M., Souza, F. M., Santos, M. C., Hammer, P., Rodrigo, M. A., Eguiluz, K. I. B. & Salazar-Banda, G. R. 2020 Microwave synthesis of Ti/$(RuO_2)_{0.5}(IrO_2)_{0.5}$ anodes: improved electrochemical properties and stability. *J. Electroanal. Chem.* **874**, 114460. doi:10.1016/j.jelechem.2020.114460.

Heidari, M., Vosoughi, M., Sadeghi, H., Dargahi, A. & Mokhtari, S. A. 2020 Degradation of diazinon from aqueous solutions by electro-Fenton process: effect of operating parameters, intermediate identification, degradation pathway, and optimization using response surface methodology (RSM). *Sep. Sci. Technol.* **56** (13), 2287–2299. doi:10.1080/01496395.2020.1821060.

Heng, G. C., Isa, M. H., Lock, S. S. M. & Ng, C. A. 2021 Process optimization of waste activated sludge in anaerobic digestion and biogas production by electrochemical pre-treatment using ruthenium oxide coated titanium electrodes. *Sustainability* **13** (9), 4874. doi:10.3390/su13094874.

Hu, X., Dong, H., Zhang, Y., Fang, B. & Jiang, W. 2021 Mechanism of *N,N*-dimethylformamide electrochemical oxidation using a Ti/RuO_2 – IrO_2 electrode. *RSC Adv.* **11** (13), 7205–7213. doi:10.1039/D0RA10181H.

Huang, H., Zeng, Q., Heynderickx, P. M., Chen, G.-H. & Wu, D. 2021 Electrochemical pretreatment (EPT) of waste activated sludge: extracellular polymeric substances matrix destruction, sludge solubilisation and overall digestibility. *Bioresour. Technol.* **330**, 125000. doi:10.1016/j.biortech.2021.125000.

Isam, M., Baloo, L., Kutty, S. R. M. & Yavari, S. 2019 Optimisation and modelling of Pb(II) and Cu(II) biosorption onto red algae (*Gracilaria changii*) by using response surface methodology. *Water* **11** (11), 2325. doi:10.3390/w11112325.

Iskurt, C., Keyikoglu, R., Kobya, M. & Khataee, A. 2021 Treatment of coking wastewater by aeration assisted electrochemical oxidation process at controlled and uncontrolled initial pH condition. *Sep. Purif. Technol.* **248**, 117043. doi:10.1016/j.sep.pur.2020.117043.

Jiang, Y., Zhao, H., Liang, J., Yue, L., Li, T., Luo, Y., Liu, Q., Lu, S., Asiri, A. M., Gong, Z. & Sun, X. 2021 Anodic oxidation for the degradation of organic pollutants: anode materials, operating conditions and mechanisms. A mini review. *Electrochem. Commun.* **123**, 106912. doi:10.1016/j.elecom.2020.106912.

Johnson, I. & Kumar, M. 2020 Electrochemical oxidation of distillery wastewater by dimensionally stable Ti-RuO_2 anodes. *Environ. Technol. Inno.* **20**, 101181. doi:10.1016/j.eti.2020.101181.

Kamyab, H., Chelliapan, S., Din, M., Rezania, S., Khademi, T. & Kumar, A. J., 2018 Palm Oil Mill Effluent as an Environmental Pollutant. In: *Palm Oil* (Waisundara, V. ed.). IntechOpen. doi:10.5772/INTECHOPEN.75811.

Kang, J., Gwon, Y. R. & Cho, S. K. 2020 Photoelectrochemical water oxidation on $PbCrO_4$ thin film photoanode fabricated via Pechini method: various solution-processes for $PbCrO_4$ film synthesis. *J. Electroanal. Chem.* **878**, 114601. doi:10.1016/j.jelechem.2020.114601.

Ken, D. S. & Sinha, A. 2021 Dimensionally stable anode (Ti/RuO_2) mediated electro-oxidation and multi-response optimization study for remediation of coke-oven wastewater. *J. Environ. Chem. Eng.* **9** (1), 105025. doi:10.1016/j.jece.2021.105025.

Moteshaker, P. M., Rokni, S. E., Farnoodian, N., Akhlaghi, N. M., Saadi, S., Ahmadidoust, G. & Yousefi, A. 2020 Application of response surface methodology for optimization of electrochemical process in metronidazole (MNZ) removal from aqueous solutions using stainless steel 316 (SS316) and lead (Pb) anodes. *Int. J. Chem. React. Eng.* **18** (8), 20200055. doi:10.1515/ijcre-2020-0055.

Nurhayati, E., Bagastyo, A. Y. & Hartatik, D. D. 2020 Effect of flow rate on electrochemical oxidation of landfill leachate using DSA and BDD anode. *IOP Conf. Ser. Earth Environ. Sci.* **506**, 012038.

Pasalari, H., Esrafili, A., Rezaee, A., Gholami, M. & Farzadkia, M. 2021 Electrochemical oxidation pretreatment for enhanced methane potential from landfill leachate in anaerobic co-digestion process: Performance, Gompertz model, and energy assessment. *Chem. Eng. J.* **422**, 130046. doi:10.1016/j.cej.2021.130046.

Pérez-Rodríguez, M., Cano, A., Durán, U. & Barrios, J. A. 2019 Solubilization of organic matter by electrochemical treatment of sludge: influence of operating conditions. *J. Environ. Manage.* **236**, 317–322. doi:10.1016/j.jenvman.2019.01.105.

Rodriguez, F. A., Pivero, E. P. & Gonzalez, I. 2018 Adapted Pechini method to prepare DSA type electrodes of RuO_2-ZrO_2 doped with Sb_2O_5 over titanium plates. *MethodsX* **5**, 1613–1617. doi:10.1016/j.mex.2018.11.020.

Ye, C., Yuan, H., Lou, Z. & Zhu, N. 2016 Combined electrochemical and hypochlorite pretreatment for improving solubilization and anaerobic digestion of waste-activated sludge: effect of hypochlorite dosage. *Energy Fuels* **30** (4), 2990–2996. doi:10.1021/acs.energyfuels.5b02884.

Zhu, X., Hu, W., Feng, C., Chen, H., Chen, N. & Li, R. 2019 Electrochemical oxidation of aniline in sodium chloride solution using Ti/RuO_2 anode. *Int. J. Electrochem. Sci.* **14**, 7516–7528. doi:10.20964/2019.08.36.

First received 17 June 2022; accepted in revised form 5 August 2022. Available online 16 August 2022

doi: 10.2166/wst.2022.407

Electron beam treatment for the removal of 1,4-dioxane in water and wastewater

Robert Pearce[a,b,*], Xi Li[c,d], John Vennekate[d], Gianluigi Ciovati[d] and Charles Bott[b]

[a] Department of Civil and Environmental Engineering, Virginia Tech, 200 Patton Hall, 750 Drillfield Dr, Blacksburg, VA 24060, USA
[b] Hampton Roads Sanitation District, 1434 Air Rail Ave, Virginia Beach, VA 23455, USA
[c] Department of Electrical and Computer Engineering, Old Dominion University, 231 Kaufman Hall, Norfolk, VA 23529, USA
[d] Thomas Jefferson National Accelerator Facility, 12000 Jefferson Ave, Newport News, VA 23606, USA
*Corresponding author. E-mail: probe93@vt.edu

RP, 0000-0003-2926-7684; XL, 0000-0002-3559-4537; JV, 0000-0002-4911-2121; CB, 0000-0002-3207-3648

ABSTRACT

Electron beam (e-beam) treatment uses accelerated electrons to form oxidizing and reducing radicals when applied to water without the use of external chemicals. In this study, electron beam treatment was used to degrade 1,4-dioxane in several water matrices. Removal improved in the progressively cleaner water matrices and removals as high as 94% to 99% were observed at a dose of 2.3 kGy in secondary effluent. 1,4-dioxane removal was confirmed to be primarily through hydroxyl radical oxidation. The calculated electrical energy per order was found to be 0.53, 0.26, and 0.08 kWh/m^3/order for secondary effluent (Avg. total organic carbon (TOC) 9.25 mg/L), granular activated carbon effluent (TOC 3.46 mg/L), and ultrapure water, respectively, with a 70% generation and transfer efficiency applied.

Key words: advanced oxidation, 1,4-dioxane, electron beam, wastewater treatment, water treatment

HIGHLIGHTS

- Electron beam treatment was used to successfully remove 1,4-dioxane from water and wastewater.
- No bromate was formed despite high bromide concentrations and bromate removal was observed.
- Energy use was comparable to other conventional AOPs without requiring additional chemicals.

GRAPHICAL ABSTRACT

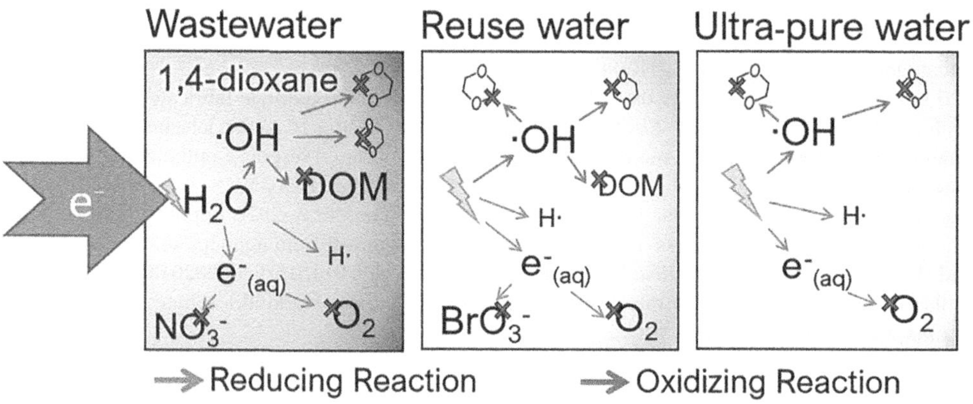

1. INTRODUCTION

1,4-dioxane is a prevalent trace contaminant in water and wastewater treatment and a potential human carcinogen with an EPA estimated 10^{-6} lifetime cancer risk associated with a concentration of 0.35 μg/L (USEPA 2013). It was historically used as a solvent stabilizer for 1,1,1-trichloroethane (TCE) and is a common groundwater and surface water contaminant and was detected in 21% of US drinking water supplies (Adamson *et al.* 2017). It is also commonly found in household detergents and cleaning products which end up in wastewater treatment plant influents (Tanabe & Kawata 2008). Because it is not readily biodegraded, adsorbed, or oxidized by traditional oxidants, it can be very difficult to treat (Otto & Nagaraja 2007).

Advanced oxidation processes (AOPs) are typically used to remove 1,4-dioxane (Otto & Nagaraja 2007; McElroy *et al.* 2019). In these processes, powerful oxidizing hydroxyl radicals (·OH) are generated in-situ to degrade contaminants. Technologies include UV advanced oxidation (UVAOP) where UV light is used for photolysis of hydrogen peroxide (H_2O_2) into ·OH. Likewise, UV-chlorine AOP generates ·OH from photolysis of free chlorine. Ozone (O_3) is applied as an oxidant and disinfectant in water and wastewater treatment and leads to the generation of ·OH through its autocatalytic decomposition and reactions with organics which can also be catalyzed by H_2O_2 (von Sonntag & von Gunten 2012; Stefan 2018). In these processes, large chemical doses are often needed. While the study of AOPs for 1,4-dioxane removal in wastewater effluent is limited, ozone is generally thought to be the best AOP for wastewater effluents. This is due to the high levels of dissolved organic matter which scavenges hydroxyl radicals and low transmittance of UV light (Katsoyiannis *et al.* 2011; Lee *et al.* 2016; Sgroi *et al.* 2021). However, ozone (and UV-Cl_2) can form bromate as a regulated disinfection byproduct if bromide (Br^-) is present.

Electron beam treatment uses accelerated electrons to generate large quantities of both oxidizing and reducing radicals through the radiolysis of water as shown in Equation (1) (Cooper *et al.* 1992; Wang *et al.* 2016):

$$0.7H_2O \sim\sim irradiation \sim\sim> 0.28 \cdot OH + 0.27e^-_{(aq)} + 0.06H \cdot + 0.07H_2O_2 + 0.27H_3O^+ + 0.05H_2 \quad (1)$$

This combination of oxidation and reduction is a major benefit over traditional advanced oxidation processes. The process has been used to treat pharmaceuticals such as carbamazepine and diclofenac in water (Tominaga *et al.* 2018; Zheng *et al.* 2020), industrial wastewaters (Han *et al.* 2002) and sludge (Farooq *et al.* 1993; Kim *et al.* 2020). Aqueous or hydrated electrons, $e^-_{(aq)}$, can also reduce bromate. Electricity is the only major input and no additional chemicals are required to generate the oxidizing and reducing species (Ponomarev & Ershov 2020). Sixty to 80% electrical conversion and transfer efficiency is typical (Stefan 2018). Recently, the first full-scale application of e-beam treatment was applied at 30,000 m^3/day for aniline and color removal from an industrial wastewater (Wang *et al.* 2022). Doses are typically given in kilogray (kGy), where 1 kGy equals 1 kJ absorbed/kg or approximately 0.278 kWh/m^3 in water.

The objectives of this study were to (1) evaluate the performance of electron beam treatment for 1,4-dioxane removal in several real water matrices used for water and wastewater treatment and indirect potable reuse with high bromide, (2) determine the mechanisms of 1,4-dioxane degradation and the effects on other water quality parameters of interest, and (3) determine the electrical energy per order (EEO) for 1,4-dioxane removal by e-beam treatment.

2. MATERIALS AND METHODS

The Upgraded Injector Test Facility (UITF) is a superconducting radio-frequency accelerator at the Thomas Jefferson National Accelerator Facility (TJNAF, i.e. JLab) capable of accelerating an electron beam up to an energy 10 MeV and a current of 100 nA, in continuous-wave mode. A dedicated beamline was added to investigate electron beam treatment on water samples. In this beamline, the water targets serve as a termination of the electron beam (Li *et al.* 2022). The sample target containers were constructed from solid aluminum blocks with a cylindrical hole to hold a sample volume of 60 mL. Figure 1 shows a schematic layout of the end of the beamline and the target sample holder. Four containers were placed on the target rail at the end of the irradiation beamline, the central position was taken by an X-ray screen on a dummy target (Figure S1). Each target was then moved in front of the beam remotely.

For the irradiation studies, the UITF was operated at an electron beam energy of 8 MeV and an average beam current of 108 nA, yielding an average beam power of 808 mW. The uncertainty in the beam power is ~1%. During commissioning of the setup, the irradiation dose rate was calibrated by the use of dedicated dosimeter rods placed inside and in front of the sample holders, as well as corresponding simulation studies. During the water irradiation, the dose was varied from 0.25 kGy to 20 kGy for the individual samples by varying the beam exposure time from 30 seconds to 40 min. A detailed description of the beamline, the relevant electromagnetic elements, dosimetry measurements and simulations can be found in Li *et al.* (2022).

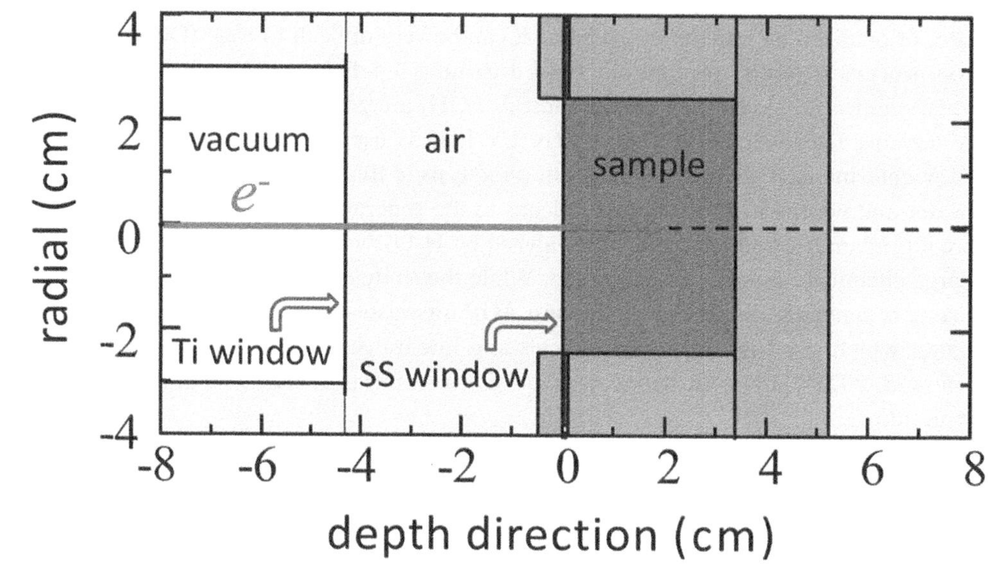

Figure 1 | Schematic layout of the water irradiation target. The electrons travel from left to right, out of the beamline vacuum, through a Ti-foil window, air, and a stainless steel window into the water sample.

2.1. Sample water collection and analysis

Secondary effluent was collected from the Hampton Roads Sanitation District Nansemond Treatment Plant (Suffolk, VA, USA). The plant consists of a five-stage bardenpho process with methanol addition for denitrification. For a sample with lower total organic carbon and further treatment representing drinking water, granular activated carbon (GAC) effluent was collected from the SWIFT Research Center at the Nansemond Treatment Plant. In this process, secondary effluent is treated through coagulation-flocculation and sedimentation, ozonation and biofiltration, followed by GAC adsorption and UV disinfection for indirect potable reuse via managed aquifer recharge. The design and operation of this plant has been described elsewhere (Gonzalez *et al.* 2021; Hogard *et al.* 2021). Ultra-pure water (UPW), resistivity >16 MΩ, was produced by a Smart2Pure UV Ultrapure Water System (Thermo-Fisher Scientific, Waltham, MA, USA).

Secondary effluent samples were collected on three separate occasions, filtered, spiked with 1,4-dioxane (when applicable), refrigerated, and taken in a cooler to Jefferson Lab where they were refrigerated until use. Samples were then allowed to come up to room temperature, 22 °C, before irradiation. The secondary effluent and GAC effluent samples were filtered through a 0.45 μm nylon filter in order to remove turbidity and pathogens and limit the potential for biodegradation of the water matrix constituents during storage. Total organic carbon (TOC) ranged from 8.98 to 10.74 mg/L in the secondary effluent, pH ranged from 7.0 to 7.2. pH was not measured at the time of testing; however, it is unlikely to have changed significantly over the course of sample storage. GAC samples were collected on two separate days, TOC was 3.47 and 3.45 mg/L, respectively. pH was 6.82 in the second sample at the time of sample collection and preparation

For most tests, a stock solution of reagent grade 1,4-dioxane diluted in ultra-pure water was spiked into the sample water to allow high enough concentrations for extraction and detection. For the test with the background level of 1,4-dioxane, two samples were irradiated at the desired dose and then combined into one 125 mL bottle. 1,4-dioxane samples were collected in 40 mL or 125 mL bottles preserved with 1 g/L sodium bisulfate depending on the desired detection limit. Samples were extracted by solid phase extraction and analyzed by a modified version of EPA method 522 by HRSD's Central Environmental Laboratory. Due to the limited sample volumes irradiated, replicate samples were not taken. Duplicate initial samples were taken on two occasions and measured within 6.5% difference (Table S1).

UV absorbance was measured on a Genesys UV-Vis spectrophotometer (Thermo-Fisher Scientific, Waltham, MA, USA). TOC and total nitrogen (TN) were measured on a Shimadzu TOC 4200 analyzer. Chemical oxygen demand (COD), nitrite and nitrate were measured by TNT methods 821, 839, and 835, respectively, with a Hach DR3900 Spectrophotometer. Dissolved oxygen (DO) was measured using a Hach LDO handheld DO probe (Hach Co, Loveland, CO, USA). Bromate was analyzed by ion chromatography by EPA method 300.1. Bromide was measured by EPA method 300.

3. RESULTS AND DISCUSSION

3.1. Removal of 1,4-dioxane

Samples of secondary effluent, GAC effluent, and ultrapure water spiked with 1,4-dioxane were irradiated at varying doses as shown in Figure 2. The reported doses represent the average dose or total energy absorbed by the sample. The relative uncertainty in the dose measurement was in the range 3–5% for all samples. Removal improved with increasing e-beam dose and it was greater in the progressively cleaner water matrices. For the secondary effluent samples, lesser removal was observed for the samples with the background concentration of 1,4-dioxane (0.48 μg/L) while the removal was quite similar between the spiked samples at high and low concentrations. Because of the low concentration, the background samples had the greatest possibility for error, as they required two irradiation runs to generate sufficient sample volume for extraction and analysis.

In some cases, moderate increases in dose did not continue to remove 1,4-dioxane as shown in the UPW results in Figure 2, though it could be decreased to below the method reporting limit (MRL) with doses >5 kGy (Table S1). There was a non-linear relationship between log-removal and dose (Figure S2) which could be due to the non-uniform distribution of dose through the sample due to the shallow penetration depth of the accelerated electrons and narrow spread of the beam over the sample. Dosimetry measurements and simulated dose distributions showed that the dose was concentrated toward the center of the sample holders and that ~10% of the sample volume at the back of the sample holders received less than 15% of the peak dose (Li *et al.* 2022). On the other hand, once the major $e^-_{(aq)}$ scavengers are consumed, $e^-_{(aq)}$ may react with ·OH and decrease 1,4-dioxane removal, though this seems less likely.

In one test shown in Figure 3, 25 mM tert-butanol (t-BuOH) was added as a hydroxyl radical scavenger and resulted in no removal of 1,4-dioxane, except for at the highest dose of 2.3 kGy where ~10% reduction occurred. Without t-BuOH addition at the same dose, 94% removal of 1,4-dioxane was observed. At the highest doses, hydrogen atom reduction may play a small part in the degradation of 1,4-dioxane. However, with a second order rate constant of $1{*}10^7\ M^{-1}s^{-1}$ (Buxton *et al.* 1988), which is two orders of magnitude slower than ·OH ($k_{\cdot OH} = 3.0{*}10^9\ M^{-1}s^{-1}$), along with the lower yields of H· and H· scavenging by dissolved oxygen, it is unlikely to play a significant role. 25 mM t-BuOH is a concentration typically used in ozonation experiments at low doses (Buffle *et al.* 2006). However, others have used concentrations as high as 0.5 M to completely scavenge ·OH in irradiation experiments (Trojanowicz *et al.* 2021). As such, it can be concluded that 1,4-dioxane removal is still primarily occurring by ·OH oxidation and not by the other oxidizing/reducing species involved in e-beam treatment.

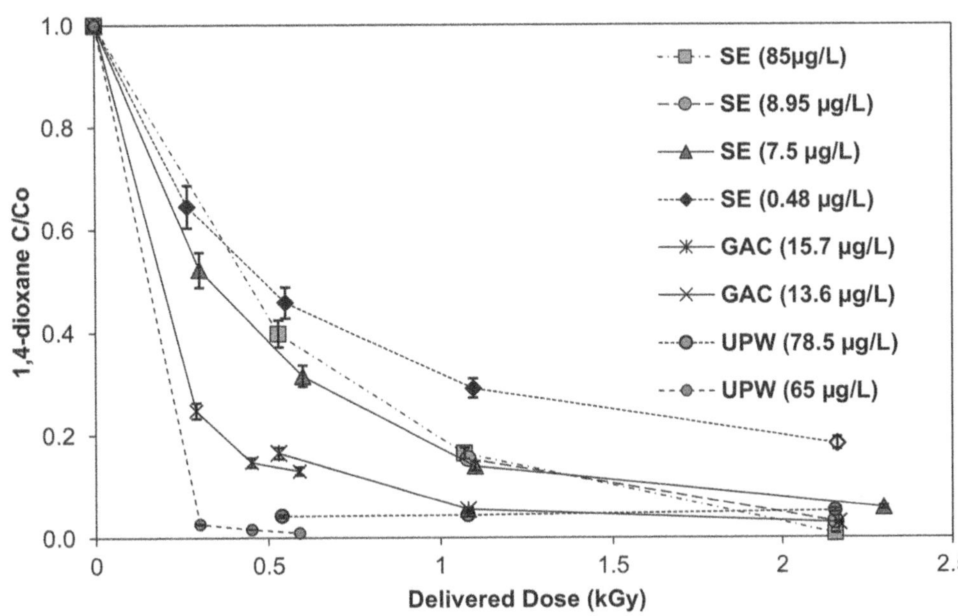

Figure 2 | 1,4-dioxane concentration vs delivered dose in secondary effluent (SE), GAC effluent (GAC), ultrapure water (UPW). Initial concentrations in parentheses. The open diamond represents a value below the method reporting limit. Error bars represent the expected analytical error of ±6.5% applied to the final concentration.

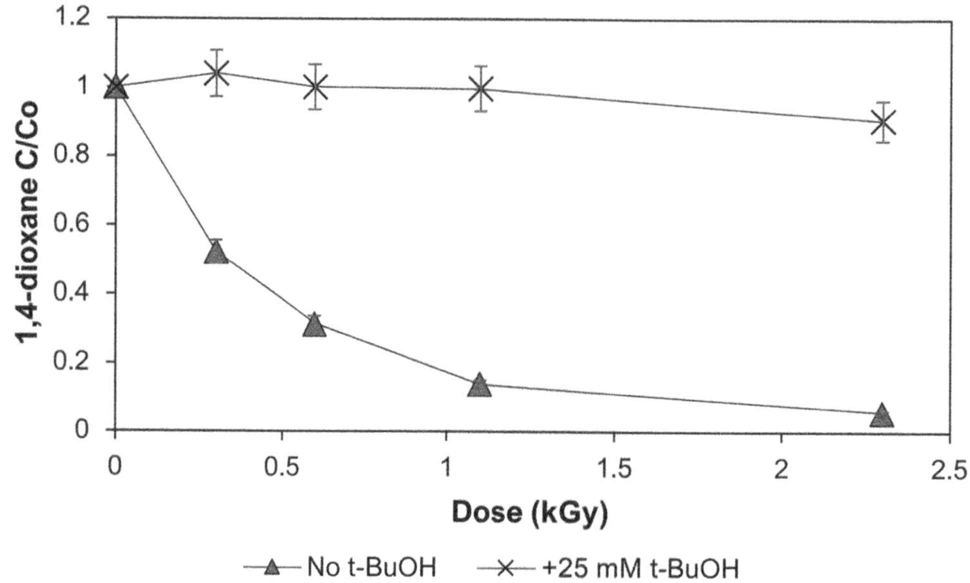

Figure 3 | 1,4-dioxane removal with and without tert-butanol. 7.5 μg/L initial 1,4-dioxane. Error bars represent the expected analytical error of ±6.5% applied to the final concentration.

3.2. Bromate and other water quality data

Bromate samples were collected during some of the secondary effluent tests. Despite high bromide concentrations, 299 μg/L, and high ·OH radical exposures as shown above, no bromate formation was observed and all values were below detection. In the GAC effluent samples, downstream of ozonation, bromate removal of up to 82% was observed at a dose of 10.8 kGy with an initial concentration of 1.99 μg/L (Figure 4). Wang *et al.* (2016) showed that bromate removal was significantly lower with dissolved oxygen and nitrate present which is consistent with $e^-_{(aq)}$ scavenging by Equations (2) and (3) (Buxton *et al.* 1988).

$$O_2 + e^-_{(aq)} \rightarrow O_2^- \qquad k = 1.1*10^9 M^{-1}s^{-1} \tag{2}$$

$$NO_3^- + e^-_{(aq)} \rightarrow NO_3^{2-} \quad k = 9.7*10^9 M^{-1}s^{-1} \tag{3}$$

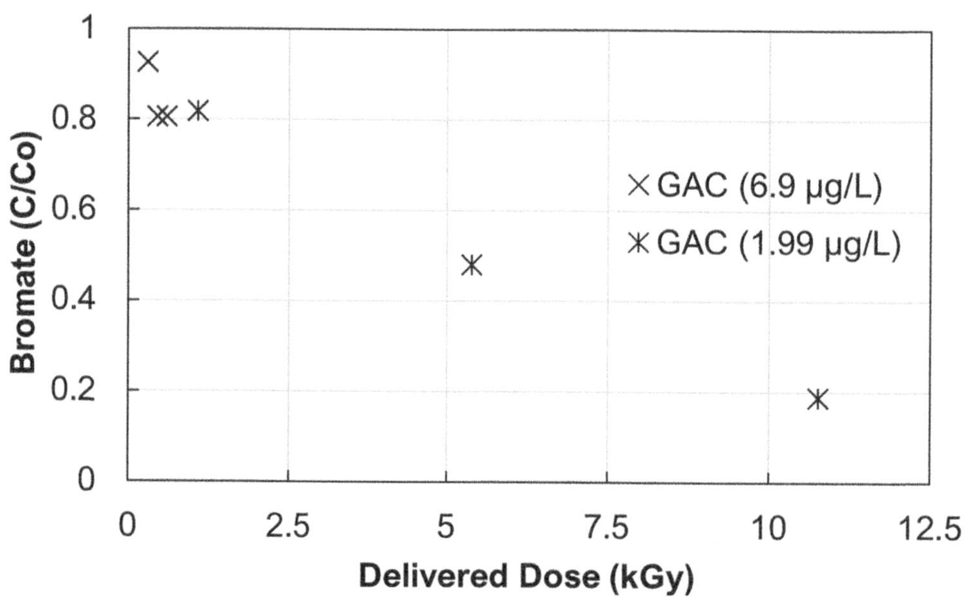

Figure 4 | Bromate removal in GAC effluent samples. Initial concentrations in parentheses.

For one test, dissolved oxygen was measured in the sample holder before and after irradiation (Figure 5(a)). It is assumed that all tests were conducted with initial saturation concentrations of DO. Figure 5(b) shows the change in total and inorganic nitrogen species through e-beam irradiation. Nitrate concentrations increased at low doses and then decreased with increasing e-beam dose above 1.16 kGy. Doses much higher than those required for 1,4-dioxane were needed for significant nitrate reduction. Dissolved oxygen may have limited nitrate reduction as it would be the predominant $e^{-}_{(aq)}$ scavenger initially due to its higher initial concentration and slightly higher second order rate constant. Nitrite formation was observed at the lower doses but then also decreased at higher doses. Nitrate and nitrite were likely reduced to another inorganic nitrogen species which was not measured as the TN concentration did not decrease through irradiation. NO and N_2O formation seem unlikely as these two species are volatile and would likely be lost between sample transfer and storage as well as the air sparging step in the TOC/TN analysis. Ammonia concentrations increased to up to 0.75 mg/L-N; this is most likely released from dissolved organic nitrogen (rather than cell lysis as the samples were pre-filtered), which has been observed due to e^- transfer in ozonation (de Vera *et al.* 2017). A similar trend in nitrite and nitrate was observed in a second test shown in the SI at doses up to 2 kGy though ammonia formation was not as consistent and is worthy of further investigation.

Bulk organic matter was measured as dissolved organic carbon (DOC), chemical oxygen demand, and UV_{254} absorbance (Figure 5(c)). Overall, there was little change in the organics indicating that, similar to other advanced oxidation processes, complete mineralization of contaminants was not occurring. The slight increase in COD and then subsequent decrease could be due to reduction products being formed and then oxidized at higher doses. However, the slight increase in DOC at the lowest dose compared to the initial sample indicates that there may be some sample contamination, perhaps leaching of organics from the cork gasket used to seal the sample holders.

3.3. Applications and engineering significance

Electrical energy per order values (EEO) were calculated from the results above using Equation (4) (Londhe *et al.* 2021).

$$EEO_{e-beam} = \frac{Dose\ (kGy)}{3.6* - \log_{10}\left(\frac{C}{Co}\right)} \tag{4}$$

Due to the non-linear relationship between log-removal and dose, EEOs were calculated using the results closest to 1-log removal in each water matrix. For secondary effluent, the average EEO value was reported from the different sample runs, excluding the samples with the background levels of 1,4-dioxane. The EEO values for 1,4-dioxane by electron beam were 0.37, 0.19, and 0.05 kWh/m^3/order for secondary effluent, GAC effluent, and UPW, respectively. With a typical 70% transfer and generation efficiency applied the values were 0.53, 0.26, and 0.08 kWh/m^3/order.

Ozonation has repeatedly been shown to be the most efficient AOP in wastewater effluent (Katsoyiannis *et al.* 2011; Lee *et al.* 2016; Sgroi *et al.* 2021). However, there is currently no available EEO data for 1,4-dioxane in secondary effluent. Lee

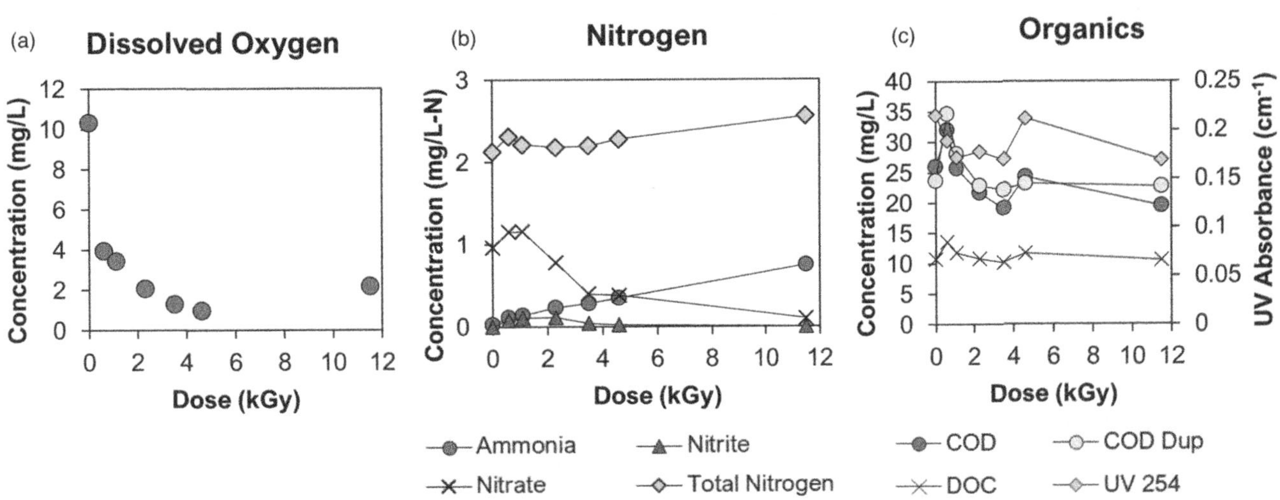

Figure 5 | Effects of e-beam treatment on (a) dissolved oxygen, (b) nitrogen species, and (c) bulk organics in secondary effluent.

et al. (2016) calculated EEO values of 0.15 kWh/m^3/order for meprobamate and atrazine, which have similar second order rate constants with ozone and hydroxyl radicals to 1,4-dioxane. While this is significantly lower than the value here, it does not account for oxygen, or any chemicals required to suppress bromate formation, which was shown to be a significant challenge in this water (Pearce *et al.* 2022). UVAOP using hydrogen peroxide is the most common technique for 1,4-dioxane removal in drinking water treatment. An EEO of 0.13 kWh/m^3/order was calculated from a bench scale UV reactor treating 1,4-dioxane in deionized water with 10 mg/L H_2O_2, and 0.20 kWh/m^3/order on average for four groundwaters treated under the same conditions (Lee *et al.* 2020). Here e-beam treatment is comparable to UVAOP and again does not require any external reagents which may need to be quenched. Additionally, as discussed above, the non-uniform dose distribution of this experimental configuration may limit the efficiency of e-beam, especially when targeting removal of 90% or greater. The configuration of the sample holders used in this study did not allow for mixing of the samples which may have improved the dose distribution. Simply as an advanced oxidation process, e-beam treatment may be comparable in terms of operating costs to other conventionally used processes as discussed above. However, the high capital costs of accelerators are an obstacle towards wider adoption of electron beam treatment (Miklos *et al.* 2018).

The clear benefit of e-beam treatment is the added reductive capacity when compared to traditional advanced oxidation processes. Certain compounds are more susceptible to degradation by reduction than oxidation. Acetone is primarily degraded by hydrated electrons (Yoon *et al.* 2009), whereas carbamazepine, bisphenol A (BPA), and atrazine are susceptible to both ·OH oxidation and $e^-_{(aq)}$/H· reduction (Basfar *et al.* 2009; Zheng *et al.* 2020; Trojanowicz *et al.* 2021). Certain halogenated contaminants and disinfection byproducts have been shown to be susceptible to radiolytic treatment and hydrated electron degradation such as trichloroethylene (TCE) (Nickelsen *et al.* 2002), and carbon tetrachloride (Mak *et al.* 1997). The removal of PFAS by e-beam is the most promising. The higher doses required for PFAS degradation would guarantee that most other pollutants are well degraded, too. If e-beam treatment could be used for PFAS removal at doses <5 kGy then it may be a cost-effective alternative to the combination of ozonation and GAC in a wastewater treatment or water reuse context (Wang *et al.* 2016; Londhe *et al.* 2021).

While e-beam has been used in a large-scale, continuous flow application (Wang *et al.* 2022), there is limited water industry experience and few equipment vendors available for these uses. The development of high-power (several hundreds of kW), high-efficiency accelerators would allow the treatment of a higher water flow (tens of thousands of m^3/day), thereby reducing the impact of their capital cost to the overall treatment cost. R&D is ongoing at National Laboratories to develop this kind of accelerators (Ciovati *et al.* 2018; Dhuley *et al.* 2022). The results presented show that electron beam treatment is effective as an advanced oxidation process and for cost effective 1,4-dioxane removal, warranting further studies on the removal of other chemicals from water and wastewater, such as PFAS.

4. CONCLUSIONS

Electron beam treatment was evaluated for 1,4-dioxane removal in water and wastewater treatment applications. Greater than 90% removal was achievable at moderate doses and removal improved in the progressively cleaner water matrices. Using tert-butanol it was shown that 1,4-dioxane degradation primarily occurs through hydroxyl radical oxidation. Water radiolysis did not result in significant reductions in TOC or COD, indicating that complete mineralization of contaminants was unlikely to have occurred. Bromate removal was observed in several samples and there was no bromate formation, despite the high bromide concentrations and ·OH exposures. The calculated EEO values for 1,4-dioxane removal in secondary wastewater effluent and GAC effluent were higher when compared to established treatment techniques in similar conditions, though electron beam treatment would result in comparable operating costs in several cases. Some of this inefficiency may be due to experimental factors which warrants further evaluation at larger scales. Electron beam treatment can also be a promising method for treating PFAS compounds in wastewater and research on this topic has begun through the collaboration between HRSD and Jefferson Lab.

ACKNOWLEDGEMENTS

The authors would like to acknowledge the work done by the HRSD Central Environmental Laboratory analysts and many colleagues in the Engineering and Accelerator Divisions at Jefferson Lab, particularly M. Poelker and M. McCaughan for helping with the installation, commissioning, and operation of the irradiation beamline.

FUNDING

This work was funded in part by Jefferson Lab's Laboratory Directed Research and Development program and Hampton Roads Sanitation District.

DATA AVAILABILITY STATEMENT

All relevant data are included in the paper or its Supplementary Information.

CONFLICT OF INTEREST

The authors declare there is no conflict.

REFERENCES

Adamson, D. T., Piña, E. A., Cartwright, A. E., Rauch, S. R., Hunter Anderson, R., Mohr, T. & Connor, J. A. 2017 1,4-Dioxane drinking water occurrence data from the third unregulated contaminant monitoring rule. *Sci. Total Environ.* **596–597**, 236–245. https://doi.org/10.1016/j.scitotenv.2017.04.085.

Basfar, A. A., Mohamed, K. A., Al-Abduly, A. J. & Al-Shahrani, A. A. 2009 Radiolytic degradation of atrazine aqueous solution containing humic substances. *Ecotoxicol. Environ. Saf.* **72**, 948–953. https://doi.org/10.1016/j.ecoenv.2008.05.006.

Buffle, M. O., Schumacher, J., Meylan, S., Jekel, M. & von Gunten, U. 2006 Ozonation and advanced oxidation of wastewater: effect of O3dose, pH, DOM and HO.-scavengers on ozone decomposition and HO.generation. *Ozone Sci. Eng.* **28**, 247–259. https://doi.org/10.1080/01919510600718825.

Buxton, G. V., Greenstock, C. L., Helman, W. P. & Ross, A. B. 1988 Critical review of rate constants for reactions of hydrated electrons, hydrogen atoms and hydroxyl radicals ($\cdot OH/\cdot O^-$ in aqueous solution. *J. Phys. Chem. Ref. Data* **17**, 513–886. https://doi.org/10.1063/1.555805.

Ciovati, G., Anderson, J., Coriton, B., Guo, J., Hannon, F., Holland, L., Lesher, M., Marhauser, F., Rathke, J., Rimmer, R., Schultheiss, T. & Vylet, V. 2018 Design of a cw, low-energy, high-power superconducting linac for environmental applications. *Phys. Rev. Accel. Beams* **21**, 91601. https://doi.org/10.1103/PhysRevAccelBeams.21.091601.

Cooper, W. J., Nickelsen, M. G., Meacham, D. E., Waite, T. D. & Kurucz, C. N. 1992 High energy electron beam irradiation: an advanced oxidation process for the treatment of aqueous based organic hazardous wastes. *Water Qual. Res. J.* **27**, 69–95. https://doi.org/https://doi.org/10.2166/wqrj.1992.005.

de Vera, G. A., Gernjak, W., Weinberg, H., Farré, M. J., Keller, J. & von Gunten, U. 2017 Kinetics and mechanisms of nitrate and ammonium formation during ozonation of dissolved organic nitrogen. *Water Res.* **108**, 451–461. https://doi.org/10.1016/j.watres.2016.10.021.

Dhuley, R. C., Gonin, I., Kazakov, S., Khabiboulline, T., Sukhanov, A., Yakovlev, V., Saini, A., Solyak, N., Sauers, A., Thangaraj, J. C. T., Zeller, K., Coriton, B. & Kostin, R. 2022 Design of a 10 MeV, 1000 kW average power electron-beam accelerator for wastewater treatment applications. *Phys. Rev. Accel. Beams* **25**, 41601. https://doi.org/10.1103/PhysRevAccelBeams.25.041601.

Farooq, S., Kurucz, C. N., Waite, T. D. & Cooper, W. J. 1993 Disinfection of wastewaters: high-energy electron vs gamma irradiation. *Water Res.* **27**, 1177–1184. https://doi.org/10.1016/0043-1354(93)90009-7.

Gonzalez, D., Thompson, K., Quiñones, O., Dickenson, E. & Bott, C. 2021 Granular activated carbon-based treatment and mobility of per- and polyfluoroalkyl substances in potable reuse for aquifer recharge. *AWWA Water Sci.* **3**, 1–16. https://doi.org/10.1002/aws2.1247.

Han, B., Ko, J., Kim, J., Kim, Y., Chung, W., Makarov, I. E., Ponomarev, A. V. & Pikaev, A. K. 2002 Combined electron-beam and biological treatment of dyeing complex wastewater. Pilot plant experiments. *Radiat. Phys. Chem.* **64**, 53–59. https://doi.org/10.1016/S0969-806X(01)00452-2.

Hogard, S., Salazar-Benites, G., Pearce, R., Nading, T., Schimmoller, L., Wilson, C., Heisig-Mitchell, J. & Bott, C. 2021 Demonstration-scale evaluation of ozone–biofiltration–granular activated carbon advanced water treatment for managed aquifer recharge. *Water Environ. Res*, 1–16. https://doi.org/10.1002/wer.1525.

Katsoyiannis, I. A., Canonica, S. & von Gunten, U. 2011 Efficiency and energy requirements for the transformation of organic micropollutants by ozone, O_3/H_2O_2 and UV/H_2O_2. *Water Res.* **45**, 3811–3822. https://doi.org/10.1016/j.watres.2011.04.038.

Kim, Y., Ershov, B. G. & Ponomarev, A. V. 2020 Features and ways to upgrade electron-beam wastewater treatment. *High Energy Chem.* **54**, 462–468. https://doi.org/10.1134/S0018143920060089.

Lee, Y., Gerrity, D., Lee, M., Gamage, S., Pisarenko, A., Trenholm, R. A., Canonica, S., Snyder, S. A. & Von Gunten, U. 2016 Organic contaminant abatement in reclaimed water by UV/H_2O_2 and a combined process consisting of O_3/H_2O_2 followed by UV/H_2O_2: prediction of abatement efficiency, energy consumption, and byproduct formation. *Environ. Sci. Technol.* **50**, 3809–3819. https://doi.org/10.1021/acs.est.5b04904.

Lee, C. S., Venkatesan, A. K., Walker, H. W. & Gobler, C. J. 2020 Impact of groundwater quality and associated byproduct formation during UV/hydrogen peroxide treatment of 1,4-dioxane. *Water Res.* **173**, 1–10. https://doi.org/10.1016/j.watres.2020.115534.

Li, X., Baumgart, H., Bott, C., Ciovati, G., Gregory, S., Hannon, F., Mccaughan, M., Pearce, R., Poelker, M., Vennekate, H. & Wang, S. 2022 Design and commissioning of an e-beam irradiation beamline at the upgraded injector test facility at Jefferson Lab. *Nucl. Inst. Methods Phys. Res. A* **1039**. https://doi.org/10.1016/j.nima.2022.167093.

Londhe, K., Lee, C.-S., Zhang, Y., Grdanovska, S., Kroc, T., Cooper, C. A. & Venkatesan, A. K. 2021 Energy evaluation of electron beam treatment of perfluoroalkyl substances in water: a critical review. *ACS ES&T Eng.* **1**, 827–841. https://doi.org/10.1021/acsestengg.0c00222.

Mak, F. T., Zele, S. R., Cooper, W. J., Kurucz, C. N., Waite, T. D. & Nickelsen, M. G. 1997 Kinetic modeling of carbon tetrachloride, chloroform and methylene chloride removal from aqueous solution using the electron beam process.pdf. *Water Res.* **31**, 219–228.

McElroy, A. C., Hyman, M. R. & Knappe, D. R. U. 2019 1,4-Dioxane in drinking water: emerging for 40 years and still unregulated. *Curr. Opin. Environ. Sci. Health.* **7**, 117–125. https://doi.org/https://doi.org/10.1016/j.coesh.2019.01.003.

Miklos, D. B., Remy, C., Jekel, M., Linden, K. G., Drewes, J. E. & Hübner, U. 2018 Evaluation of advanced oxidation processes for water and wastewater treatment – A critical review. *Water Res.* **139**, 118–131. https://doi.org/10.1016/j.watres.2018.03.042.

Nickelsen, M., Cooper, W., Secker, D., Rosocha, L., Kurucz, C. & Waite, T. 2002 Kinetic modeling and simulation of PCE and TCE removal in aqueous solutions by electron-beam irradiation. *Radiat. Phys. Chem.* **65**, 579–587.

Otto, M. & Nagaraja, S. 2007 Treatment technologies for 1,4-Dioxane: fundamentals and field applications. *Remediation* **17**, 81–88. https://doi.org/10.1002/rem.20135.

Pearce, R., Hogard, S., Buehlmann, P., Salazar-Benites, G., Wilson, C. & Bott, C. 2022 Evaluation of preformed monochloramine for bromate control in ozonation for potable reuse. *Water Res.* **211**, 118049. https://doi.org/10.1016/j.watres.2022.118049.

Ponomarev, A. V. & Ershov, B. G. 2020 The green method in water management: electron beam treatment. *Environ. Sci. Technol.* **54**, 5331–5344. https://doi.org/10.1021/acs.est.0c00545.

Sgroi, M., Snyder, S. A. & Roccaro, P. 2021 Comparison of AOPs at pilot scale: energy costs for micro-pollutants oxidation, disinfection by-products formation and pathogens inactivation. *Chemosphere* **273**, 128527. https://doi.org/10.1016/j.chemosphere.2020.128527.

Stefan, M. I., 2018 Advanced oxidation processes for water treatment – fundamentals and applications. In: *Advanced Oxidation Processes for Water Treatment* (Stefan, M. I., ed.). IWA Publishing, London, UK, pp. 241–295. https://doi.org/10.2166/9781780407197.

Tanabe, A. & Kawata, K. 2008 Determination of 1,4-dioxane in household detergents and cleaners. *J. AOAC Int.* **91**, 439–444. https://doi.org/10.1093/jaoac/91.2.439.

Tominaga, F. K., Dos Santos Batista, A. P., Silva Costa Teixeira, A. C. & Borrely, S. I. 2018 Degradation of diclofenac by electron beam irradiaton: toxicitiy removal, by-products identification and effect of another pharmaceutical compound. *J. Environ. Chem. Eng.* **6**, 4605–4611. https://doi.org/10.1016/J.JECE.2018.06.065.

Trojanowicz, M., Bojanowska-Czajka, A., Szreder, T., Męczyńska-Wielgosz, S., Bobrowski, K., Fornal, E. & Nichipor, H. 2021 Application of ionizing radiation for removal of endocrine disruptor bisphenol A from waters and wastewaters. *Chem. Eng. J.* **403**, 126169. https://doi.org/10.1016/j.cej.2020.126169.

USEPA 2013 *Summary on 1,4-Dioxane (123-91-1).*

von Sonntag, C. & von Gunten, U. 2012 *Chemistry of Ozone in Water and Wastewater Treatment: From Basic Principles to Applications.* https://doi.org/10.2166/9781780400839.

Wang, L., Batchelor, B., Pillai, S. D. & Botlaguduru, V. S. V. 2016 Electron beam treatment for potable water reuse: removal of bromate and perfluorooctanoic acid. *Chem. Eng. J.* **302**, 58–68. https://doi.org/10.1016/j.cej.2016.05.034.

Wang, S., Wang, J., Chen, C., He, S., Hu, J. & Zhang, Y. 2022 First full-scale application of electron beam technology for treating dyeing wastewater (30,000 m^3 /d) in China. *Radiat. Phys. Chem.* **196**, 110136. https://doi.org/10.1016/j.radphyschem.2022.110136.

Yoon, Y. J., Jung, Y. J., Han, B. S. & Kang, J. W. 2009 Performance of electron beam irradiation for treatment of groundwater contaminated with acetone. *Water Sci. Technol.* **59**, 2475–2483. https://doi.org/10.2166/wst.2009.234.

Zheng, M., Bao, Y., Huang, Z., Qiu, W., Xu, G. & Wang, Z. 2020 Radiolysis of carbamazepine by electron beam: roles of transient reactive species and biotoxicity of final reaction solutions on rotifer *Philodina* sp. *Sci. Total Environ.* **703**, 135013. https://doi.org/10.1016/j.scitotenv.2019.135013.

First received 1 October 2022; accepted in revised form 3 December 2022. Available online 10 December 2022

doi: 10.2166/wst.2023.119

Phthalate esters: occurrence, toxicity, bioremediation, and advanced oxidation processes

Madhu Kumari and Mrudula Pulimi *

Centre of Nanobiotechnology, Vellore Institute of Technology, Vellore, Tamil Nadu, India
*Corresponding author. E-mail: pmrudula@vit.ac.in

MP, 0000-0003-1564-3626

ABSTRACT

Phthalic acid esters are emerging pollutants, commonly used as plasticizers that are categorized as hazardous endocrine-disrupting chemicals (EDCs). A rise in anthropogenic activities leads to an increase in phthalate concentration in the environment which leads to various adverse environmental effects and health issues in humans and other aquatic organisms. This paper gives an overview of the research related to phthalate ester contamination and degradation methods by conducting a bibliometric analysis with VOS Viewer. Ecotoxicity analysis requires an understanding of the current status of phthalate pollution, health impacts, exposure routes, and their sources. This review covers five toxic phthalates, occurrences in the aquatic environment, toxicity studies, biodegradation studies, and degradation pathways. It highlights the various advanced oxidation processes like photocatalysis, Fenton processes, ozonation, sonolysis, and modified AOPs used for phthalate removal from the environment.

Key words: advanced oxidation processes, bibliometric analysis, contaminants of emerging concern (CECs), degradation, endocrine-disrupting chemicals (EDCs), phthalates

HIGHLIGHTS

- Plastic waste from urban/industrial sources is the major cause of phthalate ester occurrence in water systems.
- DEHP is the major toxic phthalate found extensively in water bodies.
- AOPs are an emerging area of treatment for the CECs.
- Bibliometric analysis of PAE degradation indicates major research focus on UV-based, ozone, and Fenton processes.
- Synergistic methods have shown better results for PAE degradation.

GRAPHICAL ABSTRACT

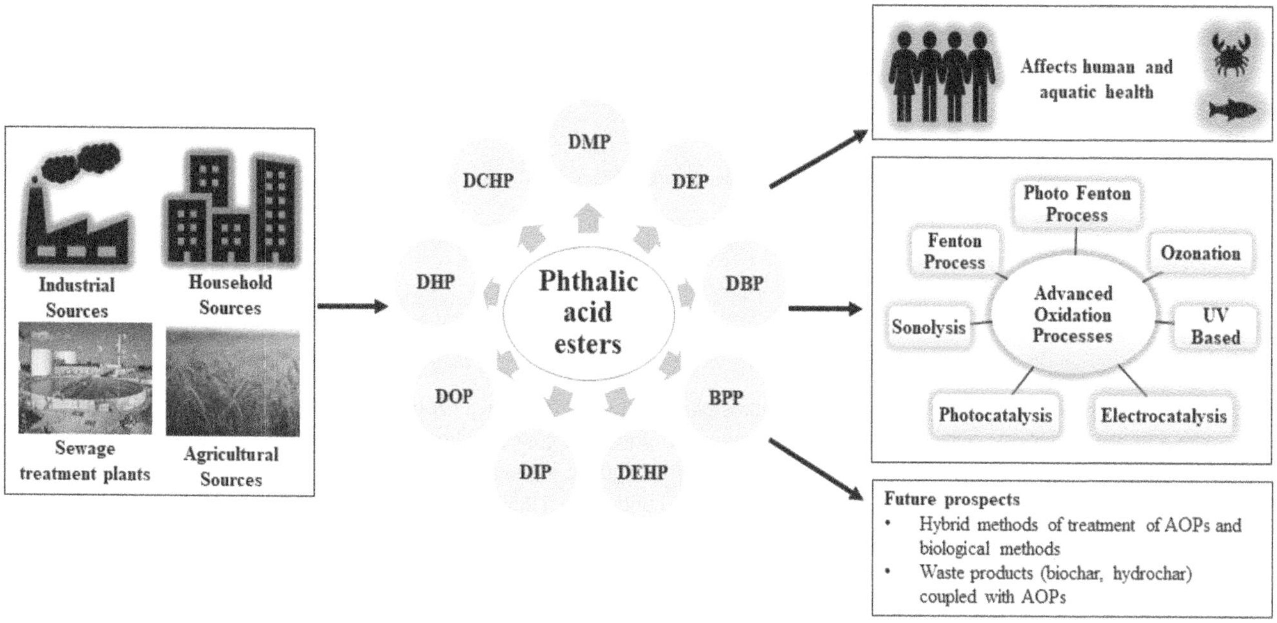

1. INTRODUCTION

Contaminants of emerging concern (CECs) also known as emerging contaminants/pollutants are the miscellaneous group of compounds which are neither monitored routinely nor regulated properly. Emerging contaminants are significant nowadays because there is a recent increase in the presence of these contaminants, and it poses health risks to humans and animals (Sengupta *et al.* 2014). They include personal care products (i.e., detergents, soaps, cosmetics, nail paints, shampoos); plastic-derived compounds (i.e., phthalates, bisphenol); medicines (i.e., prescribed and non-prescribed); pesticides; and chemical products (i.e., preservatives, additives) which make its way into the aqueous system via wastewater treatment plants, sewage, urban land surface, and agricultural runoff, septic systems, etc. (Pironti *et al.* 2021). It will remain a hot topic of interest because new compounds are continuously evolving and are produced, let it be chemical pollutants (pharmaceuticals, phthalates, and bisphenols) and its by-products or biological pollutants (SARS-CoV-2 virus). Some of these CECs act as endocrine-disrupting factors (EDFs) which change the hormone level in organisms leading to reproductive disorders. There are other health issues related to CECs which include respiratory disorders, neurological disorders, and thyroid disorders (Pastorino & Ginebreda 2021).

Diesters of phthalic acids are ordinarily known as phthalic acid esters, phthalates or phthalate esters. Phthalates adversely affect human health because of their exposure during regular activities and are represented as a worldwide issue (Giuliani *et al.* 2020). They can be present in the surrounding environment (air, water, soil) in a concentration ranging from ng L^{-1} to μg L^{-1}. PAEs are among the CECs because they act as endocrine-disrupting chemicals leading to various developmental disorders, reproductive problems, and cellular toxicity. Phthalates are used as plasticizers, which are additives in the plastic industry to increase the durability and flexibility of plastics. With the increase in the utilization of plastics, as it ages and breaks down, there has been a concurrent increase in the PAE pollution. Phthalic acid esters are colorless sticky liquids that are frequently detected as organic pollutants in the environment. Phthalate exposure is an unavoidable concern for aquatic organisms, domestic life forms and various other fauna because of their extensive use, elevated manufacturing quantity and omnipresent existence in the habitat. PAEs are lightly bound to enhance flexibility of the plastic matrix, due to which it leaches out very easily from the plastics when surface contact or mechanical pressure is enforced. Figure 1 shows the sources of phthalate esters pollution which leads directly or indirectly via soil sediment to water pollution and toxicity in human organs.

The production of phthalate all over the world has grown from 2.7 to approximately 6 million tons per annum from 2007 to 2017 (Gao *et al.* 2018). Understanding the origin of phthalates in the environment as well as their distribution is important to

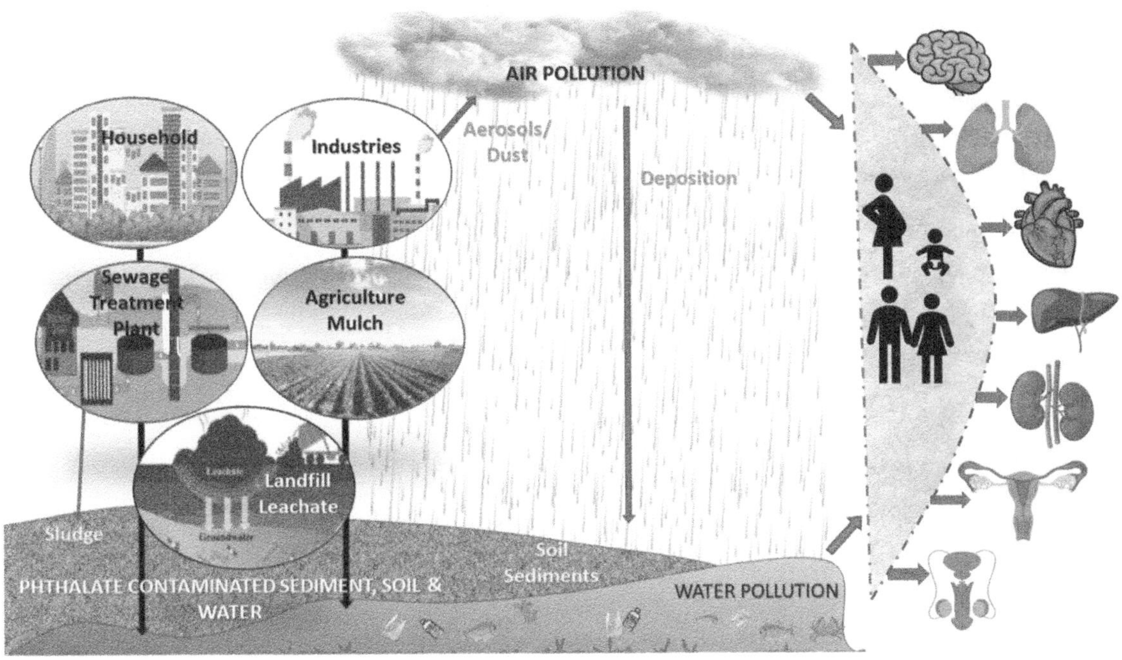

Figure 1 | Sources of phthalate pollution and different organs affected in humans by phthalates. Illustrates the households (PPCPs, toys, cosmetics, etc.), industries, sewage treatment plants, agricultural mulch and landfill leachate that are responsible for phthalate pollution in the water bodies.

evaluate the risk posed to the ecosystems. They are also used in personal care products and household plastics, which leads to higher levels of phthalates in urban wastewater plants.

Phthalates are intractable in conventional wastewater treatment, therefore making their way into the surrounding lakes and rivers where it gets accumulated in the environment and the system of fishes and other aquatic species, later entering the food web which leads to noxious effects on humans and other biotas (Kasonga *et al.* 2021). The conventional treatment methods like adsorption, and flocculation, biodegradation are not effective in the PAE remediation process as the issue arises for the disposal of the adsorbate, large amounts of sludge production, non-biodegradable compounds, etc. Therefore, it is important to understand the upcoming remediation methods used. Advanced oxidation processes have been proved to be an efficient and emerging method for the degradation of phthalates, they include a wide range of processes like photocatalysis, photolysis, Fenton processes, ozonation, UV-based methods, sonolysis, electrochemical methods, radical-based methods and coupled AOPs. Coupling together various AOPs like photo-Fenton processes, sono-Fenton, O_3/UV process, O_3/H_2O_2, and PMS/O_3 has also provided efficient degradation of the phthalate esters. AOPs have been suggested to be used in conjunction with secondary wastewater treatment methods.

The objective of the present review is to (i) understand the types of PAEs and their occurrence globally, (ii) presents a summary of the PAE toxicity and biodegradation study and (iii) document the advanced oxidation processes used for phthalate acid esters (PAEs) removal. This information will provide a basis for effective framing of regulations to limit the level of phthalates in the water and soil systems as well as an overview of the state-of-the-art methods evaluated recently for the degradation of phthalates.

2. BIBLIOMETRIC ANALYSIS

2.1. Phthalate findings

Scopus database was used to search for the literature on PAEs, the research articles were retrieved on 17/08/2022 by using the subject **((TITLE-ABS-KEY (Phthalic AND acid AND esters) OR (Phthalate AND esters) OR (Phthalate) AND (Degradation) OR (Removal) OR (Treatment)) AND PUB YEAR > 1974) AND (wastewater) AND DOC TYPE 'article').** The time duration was set from 1974 to 2022. Total of 1,739 publications were drawn out from Scopus where the top five subject categories were environmental sciences, chemistry, chemical engineering, engineering and biochemistry, genetics and molecular biology and had contributed to 74.1, 26.4, 18.1, 12.6, and 9.3% of all phthalate publications (1,739), respectively. Phthalate articles were infrequent from 1974 up to 2000. From the year 2000, a noticeable increase in the number of papers was observed. 2020 ranked as the most prolific year with 199 papers, followed by the years 2021 and 2019 with 172 and 142 papers, respectively (Figure 2(a)). With respect to the countries of phthalate articles, China was the highest PAE research publishing country (819 out of 1,739; 47.1%), followed by the United States (156; 9%) and then India (108; 6.2%) (Figure 2(b)) (Elgarahy *et al.* 2021).

2.2. Vos Viewer analysis

VOS viewer, Cite space, and Hist cite are some common software used for bibliometric analysis. VOS Viewer (version 1.6.18) is an effective software tool for assembling and analyzing bibliometric networks from literature data. Network visualizations

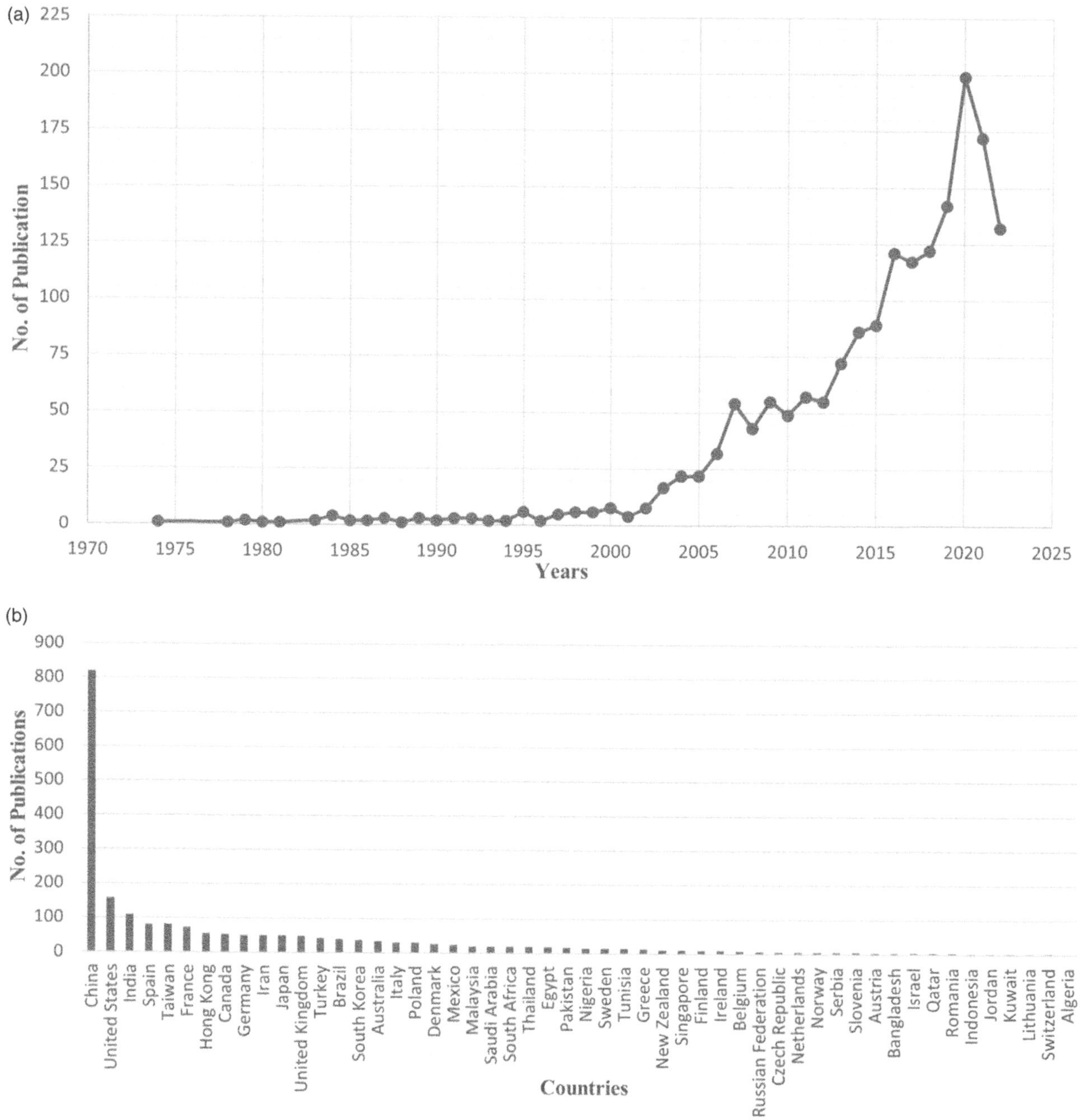

Figure 2 | (a) Publications on phthalate esters in water bodies from 1974 to 2022. (b) Publications on phthalate esters from different countries (1973–2022).

are convenient for the researchers to visualize and interpret. The network visualization of 'co-occurring keywords' plays an important role as it reveals the important area of the research field, more the frequency observed for a keyword indicates that more research is being carried out in that particular area. This software constructs maps between related data. The networks quantitatively analyze the title, keywords, and word frequency.

The keywords used in SCOPUS database were similar to that of phthalate finding with publication year from 2012 to 2022. A total of 1,360 articles were gathered from the Scopus database from the period 2012–2022. The data are extracted in .csv format and then it is imported to VOS Viewer for network visualization. The type of analysis was set to 'co-occurring' and 'indexed keywords'. The threshold value selected was 5, which means the number of times the word is reoccurring. The visualization density for 'phthalate esters' and 'degradation' can be observed in Figure 3(a). The main keywords are 'phthalate', 'biodegradation', 'bioremediation', 'wastewater treatment', 'degradation', 'toxicity', 'pollutant removal', 'adsorption', and 'oxidation'. The size is proportional to the recurrence of the keywords and the color corresponds to the different clusters present. The link thickness between two keywords indicates the number of interactions between them. There is a total of six clusters observed in Figure 3(a) which concludes the main focus of the articles. Cluster 1 (dark blue) covers phthalate studies with respect to biodegradation, and bioremediation aspects and indicates that more studies are performed on DBP. Cluster 2 (red) covers phthalate studies with respect to degradation studies like adsorption, oxidation, and photodegradation. Cluster 3 (green) covers phthalate studies with respect to wastewater treatments, environmental monitoring, and effluent studies. Cluster 4 (purple) covers phthalate study with respect to the toxicity, EDFs in humans and animals. Cluster 5 (yellow) covers phthalate studies related to water treatment and purification. Cluster 6 (blue) is the least related cluster covering pollutant, and organic compound removal. According to the cluster analysis, a lot of studies have been done with respect to biodegradation, wastewater treatment and degradation. A further focus is required in the field of toxicity and the application of AOPs.

Total of 77 articles were extracted using the keywords **(TITLE-ABS-KEY (advanced AND oxidation AND processes) OR TITLE-ABS-KEY (advanced AND oxidation) AND (phthalate AND esters) OR (phthalic AND acid AND esters)) AND PUB YEAR > 2011 AND (LIMIT-TO (DOCTYPE, 'ar'))**. The visualization density for 'phthalate esters' and 'advanced oxidation processes' can be observed in Figure 3(b). The main keywords are 'phthalate', 'advanced oxidation processes', 'ultraviolet radiation', 'photodegradation', 'ozone', 'Fenton', 'catalysts', etc. There is a total of six clusters observed in Figure 3(a) which concludes the main focus of the articles. Cluster 1 (dark blue) covers phthalate studies related to advanced oxidation processes with respect to UV, ozone, and activated carbon (AC). Cluster 2 (red) covers phthalate study with respect to AOPs using catalysts, PMS and iron compounds, and hydroxyl radicals. Cluster 3 (green) covers phthalate studies related to photocatalysis and the last cluster is a mixture of Fenton reaction related and toxicity related. With respect to AOPs and phthalate esters, major research can be observed under UV, ozone, oxidation–reduction. Further studies can be focused on sulfate radical-based, peroxide-based, and other hybrid AOPs.

3. TYPES OF PHTHALATE ESTERS

The diesters of 1,2-benzenedicarboxylic acid are frequently introduced as phthalic acid esters (PAEs) or phthalates. Toxic PAEs present in the environment are dimethyl phthalate (DMP), diethyl phthalate (DEP), di-isobutyl phthalate (DIBP), dibutyl phthalate (DnBP), butyl benzyl phthalate (BBP), and di-(2-ethylhexyl) phthalate (DEHP). Ortho, meta, and para are the isomeric forms of phthalic acid based on the orientation of the carboxylic acid with one another. However, the ortho form of benzene dicarboxylic acid, which is developed when a particular alcohol reacts with phthalic acid to form the desired ester, is used mostly as the plasticizer and brings about the significant contribution of all phthalate esters manufactured globally. They are widely utilized for manufacturing plastic products, polyvinyl chloride and other polymers to increase longevity and flexibility (Gao & Wen 2016;Yang *et al.* 2018). Nearly 60 discrete phthalate esters and its by-products are used for numerous products, i.e., cosmetics, adhesives, elasticity, lubricants, plastic manufacturing, packaging material, insecticides, and additives in paints to increase the longevity and flexibility of the plastic product (Eichler *et al.* 2019). A list of toxic phthalates with its molecular weight, chemical structure, toxic effects, and applications in the industrial sector are mentioned in Table 1.

DMP, diethyl phthalate (DEP), and di-(n-butyl) phthalate (DnBP) have low-molecular-weight (LMW) PAEs. They are utilized in industrial solvents, personal care products and cosmetics, pharmaceutical products, insecticide, perfume solvents, adhesives, waxes, and inks (Koniecki *et al.* 2011). To evaporate the perfume fragrance very slowly, DMP and DEP are used, increasing the time span for the aroma of the scent, and in nail polish to develop chip-resistant properties which is

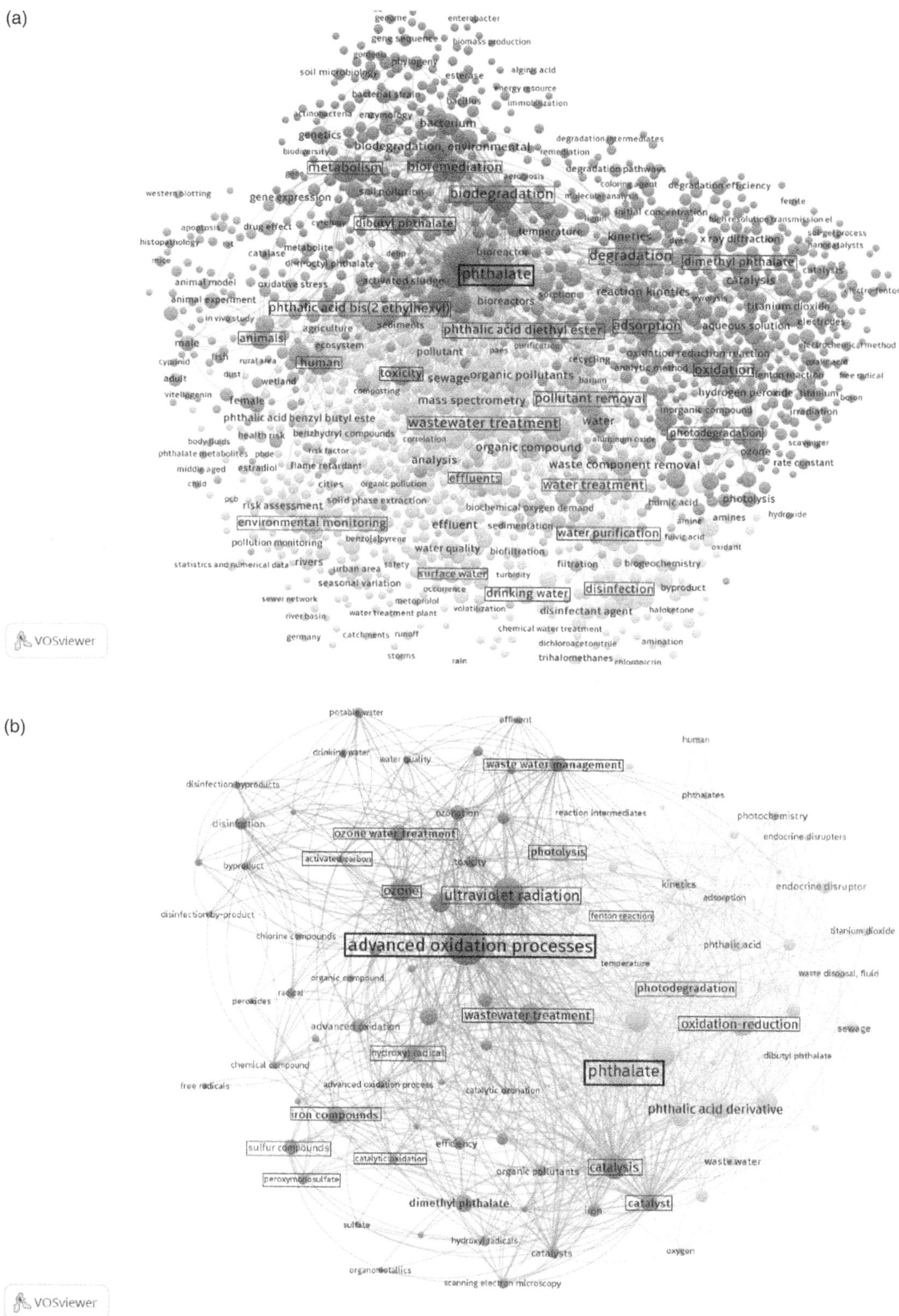

Figure 3 | (a) Co-occurrence network analysis for keywords using VOS viewer software for 'phthalate esters' and 'degradation' research. (b) Co-occurrence network analysis for keywords using VOS viewer software for 'phthalate esters' and 'advanced oxidation processes' research. Please refer to the online version of this paper to see this figure in color: http://dx.doi.org/10.2166/wst.2023.119.

Table 1 | Molecular weight, chemical structure, and application of some toxic phthalates

Phthalate esters	Mol. weight and CAS	Chemical structure	Solubility in water (mg L^{-1})	Toxic effects	Applications
Dimethyl phthalate (DMP)	194.2 and 131-11-3	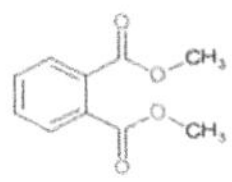	2,014	Liver problems, reproductive difficulties	Air freshener, shampoo, insecticides, etc.
Diethyl phthalate (DEP)	222.2 and 84-66-2	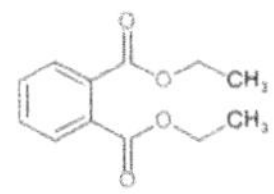	287	Male reproductive issues, liver problems	Plasticizers, aerosol sprays, cosmetics fragrances, nail polish, etc.
Dibutyl phthalate (DnBP)	278.3 and 84-74-2	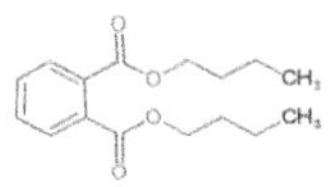	2.35	Teratogenic, testicular damage, carcinogenic	Cosmetics, anti-foaming agents, printing inks, wrapping materials.
Di-(2-ethylhexyl) phthalate (DEHP)	390.6 and 117-81-7	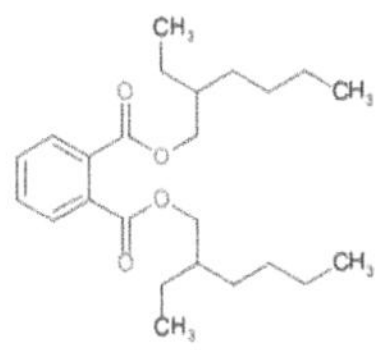	0.00	EDCs, Cytotoxicity; liver problems; elevated cancer risk.	Medical tubes, blood storage bags, gloves, furniture, diapers, children's dolls, packaging films, dialysis bags, and air tubes.
Butyl benzyl phthalate (BBP)	312.36 and 85-68-7	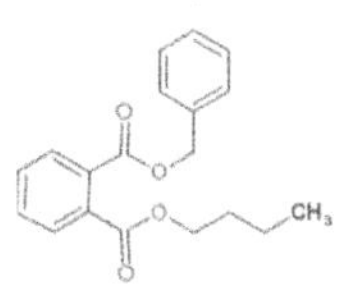	0.950	Cellular toxicity	Vinyl flooring, wrapping materials, carpet tiles, conveyor belts, automotive, adhesives, artificial leather.

Mankidy *et al.* (2013); Zhang *et al.* (2015); Giuliani *et al.* (2020); Das *et al.* (2021).

provided by a small amount of DnBP. DEHP comes under high molecular weight (HMW) PAEs. About 80% of alkyl groups with long chains are the phthalate esters utilized as plasticizers in the PVC industry which upgrades the elasticity, flooring, wall-coverings, and basic handling properties. The binding of PAE with the polymeric matrix is not strong, which leads to the gradual release of this EDC into the environment (Net *et al.* 2015). Due to the manufacturing process, leaching, and weathering, there is constant release of this pollutant into the environment.

Phthalate esters have now become an omnipresent environmental contaminant because of their excessive application in various sectors and its easy release into the environment. The leading phthalate esters used in personal care products and cosmetics are DEP followed by DBP and DEHP. DEP exposure sources are fragrances and lotions followed by deodorants, whereas the main source of DnBP are nail polishes (Koniecki *et al.* 2011). Mostly the esters have low volatility, are colorless liquids and are poorly soluble in water. These esters are soluble in oils and organic solvents. PAEs have been used as elastomers and plasticizing agents in cellulosics. There is a desperate requirement for the observation of the sources, toxicity, and ecological risk of phthalate esters in various sources in different countries, particularly in urban areas with more phthalate-based companies (Zhang *et al.* 2015).

4. OCCURRENCE AND SOURCES OF PHTHALATE ESTERS

Phthalate esters are observed in surface water, fresh water, drinking water, air, sediments, and soil worldwide. Various environmental agencies like the United States Environmental Protection Agency (USEPA), Environment Canada, European Union, and China National Environmental Monitoring (CNEM) have classified DEHP as a priority pollutant (Zolfaghari *et al.* 2014). The European Union (2013) gave guidelines for the DEHP concentration and proposed limits in marine water to be 1.3 μg L^{-1}. DEHP (0.006 mg L^{-1}) is one among the 90 contaminants listed among the legal limits set by the US EPA (US EPA 2013). Phthalates are widely discharged from rubber and chemical factories. Phthalate esters enter the surroundings very easily as it is present in a leachable form inside the PVC and they are not covalently attached to the polymeric matrix. During the manufacturing process of PVC, phthalates are consistently released into the environment by weathering, evaporating. or leaching from the final products (Gao *et al.* 2018).

In the urban areas, due to industrialization and urbanization, phthalates and their esters in different forms are released in high concentrations from the plastic materials which result in higher phthalate concentration in urban areas as compared to rural areas. They are leached out from the plastics present in the environment and objects used in homes and house dust, and also in a particulate matter which is suspended in the surroundings and contaminates the environment. Continuous bioaccumulation studies in the environments are required to explain the stress generated by the elevated concentration of phthalate esters (Sun *et al.* 2013).

Various pathways lead to the release of phthalates into marine sources including wastewater treatment units at the industrial and municipal levels, agricultural and storm runoff, deposition in the atmosphere and transportation in the river resulting via the application of sewage sludge (Net *et al.* 2015). In a study performed in the Marne Aval wastewater treatment station, a total of six phthalates were detected and DEHP (9–44 $\mu g\ L^{-1}$) was found to be the major pollutant followed by DEP (1.6–25 $\mu g\ L^{-1}$) (Dargnat *et al.* 2009). The secondary treatment process in a wastewater treatment plant removes the maximum amount of DEHP. The average concentration of DEHP in wastewater gets reduced from 30.08 to 8.13 $\mu g\ L^{-1}$ (Takdastan *et al.* 2021). Other PAEs are intractable in conventional wastewater treatment, therefore making their way into the surrounding lakes and rivers where it gets accumulated in the environment and the system of fishes and other aquatic species, later entering the food web which leads to noxious effects on humans and other biotas (Kasonga *et al.* 2021).

The occurrence and estimated concentration of phthalate esters vary between 0.03 ng m^{-3} and 24.19 $\mu g\ m^{-3}$ in air, 313 and 4,640 ng L^{-1} in water, and 40 and 348 ng g^{-1} in soil. The DMP concentration in different matrices was found to be N.D. –10.4 ng m^{-3} in atmospheric particulate matter, N.D. –31.7 $\mu g\ L^{-1}$ in fresh water, N.D. –316 $\mu g\ kg^{-1}$ in sediments (dry weight), and N.D. –43.27 $\mu g\ L^{-1}$ in landfills (Gao & Wen 2016). About 15 phthalate esters were detected with total concentration ranging from 0.386 to 3.184 $\mu g\ L^{-1}$ in lake samples (Zheng *et al.* 2014). The total occurrence of the six PAEs concentration analyzed from the Mediterranean Sea (Marseille Bay) ranged from 130 to 1,330 ng L^{-1} (avg. 522 ng L^{-1}) (Paluselli *et al.* 2018). Seven PAEs and their monoesters were detected across four sampling areas in water from the Okavango delta, Northern Botswana (fresh water) with DEHP metabolites like MEHHP and MEOHP at high concentrations ranging from <1 to 29.3 ng L^{-1} (Bartsch *et al.* 2019). Due to the rise in human activities, four dominant phthalates were detected in the lakes of Beijing, namely DIBP, DBP, BBP, and DEHP.

River sediments are excessively analyzed as an indicator for the environment to judge phthalate contamination since it bioaccumulates and is widely distributed. PAEs, particularly DEHP were detected in river sediments present worldwide with concentrations from 0.1 ng g^{-1} to 100 $\mu g\ g^{-1}$ (Liu *et al.* 2013). Higher DEHP concentrations (1.56 $\mu g\ L^{-1}$) were found in tap water in wet months (Cui *et al.* 2022). For transporting phthalate esters with an approximate annual flux of 0.705–29.4 kg, storm water acts as an important contributor (Cao *et al.* 2022).

PAE concentrations are detected more in low river-flowing areas, specifically in reservoirs than the high river-flowing areas. Hence, the proposed and completed artificial human-constructed dams on the river will potentially aggravate organic compounds pollution by making changes in hydrology. Translocation of pollutants is inhibited particularly by low river flow, which results in exacerbating ecological risk in the river and enrichment in small areas. The results show the river flow rate effect on the distribution of PAEs and the effects of urbanization (Sun *et al.* 2013). Table 2 summarizes the concentration of different types of phthalate esters present in different regions.

5. EXPOSURE AND TOXICITY OF PHTHALATE ESTERS

Most of the PAEs are listed as emerging chemicals which disrupt the endocrine function and lead to various developmental and reproductive toxicities affecting reproductive health, physical development, immunotoxicity, neurotoxicity, teratogenicity, mutagenicity, and carcinogenicity for environmentally relevant exposures in various wildlife animals. DEHP is the majorly observed PAE utilized in PVC production as a plasticizer (Magdouli *et al.* 2013). Exposure tophthalate esters mainly occurs through inhalation, skin contacts causing bioaccumulation in human bodies. The interaction of this contaminant affects the various physiology of the human body causing teratogenicity, mutagenicity, and carcinogenicity even at minimum concentrations (Gao & Wen 2016). PAEs pose a great health hazard to living beings due to the fact that they are unavoidably exposed to phthalates from the consumption of surface water contamination and up-taking aquatic organisms (Sun *et al.* 2013). Figure 4 illustrates the disorders caused by phthalate esters in different organs of the human body like brain, lungs, heart, etc.

Potential pathways of exposure to phthalate esters in human beings are drinking water, contaminated foodstuffs, inhalation or dermal contact with the phthalate ester used in products like cosmetics and personal care products. Humans and

Table 2 | Occurrence of phthalate esters in different sources of water

Phthalate	Source	Concentration	Highlight	Country	References
DEP, DBP, BBP and DEHP	River sediments (Kaveri)	DEHP – 1,400 μg kg^{-1}, DEP – 85 μg kg^{-1}, BBP – 7.8 μg kg^{-1}; and DBP – 664 μg kg^{-1}	Total phthalates ranged from 313 to 4,640 ng L^{-1}. High concentration of phthalates detected in the upstream with excessive human-made activity (industrial).	Tamil Nadu, India	Selvaraj *et al.* (2014)
DEHP, DnBP, and DiBP	Fresh and surface water	DEHP – 18.2 μg L^{-1}; DnBP – 3.9 μg L^{-1}; and DiBP – 0.69 μg L^{-1}	Phthalate contamination observed in surface water could be due to the opening Bonnet Carré Spillway.	Lake Pontchartrain, Louisiana, US	Liu *et al.* (2013)
DEP, DBP and DEHP	Sea water (surface water)	DEHP – 0.0717 μg L^{-1}; DEP – 0.012 μg L^{-1}; and DBP – 0.017 μg L^{-1}	Total concentration of PAEs is 0.177 μg L^{-1} and the most abundant PAEs are DEP, DBP, and DEHP showing maximum concentration in the tourist season.	Tunisian Coast, Northern Africa	Jebara *et al.* (2021)
DMP, DEP, DBP, BBP, DEHP and DnOP	Urban runoff and sewage discharge (Seine and Ogre River)	DMP – 112 ng L^{-1}, DEP – 225 ng L^{-1}, DBP – 134 ng L^{-1}, BBP – 55 ng L^{-1}, DEHP – 665 ng L^{-1}	Due to urbanization. Phthalates detected in raw water in rivers, with an elevated concentration of DEHP, followed by DEP.	Paris, France	Teil *et al.* (2013)
16 PAEs	Sea water	Σ_{16} PAEs concentration – 453–5,108 ng L^{-1}	DiBP, DBP, and DEHP are the predominant PAEs present in the water and sediment samples. The vertical distribution indicated high concentration of ΣPAEs in surface and bottom water.	Bohai andYellow Sea, China	Zhang *et al.* (2018a, 2018b)
14 PAEs	Surface water	Σ_{14} PAEs concentration n.d. – 2.29 μg L^{-1}	DEHP was the predominant phthalate (mean 0.11 μg L^{-1}) and DMP (mean 0.04 μg L^{-1})	Asan lake, Korea	Lee *et al.* (2019a, 2019b)
6 PAEs	Surface water	Concentration – 0.467 to 17.953 μg L^{-1}	Average Σ_6 PAEs value was 4.042 ± 3.929 μg L^{-1}. Predominant PAE was DBP – 65.8% concentration.	Lake Chaohu, China	He *et al.* (2013)
DMP, DEP, DnBP, DiBP, DEHP	Surface water	DMP – 36.54 ng L^{-1}, DEP – 42.64 ng L^{-1}, DnBP – 246.8 ng L^{-1}, DiBP – 524.7 ng L^{-1}, DEHP – 208.1 ng L^{-1}	MMP, MEP, MiBP, MnBP, and MEHP which are the secondary metabolites were observed. Primary and secondary DEHP monoesters were observed in effluents and influents of two sewage treatment plants near lake.	Taihu Lake, China	Jiang *et al.* (2018)

other species are subjected to risk via inhaling, ingesting, or dermal exposure to the water which is contaminated. These pollutants are present at nanograms and micrograms per liter concentration in groundwater, surface water and also drinking water as a result of anthropogenic activities. The research on the concentration of phthalates presents in the beverages, surrounding air, foodstuffs we consume and other phthalate-containing things present in the lifestyle of humans should be monitored (Fawell & Ong 2012; Net *et al.* 2015; Tijani *et al.* 2015). PAEs are eliminated as conjugate monoesters in urine, and oxidative transformation via secondary metabolism takes place for DEHP before urinary excretion. In serum samples and milk, phthalate esters and their metabolite concentration were detected under the method detection limit which is 0.12–3.0 μg L^{-1}, whereas the concentration detected and analyzed in urine was found to be 0.1–1,000 μg L^{-1}. MEP,

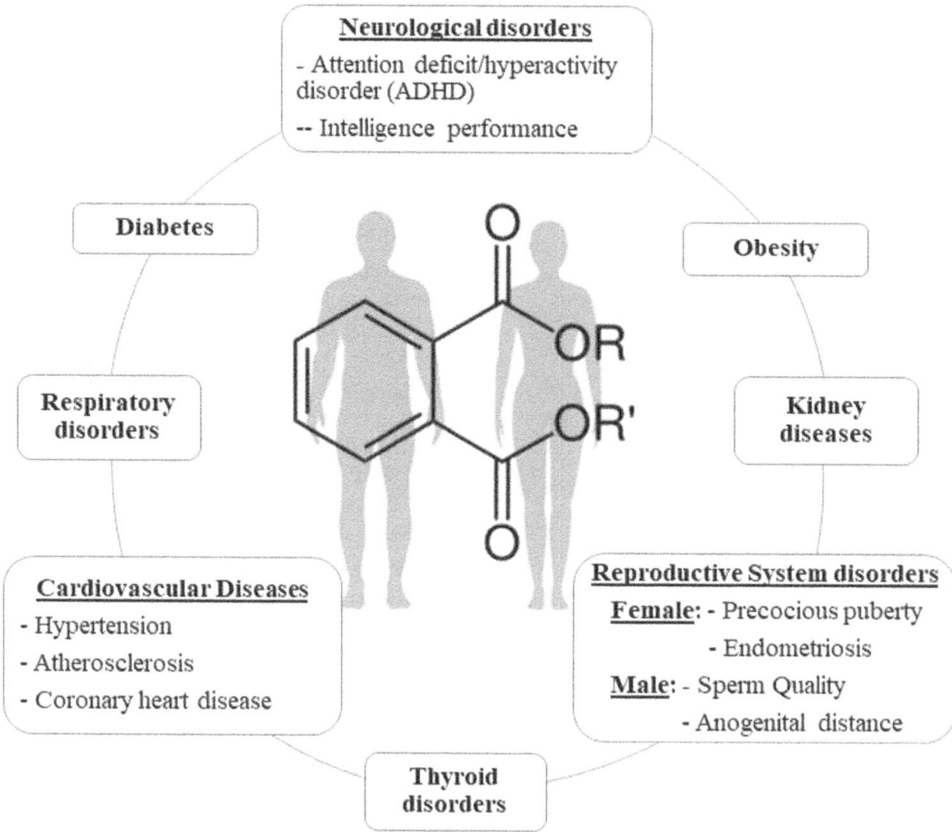

Figure 4 | Illustrates the disorders caused by phthalate esters in different organs of the human body.

MBP, and MEHP were observed in samples collected from nails, because DEP is widely used in nail polishes (Wang *et al.* 2019a, 2019b, 2019c).

Exposure to phthalates at high concentration is reported to cause pregnancy complications and miscarriage in females (Tijani *et al.* 2015). Predominant phthalate metabolite which is MEP was detected in pregnant women's urine samples from various countries like the Netherlands, USA, Canada, Norway, and Spain. In Norway, PAEs were found to be 30 $\mu g\ L^{-1}$ and 246 $\mu g\ g^{-1}$ was detected in Spanish pregnant women. It has also been reported in the saliva samples, amniotic fluids, and breast milk of lactating mothers (Buckley *et al.* 2012; Arbuckle *et al.* 2014; Sabaredzovic *et al.* 2015; Valvi *et al.* 2015).

Elevated levels of around 2.09 $\mu g\ L^{-1}$ of DEHP concentration 1.75 $\mu g\ L^{-1}$ of DBP concentration and its metabolites have been detected in the semen samples of men from Germany and USA which leads to the partition in the seminal plasma (Buck Louis *et al.* 2018; Smarr *et al.* 2018). In research conducted to observe the daily intake of PAEs through ingestion, it was observed that children (21 months old) are more exposed than adults because of household dust and dirt, their hand-to-mouth licking habit as they often lick their toys where DEHP and its metabolites were found to be present (Weiss *et al.* 2018).

In Hong Kong, the mean PAEs concentration detected in 20 fish species from fresh water varied from 1.66 to 3.14 $\mu g\ g^{-1}$ in concentration by wet-weight (ww) and ranging from 1.57 to 7.10 $\mu g\ g^{-1}$ ww in marine fish, where DBP and DEHP were the predominant phthalate esters. These PAEs are considered to be accumulated during gastrointestinal digestion (Cheng *et al.* 2013). DEHP is entering water bodies on a daily basis which is a threat to the fishes and aquatic organisms' health leading to health effects on the local surrounding people (Takdastan *et al.* 2021).

6. BIOREMEDIATION STRATEGIES FOR PHTHALATE ESTERS

There are various methods like the adsorption process (using various matrix, biochar, and AC) and biological treatments (using a microorganism, and membrane bioreactors) are used for the remediation of phthalates from the water and wastewater system. Various wastewater treatment technologies when coupled together resulted in greater efficiency for the phthalate esters remediation than using the single treatment method. When the membrane bioreactor was coupled with anaerobic wastewater remediation, the degradation of PAEs was enhanced from 65–71% to 95–97% (Gao & Wen 2016). Becky Miriyam *et al.* (2022) explained in detail about the adsorption studies and membrane technology used for PAEs removal. Chitosan beads immobilized with lipase was a cost-effective method used for DMP and DEP removal which showed 100 and 93.86% degradation, respectively (Dulazi & Liu 2011). Phytoremediation is a method where the plants and their parts are being used for the remediation of organic pollutants via degradation, uptake, and enzymatic actions. Crude enzyme extracts from *Oryza sativa L.* used for the degradation of DBP indicated that the roots of the rice plant metabolized DBP much more effectively than the stems and leaves of the plant (Zhu *et al.* 2019).

6.1. Biodegradation of phthalate esters

Aerobic, strictly anaerobic or facultatively anaerobic microorganisms are found to degrade phthalate esters from the environment. Some unfamiliar enzymes or transesterification have been reported in most of the studies of longer alkyl residues getting converted to shorter ones (Boll *et al.* 2019). DMP is more biodegradable than others since this substrate has a maximum specific growth rate and has the highest values of overall reaction rate (Ahmadi *et al.* 2017). *Glutamicibacter nicotianae* ZM05 is a DBP-degrading exogenous bacteria and *Cupriavidus metallidurans* ZM16 is a non-DBP-degrading bacterium that is effective together for co-contamination degradation. These insights assist to explicate the mutual effects of fungal and bacterial communities for co-contamination degradation of DBP-Cd and bring forth new perceptions for the fabrication of degrading consortia for the breakdown of the pollutants. It is a considerable concern for the impact that the bioaccumulation potential and toxicity of PAEs have on the environment (Gao *et al.* 2018). Through co-culture experiment, it was proven that *Pseudomonas aeruginosa* ZM03 can elevate the degradation rate of *Arthrobacter nicotianae* ZM05 for DBP removal under stressed pH conditions (Wang *et al.* 2021a, 2021b). Table 3 lists the microorganism used for the biodegradation of phthalate esters and its concentration with the removal percentage and the time duration.

Table 3 | List of phthalate-degrading microorganisms and the percentage removal

Phthalate	Organisms (bacteria, fungus, algae and enzyme)	Concentration (mg L^{-1})	Removal percentage	References
DEP	*Ralstonia pickettii*	300	100 (24 h)	Perpetuo *et al.* (2020)
DBP	*Deinococcus sp. R5*	1,000	100 (140 h)	Yang *et al.* (2014)
DBP	*Bacillus subtilis*	200	89 (120 h)	Huang *et al.* (2018)
DBP	*Rhodovulum sp. DBP07*	600	70 (96 h)	Baker *et al.* (2021)
DBP	*Pseudomonas sp. W1*-immobilization of Fe_3O_4 nanoparticle	1,000	99.88 (168 h)	Wang *et al.* (2020)
DEHP	*Pleurotus ostreatus*	100	100 (504 h)	Ahuactzin-Pérez *et al.* (2018)
DEHP	*Achromobacter denitrificans* strain SP1	10 mM	100 (96 h)	Pradeep *et al.* (2015)
DEHP	*Enterobacter spp.* Strain YC-IL1	100	86 (144 h)	Lamraoui *et al.* (2020)
DEHP	*Burkholderia pyrrocinia* B1213	500	98.05 (144 h)	Li *et al.* (2019)
DEHP	*Achromobacter sp. RX*	50–300	99.3 (96 h)	Wang *et al.* (2021a, 2021b)
DEHP	*Fusarium culmorum*	1,000	92 (36 h)	González-Márquez *et al.* (2019)
DEHP	*Rhodococcus ruber*	100–1,000	100 (72 h) 95 (144 h)	Yang *et al.* (2018)
DEHP	*Rhodococcus jostii PEVJ9* – Self assembled monolayers of silver nanoparticle	0.1	99.6 (72 h)	Annamalai and Vasudevan (2020)
DBP, DEHP	*C. oxalaticus E3*	200	100	Chen *et al.* (2021)
DMP, DEP, DPP, DBP	*Bacillus thuringiensis*	400	96, 88, 82, and 92, respectively (80 h)	Surhio *et al.* (2017)

Yang *et al.* (2013) conducted DBP degradation in synthetic and actual wastewater with enriched mixed culture in a series of batch bioreactors where 200–1,000 mg L^{-1} of initial DBP concentration was neutralized at 91–99% removal efficiency. High-performance liquid chromatography electrospray-ionization quadrupole time off light mass-spectrometry (HPLC-ESI-QTOF-MS) was used to determine some of the major PAEs metabolites when *Bacillus mojavensis* B1811 was used for the degradation (Zhang *et al.* 2018a, 2018b).

Hydrolysis of PAEs by esterase is the primary step which releases the free phthalate moiety and side chain alcohols in the degradation studies (Boll *et al.* 2019). Some by-products including mono-methyl phthalate, DMP, catechol and phthalic acid were detected in biodegradation of the specific phthalates exhibiting DAP and DEP which follows similar pathway for the degradation due to the detachment by side chain alkyl as per the MBBR effluent metabolite surveying results. During biodegradation of DEP, metabolites were observed, concluding that side chains of DEP get detached as the first step, and then mono-ethyl phthalate, mono-methyl phthalate and mono-ethyl phthalate were formed. De-methylation or de-esterification is the next step, which is followed by DMP, MMP, PA and catechol production, respectively (Ahmadi *et al.* 2015).

Degradation products/pathways of some toxic phthalate esters are explained in Figure 5. All the five phthalates mentioned in this review are DMP, DEP, DBP, BBP, and DEHP which undergo hydrolysis to form phthalate monoesters like MMP, MEP, MBP, MBzP, and MEHP, respectively. Later MBP oxidizes to form oxidation products of the monoesters 3-OH-MnBP and 4-OH-MnBP, followed by the hydrolysis of 3-OH-MnBP to MCCP. MBzP gets oxidized and forms the by-product MCPP. MEHP oxidizes and forms oxidation products of the monoesters such as MECPP, MCMHP, and MEHHP. Later MEHHP gets hydrolyzed to form MEOHP. After the oxidation of the monoesters and hydrolysis, phthalic acids are formed which undergo decarboxylation to form benzoic acid (BA). BA undergoes deoxygenation and forms protocatechuate which gets transformed into CO_2 and H_2O (Zhang *et al.* 2018a, 2018b). A combination of various methods provides efficient remediation of diesters and monoesters from the waterbodies.

7. ADVANCED OXIDATION PROCESSES

Various advanced oxidation processes like photocatalysis, electrochemical mineralization, sonolytic degradation, aqueous oxidation, Fenton oxidation, photochemical degradation, photocatalytic ozonation, and biochar facilitated degradation are applied to eliminate the phthalate esters based on the concept of adding or creating particles which are highly reactive and can oxidize molecules which are more stable (Julinová & Slavík 2012). These processes can be applied as a pre- or post-treatment method of the biological process of pollutant removal. Generation of •OH radicals play an important role in the process of pollutant degradation using AOP, which is carried out by photochemical, chemical, or sonochemical methods (Na *et al.* 2012a, 2012b).

The European Union Directive (12 August 2013) concludes Fenton-based processes to be the mostly used AOPs (31%), subsequently followed by heterogeneous photocatalysis (20%), whereas 18% of the research indicates a comparison of different advanced oxidation techniques for degradation studies in water for the contaminants present (Ribeiro *et al.* 2015). Becky Miriyam *et al.* (2022) explained electrochemical processes for phthalate degradation in detail. This review covers some important AOPs like photocatalysis, the Fenton process, ozonation, and sonolysis.

7.1. Photocatalysis process

Photocatalysis is a favorable AOP technique utilized for the treatment/removal of contaminants by mineralizing and oxidizing it under favorable conditions. Various photocatalysts and photocatalysis-based works have been investigated in water bodies for phthalate ester degradation. The basic principle of this method is that the pollutant is treated under UV radiation with or without the chemicals. DEP photodegradation was observed with Graphene quantum dots/Mn-N–TiO_2/graphene-C_3N_4 composite when photocatalytic action leads to the production of H_2 molecules (Nie *et al.* 2018). The calcination method was used to prepare g-C_3N_4/$Bi_2O_2CO_3$ and g-C_3N_4/BiOCl which considerably increased the photocatalytic activity for DnBP degradation (Shan *et al.* 2016). Research shows that DEP degradation by persulfate activated by $CuFe_2O_4$/MWCNTs magnetic nanoparticles was effective at various factors like the concentration of PS, the amount of catalyst loaded and temperature. The catalyst exhibited more catalytic performance, stability and reusability, whereas an eligible amount of metal ions was leached out from the catalysts (Zhang *et al.* 2016). PANi/CNT/TiO_2 immobilized on glass plates were represented as an effective photocatalyst for DEP degradation (Hung *et al.* 2017).

Rapid DMP breakdown and •OH radical formation was observed using hollow glass microspheres covered with photocatalytic TiO_2 (HGM–TiO_2) (Jiang *et al.* 2013). Bi_2O_3–TiO_2 composite was effective under a visible light source for mixed

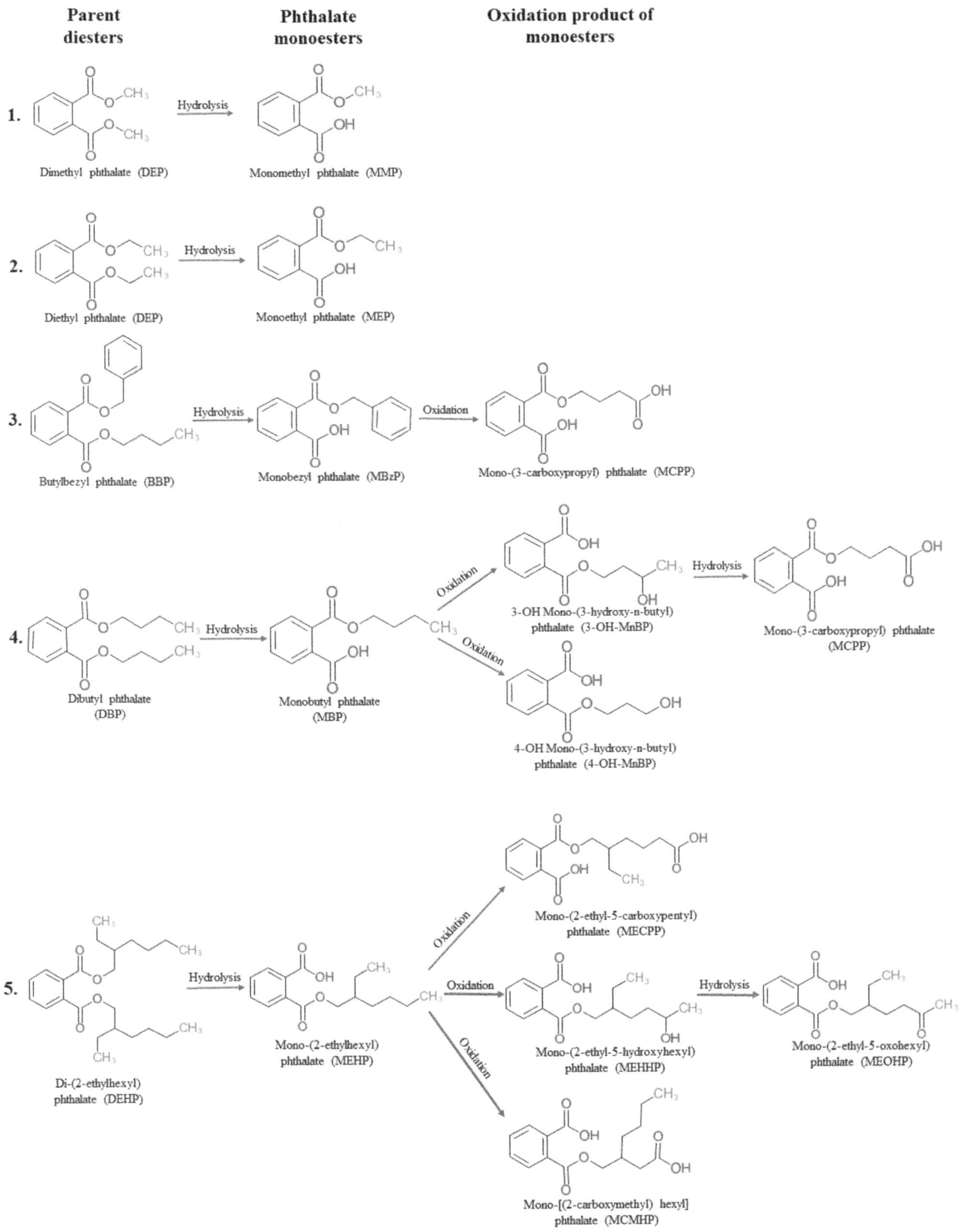

Figure 5 | Degradation products of phthalate esters – DMP, DEP, BBP, DBP, DEHP (Zhang *et al.* 2018a, 2018b).

pollutants of Pb (II) together with DBP degradation (You *et al.* 2018). In another study, Fe_3O_4@CuCr-LDH nanocomposites are used for the activation of PMS for the photocatalytic removal of DEP. This LDH-based S-scheme nano-heterojunction can be separated using magnets (Fazli *et al.* 2021). The degraded products of DMP were assessed using high-resolution (HR) Orbitrap mass-spectrometry (MS) assembled with ultra-high performance liquid chromatography (UHPLC) (Tan *et al.* 2020).

Jamil *et al.* (2017) performed the removal of total organic carbon (TOC) of DBP using Bi, Cu Co-doped $SrTiO_3$ with a visible metal halide lamp with a power of 82 µlm/W indicating increased removal. Vanadium pentoxide (V_2O_5)/molybdenum trioxide (MoO_3) composites include both doping and heterostructure which enhance the process and which is performed under a visible-light source at a wavelength greater than 420 nm using a high-pressure xenon lamp (500 W) for efficient removal of DMP (Chuai *et al.* 2015). Tungsten oxide (WO_3) deposited on the surface of TiO_2 showed a higher degradation rate than the bare TiO_2 photocatalyst under UV – LED and blue LED light sources (Ki *et al.* 2019). Nano-α-Fe_2O_3 presence with UV irradiation (300–400 nm) significantly increases the degradation rate of DEP (Shuai *et al.* 2018). The anatase phase, which is a crystalline size inhibitor, is produced when Ni is doped in TiO_2. The lower the crystalline size, the higher will be the disordering of the sample resulting in effective photocatalytic activity for the degradation of DEP (Singla *et al.* 2015).

Kaur *et al.* (2017) analyzed a DEP degradation study using transition metal-doped TiO_2 nanoparticles (Mn-, Ni-, and Co-doped TiO_2) under Hg lump (125 W) at 365 nm, where Ni-doped TiO_2 showed better results. Different ratios of reduced graphene oxide zinc oxide (rGO-ZnO) nanocomposites prepared using temperature refluxing methods showed efficient photodegradation of DEP (Kumar *et al.* 2021). Under UV radiation and magnetic separation, Fe_3O_4@TiO_2 core-shell nanoparticles which get operationalized using cyclodextrin plays an important role as a photocatalyst for DBP degradation and mineralization (Chalasani & Vasudevan 2013). Mono-methyl phthalate and phthalate are the two intermediates (aromatic) observed in the electrolysis of DMP which indicates the attack on the methyl esters group at the initial step in the oxidation process (De Souza *et al.* 2013). Organic layered double hydroxides (LHDs)/TiO_2 composites were effective in the adsorption and photodegradation of DMP and approximately 80% DMP was removed in 4 h (Huang *et al.* 2013). Table 4 shows different catalysts used for the photocatalytic process along with the light source and percentage of removal.

Table 4 | Photocatalysis method for phthalate degradation

Phthalate	Catalysts/methods	Light sources	Removal efficiency (%)	References
DMP	Multiwalled carbon nanotubes – TiO_2 composites	UV lamp (96 W)	97	Tan *et al.* (2018)
	US created N-doped TiO_2 and non-US created N-doped TiO_2	Visible light	58 (300 min)	Zhou *et al.* (2013)
	Liquid phase plasma (LPP) method with a TiO_2 photocatalyst and H_2O_2	Tungsten electrode	82.2 (180 min)	Lee *et al.* (2019a, 2019b)
	TiO_2/carbon aerogel	Xe lamp (300 W): infrared light filter	>83 (180 min)	Cui *et al.* (2016)
DEP	Nanorod ZnO/SiC nanocomposite	UV lamps (8 W); Visible-light (500W)	>90	Meenakshi and Sivasamy (2018)
	Pt/In_2O_3–TiO_2 nanotubes	Xenon lamp (350 W):	99.8(45 min)	Ma *et al.* (2012)
	Nano Fe_2O_3 embedded in montmorillonite with citric acid	Xenon light (50 W) and light irradiation	71.7	Sun *et al.* (2021)
DBP	Mesoporous TiO_2 nanotubes (m-TiO_2-NTs)	High pressure mercury lamp (125 W)	70 (60 min)	He *et al.* (2019)
	ZnO nanorods Co-doped with Fe and Ag	Visible LED lamp (7 W)	95 (60 min)	Eslami *et al.* (2017)
	Graphene – TiO_2 nanotube array photoelectrodes	Xenon lamp (150 W)	87.9 (90 min)	Wang *et al.* (2019a, 2019b, 2019c)
	α-Fe_2O_3 nanoparticles	Mercury lamp (250 W)	94	Liu *et al.* (2018)
BBP	P-doped TiO_2 (PTIO) thin-films	Xe lamp (300 W)	98 (240 min)	Mohamed and Aazam (2013)
	Chlorine-doped TiO_2	Xe lamp (300 W)	92	Wang *et al.* (2012)
DEHP	Z-scheme heterojunction catalyst of Bi_2O_3 and TiO_2	Xe lamp	89 (90 min)	Zhang *et al.* (2022a, 2022b)
	Fe-Ag@ZnO nanorods	Visible LED lamp	90 (120 min)	Eslami *et al.* (2017)

7.2. Fenton oxidation/Fenton-like processes

There are various modifications to the Fenton process like the basic Fenton Process (H_2O_2+ Fe^{2+}), Fenton-like process (H_2O_2+ Fe^{3+}), sono-Fenton process (ultrasound + H_2O_2 + Fe^{2+}), sono-photo-Fenton process, photo-Fenton process (UV + H_2O_2 + Fe^{2+}), and photo-electro-Fenton process. Fenton reagent is the mixture of H_2O_2 and Fe^{2+}, where the H_2O_2 molecule is broken down with the iron catalyst and produces hydroxyl radicals as shown in Equation (1). Whereas Equation (2) depicts the Fenton-like reaction, where ferrous ion and hydroperoxyl radicals (OOH•) are generated which are less effective than the HO• radicals in the degradation process.

$$Fe^{2+} + H_2O_2 \rightarrow Fe^{3+} + HO^- + HO\bullet \tag{1}$$

$$Fe^{3+} + H_2O_2 \rightarrow Fe^{2+} + OOH\bullet + H^+ \tag{2}$$

The UV-Fenton and Fenton processes reduced the level of DEHP and DBP, and also are considered as appropriate methods for treating concentrated solutions (Wang *et al.* 2016a, 2016b). According to Şolpan & Mehrnia (2018), phthalate degradation is affected by H_2O_2 concentration, and irradiation dose. When there is an increase in the amount of hydrogen peroxide, the degradation process is enhanced. DEP removal was enhanced when the Fe(III)/PMS process was combined with C-60 fullerenol (Zhou *et al.* 2020a, 2020b). According to Qi *et al.* 2020 combination of sono-Fenton with a photocatalytic process under visible light (Vis/P25) enhanced the mineralization process for DMP and DEP degradation, where better results were obtained for DMP degradation. Table 5 indicates the Fenton oxidation/Fenton-like processes with modifications for phthalate degradation with the removal efficiency.

The use of a gas diffusion electrode (GDE) fabricated from carbon material (CMK-3) was a much more effective electro-Fenton process for DMP degradation (Wang *et al.* 2013a, 2013b). A total of 93% of DiBP removal was observed from water using galvanic anode in electro-Fenton method due to the hydroxyl radical formation under various operating conditions, i.e., pH 5, space between the plate (5 cm), electrolyte (Na_2SO_4 with H_2O_2) (Yang *et al.* 2020a, 2020b). Homogeneous catalysts like mixtures of some transition metal ions that replaced the Fe^{2+} in the Fenton reagent achieved phthalate reduction (Ziembowicz *et al.* 2021). The PMS/MFe_2O_4 system and graphene-based $CoFe_2O_4$/PMS as sulfate and hydroxyl radical plays a crucial role in the phthalate degradation mechanism (Ren *et al.* 2015; Xu *et al.* 2015a, 2015b).

Table 5 | Fenton oxidation/Fenton-like processes for phthalate degradation

Phthalate	Catalyst/Method	Removal efficiency (%)	Type of Fenton process	References
DMP	MOF(2Fe/Co)/CA cathode	85 (120 min)	Solar photo-electro-Fenton process	Zhao *et al.* (2017)
	Quinone-like substances	100 (240 min)	Fenton-like process	Xiao *et al.* (2020)
	Graphite felt activated by KOH	100 (45 min)	Electro-Fenton cathode	Wang *et al.* (2015)
	Fe-Cu embedded Carbon aerogel	92 (240 min)	Electro-Fenton process	Zhao et al. (2018a, 2018b)
	Fe@NdFeB/AC magnetic catalyst	85.2 (120 min)	Heterogeneous Fenton-like process	Yang *et al.* (2019)
	Fe (II)/PMS-fulvic acid	85.70	Fenton-like process	Ding *et al.* (2022)
DEP	Zero valent copper (ZVC)	100 (120 min)	Fenton-like process	Wen *et al.* (2014)
	Biphase Co@C core-shell catalyst ($CaCO_3$and Co_3O_4)	100	Catalytic oxidation – Fenton-like process	Ma *et al.* (2022a, 2022b, 2022c)
	Vanadium (V) oxides/H_2O_2/oxalic acid	92 (240 min)	VO_2-Fenton-like process	Huang *et al.* (2021)
	Ferrous sulfate, Ferric oxide, Pyrite and FeF_2	100 (300 min)	Homo and heterogeneous Fenton oxidation	Bensalah *et al.* (2019)
	Pyrite (FeS_2)/CaO_2 system	78	Catalytic oxidation	Zhou *et al.* (2020a, 2020b)
DBP	U/Fe^{2+}, UV/Fe^{2+} and US/UV/Fe^{2+}	100 (75 min)	Sono-photo-Fenton treatment process	Xu *et al.* (2014)
	FeMnCu5%i/H_2O_2	95	Heterogeneous Fenton-like processes	Ziembowicz *et al.* (2021)
DEHP	Fe^{3+}/PCA/H_2O_2	92 (240 min)	Surfactant-enhanced Fenton-like system	Zhao et al. (2018a, 2018b)

Chen *et al.* (2016) found that in the presence of clay minerals, the Fe content present in the clay structure leads to an effective Fenton reaction for the degradation of DEP with the formation of six by-products majorly mono-ethyl phthalate, followed by phthalate acid, hydroxyl DEP, and many more compounds. Ferric reduction reaction–electro-Fenton increased the H_2O_2 production in the gas diffusion device which leads to efficient DMP degradation (Liang *et al.* 2021a, 2021b). 99.1% of DMP degradation is observed using the electro-Fenton process with the generation of hydroxyl radicals in 8 min (Dolatabadi *et al.* 2021).

7.3. Ozone/ultraviolet-based AOPs

Ozone (O_3) is often used in water treatment because of its high oxidation activity with the contaminants (Gucheng *et al.* 2017). In the ozonation process, two different pathways are followed: (i) direct reaction with O_3 and (ii) indirect reactions with hydroxyl radicals. According to Mansouri *et al.* (2019), the AOPs which are effective for the oxidation and mineralization of a broad range of contaminants from wastewater are ozonation, ultraviolet radiation and H_2O_2 based. AOPs like O_3/H_2O_2 and O_3/AC are effective for DEP degradation when compared to the traditional methods (O_3 and UV alone) (Medellin-Castillo *et al.* 2013). On performing the three different ozonation processes (ozone alone, O_3/H_2O_2 and O_3/ZnO) for DEP mineralization, O_3/H_2O_2 showed better results (Wen *et al.* 2011). O_3 and UV/O_3 effectively reduced DEHP as compared to the conventional UV method, where pH plays an important role in the degradation process (Yang & Lin 2012). Whereas in a study for DMP degradation, UV/O_3 process performs better than the free O_3 process (Yang *et al.* 2020a, 2020b). AOPs such as UV/H_2O_2 were effective in alkaline conditions with natural organic matter (NOM) for DBP degradation (Wang *et al.* 2016a, 2016b). Table 6 depicts the list of ozone/UV-based AOPs for phthalate degradation showing the radicals generated and the removal efficiency of the process.

Catalytic ozonation for DBP degradation was analyzed using magnetic, stable graphene-$MnFe_2O_4$ hybrids. The synergistic function between the support matrix and the nanomaterial results in efficient degradation. Also, $NiFe_2O_4$ and $Ag_{0.1}Ni_{0.95}Fe_2O_4$ formulated using sol–gel process enhanced ozone decomposition during the DBP degradation process, and during the ozonation process it concludes that the removal is done by degradation and not adsorption. Catalytic ozonation of DMP by AC was performed on the basis of hydrodynamics, adsorption, mass transfer and reaction kinetics to enhance the catalytic ozonation reactors where gas-liquid continuous flow is present (Zheng *et al.* 2020; Yu *et al.* 2021; Ma *et al.* 2022a, 2022b, 2022c).

Ozone-based AOPs study like heterogeneous catalytic ozonation with Al_2O_3 and AC showed effective results for DEP removal, whereas in the presence of O_3/N-methyl hydroxylamine (N-HA) coupled with catalytic ozonation efficiently removed DMP with increasing the N-HA concentration (Gucheng *et al.* 2017; Mansouri *et al.* 2019). Heterogeneous photocatalysis coupled with ozonation is effective for the removal of DEHP from wastewater using Nx–TiO_2x nanoparticles. Ozonation if combined with other processes like photocatalytic reaction and sonolysis enhances the degradation rate of phthalates (Anandan *et al.* 2013). Novel Fe/Mn@$\gamma - Al_2O_3/O_3$ nano catalyst showed great efficiency in real water samples for DMP degradation (Liang *et al.* 2021a, 2021b).

UV/H_2O_2 method can be enhanced tenfold in the presence of ferric ions for DMP degradation, and the first degradation intermediate is dimethyl 4-hydroxy phthalate (DMP-4OH) (Du *et al.* 2015). For the degradation of DEP, UV/H_2O_2 is much

Table 6 | Ultraviolet/ozone-based AOPs for phthalate degradation

Phthalate	Catalyst/process	Removal efficiency (%)	Radicals generated	References
DMP	$(Cu_2O)_{0.5}\cdot CuO\cdot Fe_2O_3$ nanoparticles (CFO NPs)	100 (20 min)	•OH, •CH_3 and $O_2\bullet^-$ radicals	Liu *et al.* (2019)
	Co-Mn-Mesoporous siliceous (MCM-41) catalyst	99.7 (15 min)	•OH radicals	Tang *et al.* (2017)
	Cerium-loaded SBA-15 (Ce/SBA-15)	88.7 (60 min)	•OH radicals	Yan *et al.* (2013)
DEP	Electro-peroxone with carbon – polytetrafluorethylene	99 (60 min)	•OH radicals	Hou *et al.* (2016)
DBP	Magnetic porous ferro spinel $NiFe_2O_4$	100 (60 min)	HO_2^- and O_2 radicals	Ren *et al.* (2012)
	O_3/UV process	100 (60 min)	H_2O and CO_2	Wang *et al.* (2013a, 2013b)
BBP	O_3/UV process with scavenger	91 (15 min)	Tert butanol as radical scavenger	Lovato *et al.* (2014)
DEHP	V_2O_5/TiO_2 – Ozone	58.7 (1 min)	•OH radicals	Tak *et al.* (2022)

more effective when compared with the traditional method (O_3, UV). UV/H_2O_2 and UV/TiO_2 were compared for DEP photodegradation, where UV/H_2O_2 proved to be highly effective for DEP removal (Medellin-Castillo *et al.* 2013; Song *et al.* 2015).

7.4. Sonolysis

If ultrasound (US) is applied to an organic pollutant solution, a pyrolysis mechanism is generated due to cavitation inside the cavitation bubbles which leads to the production of •OH radicals (Na *et al.* 2012a, 2012b). High-frequency ultrasonic methods in combination with H_2O_2 can enhance the radical formation to improve the DMP degradation (Xu *et al.* 2013a, 2013b). Cavitation bubbles play an important role because in the ultrasonic method, the removal ability is similar to the amount of cavitation bubbles in the water due to the introduction of mechanical agitation (Nie *et al.* 2021). The combination of US with titanium dioxide (TiO_2) catalyst and UV irradiation known as sono-photocatalysis and sono-photolysis, respectively, of DEP, exhibited synergistic effects which were advantageous (Na *et al.* 2012a, 2012b). Phthalate degradation takes place via radical oxidation in the sonolysis process (Xu *et al.* 2015a, 2015b).

According to Souza *et al.* (2014), when UV-light irradiation and US are coupled to obtain improvement in conductive-diamond electrochemical oxidation (CDEO) for the treatment of DMP in synthetic wastewater. The collaboration of high-frequency US and catalyst-free UV radiation which is known as sono-photolysis is observed to perform effectively for DMP degradation. Also, effective sonochemical degradation of DMP was performed with high-frequency ultrasonic processes (HFSU) under various operating parameters. HSFU undergoes pseudo-first order kinetics after reacting with the *OH radicals, whereas the degradation rate is enhanced with the increase in the power density of US (Xu *et al.* 2013a, 2013b).

Na *et al.* (2012a, 2012b) analyzed the degradation efficiency with sonolytic, photolytic and sonophotolytic processes using US (283 kHz) and two different types of UV lamps (UVC-254 nm) and (VUV-185 nm + 254 nm), where H_2O_2 generated via UV irradiation and application of US to generate •OH radicals. Hence, enhanced DEP degradation indicates the major role of the US in sonophotolytic processes. Similarly, US treatment was given to manganese oxide catalyst for the activation of PMS efficiently, which resulted in the complete degradation of DMP (10 mg/L) in 90 min (Yi *et al.* 2022). In a study, sono-Fenton method combined with H_2O_2 and photocatalytic process under visible light was effective in degrading DMP and DEP (Qi *et al.* 2020). Use of TiO_2 at high-frequency enhanced the sonocatalytic degradation of DEP, whereas sonolysis was not a very effective method (Cho *et al.* 2012).

7.5. Sulfate radicals-based studies

Several compounds are used to activate the peroxymonosulfate (PMS) which gives much more effective results as compared to bare PMS. Metal-organic frameworks like (Fe (II)-MOFs) altered with a particular molecular imprinting layer (Fe(II)-MOFs@MIP) can be used for DBP degradation as adsorption-catalytic degradation (Chi *et al.* 2021). The hydroxyl radicals increase the degradation step in advanced oxidation systems. Electron beam radiolysis can produce both •OH radicals and electrons which can be advantageous than the method where only •OH radicals are generated (Liu *et al.* 2012). Hydroxyl and sulfate radicals are initially responsible for DEP removal leading to the production of intermediary products (Zhou *et al.* 2020a, 2020b). Metal-free graphitic carbon nitride (g-C_3N_4) synergistically works with PMS for DMP degradation and mineralization which took place mainly due to the SO_4^- and •OH radical attack to the benzene ring and oxidation of the aliphatic chains (Xu *et al.* 2020).

Different combinations of AOPs like UV-activated PMS oxidation have shown effective results. In a study by Huang *et al.* (2017), activated PMS using UV was used for generating SO_4^- and •OH radicals which resulted in 90% DEHP degradation in 70 mins. Similarly, Lei *et al.* (2020) reported 98% removal of DEP in 10 min generating SO_4^-, •OH, Cl_2^-, CO_3 and Cl radicals. The UV/PS process produced hydroxyl radicals (OH*) and sulfate radicals (SO_4^-*) for effective DBP degradation which can be enhanced by increasing PS concentration. UV/PS is considered as an effective treatment method for phthalate contaminated water (Wang *et al.* 2018, 2019a, 2019b, 2019c). DEP degradation under alkaline conditions was effective for a UV/sulfite system with oxygen, since •OH accumulation increases the pH (Chu *et al.* 2021).

The catalytic oxidation of DMP is seen to be enhanced by α-Fe_2O_3–TiO_2 nanocomposite coupled with persulfate using non-thermal dielectric barrier discharge (NTP-DBD) plasma reactor (Ahmadi *et al.* 2020). PMS gets decomposed by a metal-free activator like amorphous boron to produce reactive oxygen species (ROS) for effective degradation of DEP (Ren *et al.* 2020). Oxidation of PMS was performed using a heterogeneous catalyst like Magnetic ferro spinel MFe_2O_4 (Metal = Co, Cu, Mn, and Zn) for DBP degradation, where MFe_2O_4 showed high PMS catalytic effect even after the seventh used (Ren *et al.*

2015). In a study, 99% DMP (20 mg L^{-1}) was removed in 90 min with PMS activated using manganese oxide nanomaterial (H_2^-OMS-2) with abundant Bronsted-acid sites by means of ion-exchange of crypto melane-type MnO_2 (OMS-2) (Zhang *et al.* 2022a, 2022b).

Greenhouse gas, CO_2 was used to prepare a core-shell (Co@C) catalyst using molten salt electrolysis which activates the PMS to degrade DEP (Ma *et al.* 2022a, 2022b, 2022c). In a study, reduced Fe-bearing smectite clays are used for the activation of PMS which contributes to rapid DEP degradation in 30 s, where the SO_4^- and •OH radicals bind to the surface of the clay rather than mixing in the bulk solution (Chen *et al.* 2020). When 0.1 gL^{-1} of $LiCoPO_4$ (LCP) was used as an activator for PMS (0.5 g L^{-1}) producing SO_4^- and •OH radicals, 98.6% OF DEP was removed in 6 min. Likewise, SO_4^- and •OH radicals are generated when nanoscale zero-valent tungsten (nZVW) is utilized for activating PMS for DMP degradation (Lin *et al.* 2017; Zhou *et al.* 2019). A $(Co,Mn)_3O_4$/PMS system, which is prepared using two metals, i.e., cobalt and manganese was considered more efficient for DEP removal than the single metal systems like Co_3O_4/PMS and Mn_5O_8/PMS. The synergistic effects between the Co and Mn oxides showed good utilization results for pollutant degradation by enhancing the PMS activity (Klu *et al.* 2022).

According to Lin *et al.* (2020) TOC estimation with the degradation of DEP showed that it can be mineralized into CO_2 and H_2O by the use of PMS activated using $Co_{0.59}Fe_{0.41}P$ nanocubes prepared from nanoscale metal-organic frameworks, which removed 95% of DEP in 60 min. Similarly, MOF core-SiO_2 shell nanomaterial (Fe-MOF-74@SiO_2) prepared using a hydrothermal process, can be used for persulfate (PS) activation effective for DMP degradation. Also, (Fe-MOF-74) can be improved by a molecular imprinting technique which provides much more efficient results (Ding *et al.* 2021a, 2021b). In a study, cage-core PMS catalysts were synthesized using electrospinning–calcination of the cobalt–zeolitic imidazole framework (ZIF-67) crystals to remove 90.4% DMP in 60 mins (Pang *et al.* 2022). Using a ball milling process, a novel sulfur-doped activated carbon (SACx) particle was prepared to activate PMS which was effective in the degradation of DEP (Huang *et al.* 2021a, 2021b). Sulfate-radical-based AOPs are much more efficient than other AOPs methods as they can be combined with other AOPs present.

7.6. Using biochar as a catalyst

Biochar as a sorbent is applied for pollutant removal, whereas some research works have been performed with biochar merged with some AOPs which is an emerging topic of interest. According to Nidheesh *et al.* (2021), catalyst which is supported by biochar performs better than the unsupported catalysts for contaminant removal.

DEP degradation with partial mineralization was accounted for using UV radiation and solar lights suspended in biochar solution, due to the formation of quinone-like formation of bio carbon minerals (BCM-Q) and BCM-enclosed with persistent free radicals (BCM-PFRs) which is the principal factor affecting the production of ROS like •OH and $O_2^{-\bullet}$. For the degradation of DEP, generation of ROS and hydroxyl radicals takes place from the biochar in the presence of oxygen. Catalase, oxide dismutase, and deferoxamine are utilized for the free-radical quenching studies to justify the mechanism. For phthalate degradation, •OH production plays an important role, where biochar carbon matrix-bound persistent free radicals' activation and photo-Fenton steps are the route cause (Fang *et al.* 2015, 2017). Nitrogen doped over biochar improves H_2O_2 production due to O_2 reduction reaction leading to 86.8% of DMP degradation (Nidheesh *et al.* 2021).

Biochar-based study was performed for the first time, where carbon nanofiber (CCNF) prepared using cellulose (biochar) combined with cobalt ferrite ($CoFe_2O_4$) that forms a nanocomposite which is used to activate PMS to enhance the catalytic activity. It increases the transfer rate of the electrons present and blocking the agglomeration of the $CoFe_2O_4$ nanoparticles, which leads to the formation of radicals, i.e., SO_4^- and •OH where $SO_4^{-\bullet}$ acts as the dominant active species (Gan *et al.* 2019). Complete DMP degradation within 60 min was observed with the PMS activated using biochar@CoFe-LDH composite (BC-LDH). The synergistic effect possessed by the biochar and CoFe-LDH results in higher catalytic activity and increased stability for the PMS activation (Ye *et al.* 2021). Xiong & Pei (2021) mentioned that persulfate activated using transition metal shows good results for PAEs degradation. Synergism of biochar and transition metal-PS can get over the disadvantages of a single approach and perform efficiently in the degradation process.

Biochar combined with PS, PDS, and PMS is a novel approach for PAE degradation. Red seaweed-derived biochar is used to activate sodium percarbonate (SPC) for DEHP degradation by undergoing carbo-catalysis by generating OH radicals (Hung *et al.* 2021). Magnetite nanoparticles (Fe_3O_4) – rice husk biochar (RHB) composite activated the sodium persulfate ($Na_2S_2O_8$, PS) to form $SO_4^{-\bullet}$ for PAE degradation in marine environments (Dong *et al.* 2018). Biochar-based advanced oxidation process is an emerging concept for the contaminant's degradation. Magnetic biochar doped with Fe exhibited photocatalytic effects in the presence of solar radiation for effective remediation of DEP because of the formation of PFRs, •OH radicals, and $\bullet O_2^-$ radicals (Yi *et al.* 2021).

8. CONCLUSION

CECs/emerging pollutants are a major concern for the society in the present scenario because of the increase in anthropogenic activities. AOPs can generally achieve higher removal efficiencies of CECs from contaminated sites than the conventional treatment methods. Data gap exists for the occurrence of phthalate esters in India, which needs to be explored. Toxicology experiments related to phthalates need to be explored to understand the level of toxicity present in various subjects. Some future trends and prospects are mentioned in the following:

- Various by-products are formed as the oxidation of pollutants takes place, which could be potentially toxic. The extent of phthalate degradation and the resultant residue toxicity need to be evaluated.
- In the present scenario, AOPs can be coupled with the biological process which makes use of reactors and waste materials like biochar and hydrochar that may produce effective results. This step can be beneficial as the AOPs can remove the recalcitrant compounds that can impair the oxidation in biological processes. Whereas the biological processes can be advantageous as it overcomes the pre-oxidized water condition which is not possible with AOPs.
- Much less research related to hydrochar-based AOPs are explored for phthalate degradation. Less literature is available for biochar-based AOPs.
- The detection level of occurrence for phthalate esters was found to be in the decreasing order of magnitude in wastewater, surface water, groundwater and drinking water. It is important to identify efficient remediation techniques for appropriate concentrations.
- Less information is available on the industrial scale studies for phthalate removal, so further studies involving scaling up can be explored.

DATA AVAILABILITY STATEMENT

All relevant data are included in the paper or its Supplementary Information.

CONFLICT OF INTEREST

The authors declare there is no conflict.

REFERENCES

Ahmadi, E., Gholami, M., Farzadkia, M., Nabizadeh, R. & Azari, A. 2015 Study of moving bed biofilm reactor in diethyl phthalate and diallyl phthalate removal from synthetic wastewater. *Bioresource Technology* **183**, 129–135.

Ahmadi, E., Yousefzadeh, S., Ansari, M., Ghaffari, H., Azari, A., Miri, M., Mesdaghinia, A., Nabizadeh, R., Kakavandi, B., Ahmadi, P., Badi, M., Gholami, M., Sharafi, K., Karimaei, M., Ghoochani, M., Brahmand, M., Mohseni, S., Sarkhosh, M., Rezaei, S., Asgharnia, H., Dehghanifard, E., Jafari, B., Mortezapour, A., Moghaddam, V., Mahmoudi, M. & Taghipour, N. 2017 Performance, kinetic, and biodegradation pathway evaluation of anaerobic fixed film fixed bed reactor in removing phthalic acid esters from wastewater. *Scientific Reports* **7** (1), 41020.

Ahmadi, E., Shokri, B., Mesdaghinia, A., Nabizadeh, R., Reza Khani, M., Yousefzadeh, S., Salehi, M. & Yaghmaeian, K. 2020 Synergistic effects of α-Fe_2O_3-TiO_2 and $Na_2S_2O_8$ on the performance of a non-thermal plasma reactor as a novel catalytic oxidation process for dimethyl phthalate degradation. *Separation and Purification Technology* **250**, 117185.

Ahuactzin-Pérez, M., Tlecuitl-Beristain, S., García-Dávila, J., Santacruz-Juárez, E., González-Pérez, M., Gutiérrez-Ruíz, M. & Sánchez, C. 2018 A novel biodegradation pathway of the endocrine-disruptor di(2-ethyl hexyl) phthalate by Pleurotus ostreatus based on quantum chemical investigation. *Ecotoxicology and Environmental Safety* **147**, 494–499.

Anandan, S., Pugazhenthiran, N., Lana-Villarreal, T., Lee, G. & Wu, J. 2013 Catalytic degradation of a plasticizer, di-ethylhexyl phthalate, using Nx–TiO_2 – x nanoparticles synthesized via co-precipitation. *Chemical Engineering Journal* **231**, 182–189.

Annamalai, J. & Vasudevan, N. 2020 Enhanced biodegradation of an endocrine disrupting micro-pollutant: di (2-ethylhexyl) phthalate using biogenic self-assembled monolayer of silver nanoparticles. *Science of The Total Environment* **719**, 137115.

Arbuckle, T., Davis, K., Marro, L., Fisher, M., Legrand, M., LeBlanc, A., Gaudreau, E., Foster, W., Choeurng, V. & Fraser, W. 2014 Phthalate and bisphenol A exposure among pregnant women in Canada – results from the MIREC study. *Environment International* **68**, 55–65.

Baker, A., Ahmad, B., Alarjani, K., Aldosri, N. & Khan, M. 2021 Biostimulation of Rhodovulum sp., for enhanced degradation of di-n-butyl phthalate under optimum conditions. *Chemosphere* **266**, 128998.

Bartsch, P., Edwards, T. & Brock, J. 2019 Prevalence of eight phthalate monoesters in water from the Okavango Delta, Northern Botswana. *Bulletin of Environmental Contamination and Toxicology* **103** (2), 274–279.

Becky Miriyam, I., Anbalagan, K. & Magesh Kumar, M. 2022 Phthalates removal from wastewater by different methods – a review. *Water Science and Technology* **85** (9), 2581–2600.

Bensalah, N., Dbira, S. & Bedoui, A. 2019 Mechanistic and kinetic studies of the degradation of diethyl phthalate (DEP) by homogeneous and heterogeneous Fenton oxidation. *Environmental Nanotechnology, Monitoring & Management* **11**, 100224.

Boll, M., Geiger, R., Junghare, M. & Schink, B. 2019 Microbial degradation of phthalates: biochemistry and environmental implications. *Environmental Microbiology Reports* **12** (1), 3–15.

Buck Louis, G., Smarr, M., Sun, L., Chen, Z., Honda, M., Wang, W., Karthikraj, R., Weck, J. & Kannan, K. 2018 Endocrine disrupting chemicals in seminal plasma and couple fecundity. *Environmental Research* **163**, 64–70.

Buckley, J., Palmieri, R., Matuszewski, J., Herring, A., Baird, D., Hartmann, K. & Hoppin, J. 2012 Consumer product exposures associated with urinary phthalate levels in pregnant women. *Journal of Exposure Science & Environmental Epidemiology* **22** (5), 468–475.

Cao, Y., Xu, S., Zhang, K., Lin, H., Wu, R., Lao, J., Tao, D., Liu, M., Leung, K. & Lam, P. 2022 Spatiotemporal occurrence of phthalate esters in stormwater drains of Hong Kong, China: mass loading and source identification. *Environmental Pollution* **308**, 119683.

Chalasani, R. & Vasudevan, S. 2013 Cyclodextrin-Functionalized Fe_3O_4@TiO_2: reusable, magnetic nanoparticles for photocatalytic degradation of endocrine-disrupting chemicals in water supplies. *ACS Nano* **7** (5), 4093–4104.

Chen, N., Fang, G., Zhou, D. & Gao, J. 2016 Effects of clay minerals on diethyl phthalate degradation in fenton reactions. *Chemosphere* **165**, 52–58.

Chen, N., Fang, G., Zhu, C., Wu, S., Liu, G., Dionysiou, D. D., Wang, X., Gao, J. & Zhou, D. 2020 Surface-bound radical control rapid organic contaminant degradation through peroxymonosulfate activation by reduced Fe-bearing smectite clays. *Journal of Hazardous Materials* **389**, 121819. https://doi.org/10.1016/j.jhazmat.2019.121819.

Chen, F., Li, X., Dong, Y., Li, J., Li, Y., Li, H., Chen, L., Zhou, M. & Hou, H. 2021 Biodegradation of phthalic acid esters (PAEs) by Cupriavidus oxalaticus strain E3 isolated from sediment and characterization of monoester hydrolases. *Chemosphere* **266**, 129061.

Cheng, Z., Nie, X., Wang, H. & Wong, M. 2013 Risk assessments of human exposure to bioaccessible phthalate esters through market fish consumption. *Environment International* **57–58**, 75–80.

Chi, H., Li, C., Huang, M., Wan, J., Zhou, X. & Yan, B. 2021 Targeted accumulation and spatial confinement effect of Fe(II)-MOFs@MIP for efficiently removing low concentration dibutyl phthalate. *Chemical Engineering Journal* **424**, 130367.

Cho, E., Park, B., Na, S. & Khim, J. 2012 Effects of power density and tio2 dose in the sonocatalytic degradation of diethyl phthalate using high frequency. *Japanese Journal of Applied Physics* **51** (7S). https://doi.org/10.1143/jjap.51.07gd09.

Chu, Y., Xu, L., Gan, L., Qiao, W., Han, J., Mei, X., Guo, H., Li, W., Pei, C., Gong, H. & Guo, X. 2021 Efficient destruction of emerging contaminants in water by UV/S(IV) process with natural reoxygenation: effect of pH on reactive species. *Water Research* **198**, 117143.

Chuai, H., Zhou, D., Zhu, X., Li, Z. & Huang, W. 2015 Characterization of V_2O_5/MoO_3 composite photocatalysts prepared via electrospinning and their photodegradation activity for dimethyl phthalate. *Chinese Journal of Catalysis* **36** (12), 2194–2202.

Cui, H., Liang, Z., Zhang, J., Liu, H. & Shi, J. 2016 Enhancement of the photocatalytic activity of a TiO_2/carbon aerogel based on a hydrophilic secondary pore structure. *RSC Advances* **6** (72), 68416–68423.

Cui, D., Ricardo, M. & Quinete, N. 2022 A novel report on phthalates levels in Biscayne Bay surface waters and drinking water from South Florida. *Marine Pollution Bulletin* **180**, 113802.

Dargnat, C., Teil, M., Chevreuil, M. & Blanchard, M. 2009 Phthalate removal throughout wastewater treatment plant. *Science of The Total Environment* **407** (4), 1235–1244.

Das, M., Kumar, S., Ghosh, P., Shah, G., Malyan, S., Bajar, S., Thakur, I. & Singh, L. 2021 Remediation strategies for mitigation of phthalate pollution: challenges and future perspectives. *Journal of Hazardous Materials* **409**, 124496.

de Souza, F., Sáez, C., Cañizares, P., de Motheo, A. & Rodrigo, M. 2013 Electrochemical removal of dimethyl phthalate with diamond anodes. *Journal of Chemical Technology & Biotechnology* **89** (2), 282–289.

Ding, S., Wan, J., Ma, Y., Wang, Y., Li, X., Sun, J. & Pu, M. 2021a Targeted degradation of dimethyl phthalate by activating persulfate using molecularly imprinted Fe-MOF-74. *Chemosphere* **270**, 128620.

Ding, S., Wan, J., Ma, Y., Wang, Y., Pu, M., Li, X. & Sun, J. 2021b Water stable SiO_2-coated Fe-MOF-74 for aqueous dimethyl phthalate degradation in PS activated medium. *Journal of Hazardous Materials* **411**, 125194.

Ding, Y., Zhang, M., Zhou, S., Xie, L., Li, A. & Wang, P. 2022 Degradation of dimethyl phthalate through Fe(II)/peroxymonosulphate heightened by fulvic acid: efficiency and possible mechanism. *Environmental Technology*, 1–13.

Dolatabadi, M., Świergosz, T. & Ahmadzadeh, S. 2021 Electro-Fenton approach in oxidative degradation of dimethyl phthalate – The treatment of aqueous leachate from landfills. *Science of The Total Environment* **772**, 145323.

Dong, C., Chen, C. & Hung, C. 2018 Persulfate activation with rice husk-based magnetic biochar for degrading PAEs in marine sediments. *Environmental Science and Pollution Research* **26** (33), 33781–33790.

Du, E., Feng, X., Guo, Y., Peng, M., Feng, H., Wang, J. & Zhang, S. 2015 Dimethyl phthalate degradation by UV/H2O2: combination of experimental methods and quantum chemical calculation. *CLEAN – Soil, Air, Water* **43** (6), 811–821.

Dulazi, A. & Liu, H. 2011 Removal of phthalate esters from water using immobilized lipase on chitosan beads. *Environmental Technology* **32** (13), 1443–1451.

Eichler, C., Cohen Hubal, E. & Little, J. 2019 Assessing human exposure to chemicals in materials, products and articles: the international risk management landscape for phthalates. *Environmental Science & Technology* **53** (23), 13583–13597.

Elgarahy, A., Akhdhar, A. & Elwakeel, K. 2021 Microplastics prevalence, interactions, and remediation in the aquatic environment: a critical review. *Journal of Environmental Chemical Engineering* **9** (5), 106224.

Eslami, A., Akbari-Adergani, B., Mohseni-Bandpei, A., Rabbani, M. & Saghi, M. 2017 Synthesis and characterization of a coated Fe-Ag@ZnO nanorod for the purification of a polluted environmental solution under simulated sunlight irradiation. *Materials Letters* **197**, 205–208.

European Union 2013 Directive 2013/39/EU of the European parliament and of the council of 12 August 2013 amending directives 2000/60/EC and 2008/105/EC as regards priority substances in the field of water policy text with EEA relevance. *Official Journal L* **226**, 1–17. Available from: http://data.europa.eu/eli/dir/2013/39/oj.

Fang, G., Zhu, C., Dionysiou, D., Gao, J. & Zhou, D. 2015 Mechanism of hydroxyl radical generation from biochar suspensions: implications to diethyl phthalate degradation. *Bioresource Technology* **176**, 210–217.

Fang, G., Liu, C., Wang, Y., Dionysiou, D. & Zhou, D. 2017 Photogeneration of reactive oxygen species from biochar suspension for diethyl phthalate degradation. *Applied Catalysis B: Environmental* **214**, 34–45.

Fawell, J. & Ong, C. 2012 Emerging contaminants and the implications for drinking water. *International Journal of Water Resources Development* **28** (2), 247–263.

Fazli, A., Brigante, M., Khataee, A. & Mailhot, G. 2021 Synthesis of a magnetically separable LDH-based S-scheme nano-heterojunction for the activation of peroxymonosulfate towards the efficient visible-light photodegradation of diethyl phthalate. *Applied Surface Science* **559**, 149906.

Gan, L., Zhong, Q., Geng, A., Wang, L., Song, C., Han, S., Cui, J. & Xu, L. 2019 Cellulose derived carbon nanofiber: a promising biochar support to enhance the catalytic performance of $CoFe_2O_4$ in activating peroxymonosulfate for recycled dimethyl phthalate degradation. *Science of The Total Environment* **694**, 133705.

Gao, D. & Wen, Z. 2016 Phthalate esters in the environment: a critical review of their occurrence, biodegradation, and removal during wastewater treatment processes. *Science of The Total Environment* **541**, 986–1001.

Gao, D., Li, Z., Wang, H. & Liang, H. 2018 An overview of phthalate acid ester pollution in China over the last decade: environmental occurrence and human exposure. *Science of The Total Environment* **645**, 1400–1409.

Giuliani, A., Zuccarini, M., Cichelli, A., Khan, H. & Reale, M. 2020 Critical review on the presence of phthalates in food and evidence of their biological impact. *International Journal of Environmental Research and Public Health* **17** (16), 5655.

González-Márquez, A., Loera-Corral, O., Santacruz-Juárez, E., Tlécuitl-Beristain, S., García-Dávila, J., Viniegra-González, G. & Sánchez, C. 2019 Biodegradation patterns of the endocrine disrupting pollutant di(2-ethyl hexyl) phthalate by Fusarium culmorum. *Ecotoxicology and Environmental Safety* **170**, 293–299.

Gucheng, Z., Jing, Z., Yongli, Z., Peng, Z., Chenmo, W., Wenshu, L. & Tao, L. 2017 Ozonation of dimethyl phthalate in water activated by N-methyl hydroxylamine. *Journal of Water Supply: Research and Technology – Aqua* **67** (1), 22–29.

He, W., Qin, N., Kong, X., Liu, W., He, Q., Ouyang, H., Yang, C., Jiang, Y., Wang, Q., Yang, B. & Xu, F. 2013 Spatio-temporal distributions and the ecological and health risks of phthalate esters (PAEs) in the surface water of a large, shallow Chinese lake. *Science of The Total Environment* **461–462**, 672–680.

He, G., Zhang, J., Hu, Y., Bai, Z. & Wei, C. 2019 Dual-template synthesis of mesoporous TiO_2 nanotubes with structure-enhanced functional photocatalytic performance. *Applied Catalysis B: Environmental* **250**, 301–312.

Hou, M., Chu, Y., Li, X., Wang, H., Yao, W., Yu, G., Murayama, S. & Wang, Y. 2016 Electro-peroxone degradation of diethyl phthalate: cathode selection, operational parameters, and degradation mechanisms. *Journal of Hazardous Materials* **319**, 61–68.

Huang, Z., Wu, P., Lu, Y., Wang, X., Zhu, N. & Dang, Z. 2013 Enhancement of photocatalytic degradation of dimethyl phthalate with nano-TiO_2 immobilized onto hydrophobic layered double hydroxides: a mechanism study. *Journal of Hazardous Materials* **246–247**, 70–78.

Huang, J., Li, X., Ma, M. & Li, D. 2017 Removal of di-(2-ethylhexyl) phthalate from aqueous solution by UV/peroxymonosulfate: influencing factors and reaction pathways. *Chemical Engineering Journal* **314**, 182–191.

Huang, Y., Huang, X., Chen, X., Cai, Q., Chen, S., Mo, C., Lü, H. & Wong, M. 2018 Biodegradation of di-butyl phthalate (DBP) by a novel endophytic bacterium bacillus subtilis and its bioaugmentation for removing DBP from vegetation slurry. *Journal of Environmental Management* **224**, 1–9.

Huang, M., Wang, X., Liu, C., Fang, G., Gao, J., Wang, Y. & Zhou, D. 2021a Facile ball milling preparation of sulfur-doped carbon as peroxymonosulfate activator for efficient removal of organic pollutants. *Journal of Environmental Chemical Engineering* **9** (6), 106536.

Huang, M., Zhu, C., Zhu, F., Fang, G. & Zhou, D. 2021b Mechanism of significant enhancement of VO2-Fenton-like reactions by oxalic acid for diethyl phthalate degradation. *Separation and Purification Technology* **279**, 119671.

Hung, C., Yuan, C. & Li, H. 2017 Photodegradation of diethyl phthalate with PANi/CNT/TiO_2 immobilized on glass plate irradiated with visible light and simulated sunlight – effect of synthesized method and pH. *Journal of Hazardous Materials* **322**, 243–253.

Hung, C., Huang, C., Chen, C. & Dong, C. 2021 The degradation of di-(2-ethylhexyl) phthalate, DEHP, in sediments using percarbonate activated by seaweed biochars and its effects on the benthic microbial community. *Journal of Cleaner Production* **292**, 126108.

Jamil, T., Abbas, H., Youssief, A., Mansor, E. & Hammad, F. 2017 The synthesis of nano-sized undoped, Bi doped and Bi, Cu co-doped srtio3 using two sol–gel methods to enhance the photocatalytic performance for the degradation of dibutyl phthalate under visible light. *Comptes Rendus Chimie* **20** (2), 97–106.

Jebara, A., Albergamo, A., Rando, R., Potortì, A., Lo Turco, V., Mansour, H. & Di Bella, G. 2021 Phthalates and non-phthalate plasticizers in Tunisian marine samples: occurrence, spatial distribution and seasonal variation. *Marine Pollution Bulletin* **163**, 111967.

Jiang, W., Joens, J., Dionysiou, D. & O'Shea, K. 2013 Optimization of photocatalytic performance of TiO_2 coated glass microspheres using response surface methodology and the application for degradation of dimethyl phthalate. *Journal of Photochemistry and Photobiology A: Chemistry* **262**, 7–13.

Jiang, J., Mu, D., Ding, M., Zhang, S., Zhang, H. & Hu, J. 2018 Simultaneous determination of primary and secondary phthalate monoesters in the Taihu Lake: exploration of sources. *Chemosphere* **202**, 17–24.

Julinová, M. & Slavík, R. 2012 Removal of phthalates from aqueous solution by different adsorbents: a short review. *Journal of Environmental Management* **94** (1), 13–24.

Kasonga, T., Coetzee, M., Kamika, I., Ngole-Jeme, V. & Benteke Momba, M. 2021 Endocrine-disruptive chemicals as contaminants of emerging concern in wastewater and surface water: a review. *Journal of Environmental Management* **277**, 111485.

Kaur, R., Singla, P. & Singh, K. 2017 Transition metals (Mn, Ni, Co) doping in TiO_2 nanoparticles and their effect on degradation of diethyl phthalate. *International Journal of Environmental Science and Technology* **15** (11), 2359–2368.

Ki, S., Park, Y., Kim, J., Lee, W., Lee, H. & Jung, S. 2019 Facile preparation of tungsten oxide doped TiO_2 photocatalysts using liquid phase plasma process for enhanced degradation of diethyl phthalate. *Chemical Engineering Journal* **377**, 120087.

Klu, P., Nasir Khan, M., Wang, C., Qi, J., Sun, X. & Li, J. 2022 Mechanism of peroxymonosulfate activation and the utilization efficiency using hollow (Co, Mn)3O4 nanoreactor as an efficient catalyst for degradation of organic pollutants. *Environmental Research* **207**, 112148.

Koniecki, D., Wang, R., Moody, R. & Zhu, J. 2011 Phthalates in cosmetic and personal care products: concentrations and possible dermal exposure. *Environmental Research* **111** (3), 329–336.

Kumar, S., Kaushik, R., Upadhyay, G. & Purohit, L. 2021 rGO-ZnO nanocomposites as efficient photocatalyst for degradation of 4-BP and DEP using high temperature refluxing method in in-situ condition. *Journal of Hazardous Materials* **406**, 124300.

Lamraoui, I., Eltoukhy, A., Wang, J., Lamraoui, M., Ahmed, A., Jia, Y., Lu, T. & Yan, Y. 2020 Biodegradation of Di (2-Ethylhexyl) phthalate by a novel Enterobacter spp. strain YC-IL1 isolated from polluted soil, Mila, Algeria. *International Journal of Environmental Research and Public Health* **17** (20), 7501.

Lee, H., Park, Y., Kim, J., Park, Y. & Jung, S. 2019a Degradation of dimethyl phthalate using a liquid phase plasma process with TiO_2 photocatalysts. *Environmental Research* **169**, 256–260.

Lee, Y., Lee, J., Choe, W., Kim, T., Lee, J., Kho, Y., Choi, K. & Zoh, K. 2019b Distribution of phthalate esters in air, water, sediments, and fish in the Asan Lake of Korea. *Environment International* **126**, 635–643.

Lei, Y., Lu, J., Zhu, M., Xie, J., Peng, S. & Zhu, C. 2020 Radical chemistry of diethyl phthalate oxidation via UV/peroxymonosulfate process: roles of primary and secondary radicals. *Chemical Engineering Journal* **379**, 122339.

Li, J., Zhang, J., Yadav, M. & Li, X. 2019 Biodegradability and biodegradation pathway of di-(2-ethylhexyl) phthalate by *Burkholderia pyrrocinia* b1213. *Chemosphere* **225**, 443–450.

Liang, C., Luo, X. & Hu, Y. 2021a Enhanced ozone oxidation by a novel $Fe/Mn@\gamma - Al_2O_3$ nanocatalyst: the role of hydroxyl radical and singlet oxygen. *Water* **14** (1), 19.

Liang, J., Xiang, Q., Lei, W., Zhang, Y., Sun, J., Zhu, H. & Wang, S. 2021b Ferric iron reduction reaction electro-Fenton with gas diffusion device: a novel strategy for improvement of comprehensive efficiency in electro-Fenton. *Journal of Hazardous Materials* **412**, 125195.

Lin, X., Ma, Y., Wan, J. & Wang, Y. 2017 LiCoPO4 (LCP) as an effective peroxymonosulfate activator for degradation of diethyl phthalate in aqueous solution without controlling pH: efficiency, stability and mechanism. *Chemical Engineering Journal* **315**, 304–314.

Lin, X., Ma, Y., Wan, J., Wang, Y. & Li, Y. 2020 Co0.59Fe0.41p nanocubes derived from nanoscale metal–organic frameworks for removal of diethyl phthalate by activation of peroxymonosulfate. *Applied Catalysis A: General* **589**, 117307. https://doi.org/10.1016/j.apcata.2019.117307.

Liu, N., Xu, G., Wu, M., He, X., Tang, L., Shi, W., Wang, L. & Shao, H. 2012 Radical-induced destruction of diethyl phthalate in aqueous solution: kinetics, spectral properties, and degradation efficiencies studies. *Research on Chemical Intermediates* **39** (8), 3727–3737.

Liu, P., Tian, T., Barreto, J. & Chou, J. 2013 Assessment and analysis of phthalate esters, in Lake Pontchartrain, by SPME combining with GC-MS. *Environmental Technology* **34** (4), 453–462.

Liu, Y., Sun, N., Hu, J., Li, S. & Qin, G. 2018 Photocatalytic degradation properties of α-Fe_2O_3 nanoparticles for dibutyl phthalate in aqueous solution system. *Royal Society Open Science* **5** (4), 172196.

Liu, Y., Wu, D., Peng, S., Feng, Y. & Liu, Z. 2019 Enhanced mineralization of dimethyl phthalate by heterogeneous ozonation over nanostructured Cu-Fe-O surfaces: synergistic effect and radical chain reactions. *Separation and Purification Technology* **209**, 588–597.

Lovato, M., Gilliard, M., Cassano, A. & Martín, C. 2014 Kinetics of the degradation of n-butyl benzyl phthalate using O3/UV, direct photolysis, direct ozonation and UV effects. *Environmental Science and Pollution Research* **22** (2), 909–917.

Ma, F., Zhang, S., Yang, X., Guo, W., Guo, Y. & Huo, M. 2012 Fabrication of metallic platinum and indium oxide codoped titania nanotubes for the simulated sunlight photocatalytic degradation of diethyl phthalate. *Catalysis Communications* **24**, 75–79.

Ma, Y., Du, K., Guo, Y., Tang, M., Yin, H., Mao, X. & Wang, D. 2022a Biphase Co@C core-shell catalysts for efficient fenton-like catalysis. *Journal of Hazardous Materials* **429**, 128287.

Ma, Y., Du, K., Guo, Y., Tang, M., Yin, H., Mao, X. & Wang, D. 2022b Electrolytic core–shell Co@C for diethyl phthalate degradation. *Chemical Engineering Journal* **431**, 134065.

Ma, Y., Wang, J., Wang, H., Wang, C., He, C. & Zhang, X. 2022c Performance and integrated mathematical simulations of catalytic ozonation of dimethyl phthalate with activated carbon in continuous gas-liquid-solid reactors. *Process Safety and Environmental Protection* **167**, 67–76.

Magdouli, S., Daghrir, R., Brar, S., Drogui, P. & Tyagi, R. 2013 Di 2-ethylhexylphtalate in the aquatic and terrestrial environment: a critical review. *Journal of Environmental Management* **127**, 36–49.

Mankidy, R., Wiseman, S., Ma, H. & Giesy, J. 2013 Biological impact of phthalates. *Toxicology Letters* **217** (1), 50–58.

Mansouri, L., Tizaoui, C., Geissen, S. & Bousselmi, L. 2019 A comparative study on ozone, hydrogen peroxide and UV based advanced oxidation processes for efficient removal of diethyl phthalate in water. *Journal of Hazardous Materials* **363**, 401–411.

Medellin-Castillo, N., Ocampo-Pérez, R., Leyva-Ramos, R., Sanchez-Polo, M., Rivera-Utrilla, J. & Méndez-Díaz, J. 2013 Removal of diethyl phthalate from water solution by adsorption, photo-oxidation, ozonation and advanced oxidation process (UV/H_2O_2, O_3/H_2O_2 and O_3/activated carbon). *Science of The Total Environment* **442**, 26–35.

Meenakshi, G. & Sivasamy, A. 2018 Nanorod ZnO/SiC nanocomposite: an efficient catalyst for the degradation of an endocrine disruptor under UV and visible light irradiations. *Journal of Environmental Chemical Engineering* **6** (3), 3757–3769.

Mohamed, R. & Aazam, E. 2013 Synthesis and characterization of P-doped TiO_2 thin-films for photocatalytic degradation of butyl benzyl phthalate under visible-light irradiation. *Chinese Journal of Catalysis* **34** (6), 1267–1273.

Na, S., Ahn, Y., Cui, M. & Khim, J. 2012a Significant diethyl phthalate (DEP) degradation by combined advanced oxidation process in aqueous solution. *Journal of Environmental Management* **101**, 104–110.

Na, S., Jinhua, C., Cui, M. & Khim, J. 2012b Sonophotolytic diethyl phthalate (DEP) degradation with UVC or VUV irradiation. *Ultrasonics Sonochemistry* **19** (5), 1094–1098.

Net, S., Sempéré, R., Delmont, A., Paluselli, A. & Ouddane, B. 2015 Occurrence, fate, behavior and ecotoxicological state of phthalates in different environmental matrices. *Environmental Science & Technology* **49** (7), 4019–4035.

Nidheesh, P., Gopinath, A., Ranjith, N., Praveen Akre, A., Sreedharan, V. & Suresh Kumar, M. 2021 Potential role of biochar in advanced oxidation processes: a sustainable approach. *Chemical Engineering Journal* **405**, 126582.

Nie, Y., Yu, F., Wang, L., Xing, Q., Liu, X., Pei, Y., Zou, J., Dai, W., Li, Y. & Suib, S. 2018 Photocatalytic degradation of organic pollutants coupled with simultaneous photocatalytic H2 evolution over graphene quantum dots/Mn-N-TiO_2/g-C_3N_4 composite catalysts: performance and mechanism. *Applied Catalysis B: Environmental* **227**, 312–321.

Nie, G., Hu, K., Ren, W., Zhou, P., Duan, X., Xiao, L. & Wang, S. 2021 Mechanical agitation accelerated ultrasonication for wastewater treatment: Sustainable production of hydroxyl radicals. *Water Research* **198**, 117124. https://doi.org/10.1016/j.watres.2021.117124.

Paluselli, A., Fauvelle, V., Schmidt, N., Galgani, F., Net, S. & Sempéré, R. 2018 Distribution of phthalates in Marseille Bay (NW Mediterranean Sea). *Science of The Total Environment* **621**, 578–587.

Pang, G., Ji, M., Li, Z., Yang, Z., Qiu, X. & Zhao, Y. 2022 Electrospinning of ZIF-67 derived Co-C-N composite efficiently activating peroxymonosulfate to degrade dimethyl phthalate. *Water* **14** (14), 2248.

Pastorino, P. & Ginebreda, A. 2021 Contaminants of emerging concern (CECs): occurrence and fate in aquatic ecosystems. *International Journal of Environmental Research and Public Health* **18** (24), 13401.

Perpetuo, E., da Silva, E., Karolski, B. & do Nascimento, C. 2020 Biodegradation of diethyl-phthalate (DEP) by halotolerant bacteria isolated from an estuarine environment. *Biodegradation* **31** (4–6), 331–340.

Pironti, C., Ricciardi, M., Proto, A., Bianco, P., Montano, L. & Motta, O. 2021 Endocrine-disrupting compounds: an overview on their occurrence in the aquatic environment and human exposure. *Water* **13** (10), 1347.

Pradeep, S., Sarath Josh, M., Binod, P., Sudha Devi, R., Balachandran, S., Anderson, R. & Benjamin, S. 2015 Achromobacter denitrificans strain SP1 efficiently remediates di(2-ethylhexyl) phthalate. *Ecotoxicology and Environmental Safety* **112**, 114–121.

Qi, L., Lu, W., Tian, G., Sun, Y., Han, J. & Xu, L. 2020 Enhancement of sono-Fenton by P25-Mediated visible light photocatalysis: analysis of synergistic effect and influence of emerging contaminant properties. *Catalysts* **10** (11), 1297.

Ren, Y., Dong, Q., Feng, J., Ma, J., Wen, Q. & Zhang, M. 2012 Magnetic porous ferrospinel $NiFe_2O_4$: a novel ozonation catalyst with strong catalytic property for degradation of di-n-butyl phthalate and convenient separation from water. *Journal of Colloid and Interface Science* **382** (1), 90–96.

Ren, Y., Lin, L., Ma, J., Yang, J., Feng, J. & Fan, Z. 2015 Sulfate radicals induced from peroxymonosulfate by magnetic ferrospinel MFe_2O_4 (M = Co, Cu, Mn, and Zn) as heterogeneous catalysts in the water. *Applied Catalysis B: Environmental* **165**, 572–578.

Ren, W., Zhou, P., Nie, G., Cheng, C., Duan, X., Zhang, H. & Wang, S. 2020 Hydroxyl radical dominated elimination of plasticizers by peroxymonosulfate on metal-free boron: kinetics and mechanisms. *Water Research* **186**, 116361.

Ribeiro, A., Nunes, O., Pereira, M. & Silva, A. 2015 An overview on the advanced oxidation processes applied for the treatment of water pollutants defined in the recently launched directive 2013/39/EU. *Environment International* **75**, 33–51.

Sabaredzovic, A., Sakhi, A., Brantsæter, A. & Thomsen, C. 2015 Determination of 12 urinary phthalate metabolites in Norwegian pregnant women by core–shell high performance liquid chromatography with on-line solid-phase extraction, column switching and tandem mass spectrometry. *Journal of Chromatography B* **1002**, 343–352.

Selvaraj, K., Sundaramoorthy, G., Ravichandran, P., Girijan, G., Sampath, S. & Ramaswamy, B. 2014 Phthalate esters in water and sediments of the Kaveri River, India: environmental levels and ecotoxicological evaluations. *Environmental Geochemistry and Health* **37** (1), 83–96.

Sengupta, A., Lyons, J., Smith, D., Drewes, J., Snyder, S., Heil, A. & Maruya, K. 2014 The occurrence and fate of chemicals of emerging concern in coastal urban rivers receiving discharge of treated municipal wastewater effluent. *Environmental Toxicology and Chemistry* **33** (2), 350–358.

Shan, W., Hu, Y., Bai, Z., Zheng, M. & Wei, C. 2016 In situ preparation of g-C_3N_4/bismuth-based oxide nanocomposites with enhanced photocatalytic activity. *Applied Catalysis B: Environmental* **188**, 1–12.

Shuai, W., Liu, C., Fang, G., Zhou, D. & Gao, J. 2018 Nano-α-Fe_2O_3 enhanced photocatalytic degradation of diethyl phthalate ester by citric acid/UV (300–400 nm): a mechanism study. *Journal of Photochemistry and Photobiology A: Chemistry* **360**, 78–85.

Singla, P., Pandey, O. & Singh, K. 2015 Study of photocatalytic degradation of environmentally harmful phthalate esters using Ni-doped TiO_2 nanoparticles. *International Journal of Environmental Science and Technology* **13** (3), 849–856.

Smarr, M., Kannan, K., Sun, L., Honda, M., Wang, W., Karthikraj, R., Chen, Z., Weck, J. & Buck Louis, G. 2018 Preconception seminal plasma concentrations of endocrine disrupting chemicals in relation to semen quality parameters among male partners planning for pregnancy. *Environmental Research* **167**, 78–86.

Şolpan, D. & Mehrnia, M. 2018 Dimethyl phthalate (DMP) degradation in aqueous solution by gamma-irradiation/H_2O_2. *Journal of Radioanalytical and Nuclear Chemistry* **317** (2), 841–851.

Song, C., Wang, L., Ren, J., Lv, B., Sun, Z., Yan, J., Li, X. & Liu, J. 2015 Comparative study of diethyl phthalate degradation by UV/H_2O_2 and UV/TiO_2: kinetics, mechanism, and effects of operational parameters. *Environmental Science and Pollution Research* **23** (3), 2640–2650.

Souza, F., Sáez, C., Cañizares, P., Motheo, A. & Rodrigo, M. 2014 Coupling photo and sono technologies to improve efficiencies in conductive diamond electrochemical oxidation. *Applied Catalysis B: Environmental* **144**, 121–128.

Sun, J., Huang, J., Zhang, A., Liu, W. & Cheng, W. 2013 Occurrence of phthalate esters in sediments in Qiantang River, China and inference with urbanization and river flow regime. *Journal of Hazardous Materials* **248–249**, 142–149.

Sun, Z., Feng, L., Fang, G., Chu, L., Zhou, D. & Gao, J. 2021 Nano Fe_2O_3 embedded in montmorillonite with citric acid enhanced photocatalytic activity of nanoparticles towards diethyl phthalate. *Journal of Environmental Sciences* **101**, 248–259.

Surhio, M., Talpur, F., Nizamani, S., Talpur, M., Amin, F., Khaskheli, A., Bhurgri, S., Afridi, H. & Rahman, S. 2017 Effective bioremediation of endocrine-disrupting phthalate esters, mediated by bacillus strains. *Water, Air, & Soil Pollution* **228** (10), 386.

Tak, H., Chung, Y., Kim, G., Kim, H., Lee, J., Kang, J., Do, Q., Bae, B. & Kang, S. 2022 Catalytic ozonation with vanadium oxide-doped TiO_2 nanoparticles for the removal of di-2-ethylhexyl phthalate. *Chemosphere* **306**, 135646.

Takdastan, A., Niari, M., Babaei, A., Dobaradaran, S., Jorfi, S. & Ahmadi, M. 2021 Occurrence and distribution of microplastic particles and the concentration of Di 2-ethyl hexyl phthalate (DEHP) in microplastics and wastewater in the wastewater treatment plant. *Journal of Environmental Management* **280**, 111851.

Tan, T., Lai, C., Hong, S. & Rashid, S. 2018 New insights into the photocatalytic endocrine disruptors dimethyl phathalate esters degradation by UV/MWCNTs-TiO_2 nanocomposites. *Journal of Photochemistry and Photobiology A: Chemistry* **364**, 177–189.

Tan, T., Lee, K., Lai, C., Hong, S. & Rashid, S. 2020 Photocatalytic degradation mechanisms of dimethyl phthalate esters by MWCNTs-anatase TiO_2 nanocomposites using the UHPLC/Orbitrap/MS technique. *Advanced Powder Technology* **31** (2), 533–547.

Tang, Y., Pan, Z. & Li, L. 2017 pH-insusceptible cobalt-manganese immobilizing mesoporous siliceous MCM-41 catalyst for ozonation of dimethyl phthalate. *Journal of Colloid and Interface Science* **508**, 196–202.

Teil, M., Tlili, K., Blanchard, M., Labadie, P., Alliot, F. & Chevreuil, M. 2013 Polychlorinated biphenyls, polybrominated diphenyl ethers, and phthalates in roach from the Seine River Basin (France): impact of densely urbanized areas. *Archives of Environmental Contamination and Toxicology* **66** (1), 41–57.

Tijani, J., Fatoba, O., Babajide, O. & Petrik, L. 2015 Pharmaceuticals, endocrine disruptors, personal care products, nanomaterials and perfluorinated pollutants: a review. *Environmental Chemistry Letters* **14** (1), 27–49.

US EPA 2013 *National Primary Drinking Water Regulations*. US EPA. Available from: https://www.epa.gov/ground-water-and-drinking-water/national-primary-drinking-water-regulations#Organic (accessed 1 September 2022).

Valvi, D., Monfort, N., Ventura, R., Casas, M., Casas, L., Sunyer, J. & Vrijheid, M. 2015 Variability and predictors of urinary phthalate metabolites in Spanish pregnant women. *International Journal of Hygiene and Environmental Health* **218** (2), 220–231.

Wang, X., Wang, C., Jiang, W., Guo, W. & Wang, J. 2012 Sonochemical synthesis and characterization of Cl-doped TiO_2 and its application in the photodegradation of phthalate ester under visible light irradiation. *Chemical Engineering Journal* **189–190**, 288–294.

Wang, L., Fu, G., Zhao, B., Zhang, Z., Guo, X. & Zhang, H. 2013a Degradation of di-n-butyl phthalate in aqueous solution by the O_3/UV process. *Desalination and Water Treatment* **52** (4–6), 824–833.

Wang, Y., Liu, Y., Li, X., Zeng, F. & Liu, H. 2013b A highly-ordered porous carbon material based cathode for energy-efficient electro-Fenton process. *Separation and Purification Technology* **106**, 32–37.

Wang, Y., Liu, Y., Wang, K., Song, S., Tsiakaras, P. & Liu, H. 2015 Preparation and characterization of a novel KOH activated graphite felt cathode for the electro-Fenton process. *Applied Catalysis B: Environmental* **165**, 360–368.

Wang, D., Duan, X., He, X. & Dionysiou, D. 2016a Degradation of dibutyl phthalate (DBP) by UV-254nm/H_2O_2 photochemical oxidation: kinetics and influence of various process parameters. *Environmental Science and Pollution Research* **23** (23), 23772–23780.

Wang, G., Lu, G., Zhao, J., Yin, P. & Zhao, L. 2016b Evaluation of toxicity and estrogenicity of the landfill-concentrated leachate during advanced oxidation treatment: chemical analyses and bioanalytical tools. *Environmental Science and Pollution Research* **23** (16), 16015–16024.

Wang, Z., Shao, Y., Gao, N. & An, N. 2018 Degradation kinetic of dibutyl phthalate (DBP) by sulfate radical- and hydroxyl radical-based advanced oxidation process in UV/persulfate system. *Separation and Purification Technology* **195**, 92–100.

Wang, G., Zhang, Q., Chen, Q., Ma, X., Xin, Y., Zhu, X., Ma, D., Cui, C., Zhang, J. & Xiao, Z. 2019a Photocatalytic degradation performance and mechanism of dibutyl phthalate by graphene/TiO_2 nanotube array photoelectrodes. *Chemical Engineering Journal* **358**, 1083–1090.

Wang, Y., Zhu, H. & Kannan, K. 2019b A review of biomonitoring of phthalate exposures. *Toxics* **7** (2), 21.

Wang, Z., Shao, Y., Gao, N., Lu, X. & An, N. 2019c Degradation kinetic of phthalate esters and the formation of brominated byproducts in heat-activated persulfate system. *Chemical Engineering Journal* **359**, 1086–1096.

Wang, Q., Wu, X., Jiang, L., Fang, C., Wang, H. & Chen, L. 2020 Effective degradation of Di-n-butyl phthalate by reusable, magnetic Fe_3O_4 nanoparticle-immobilized Pseudomonas sp. W1 and its application in simulation. *Chemosphere* **250**, 126339.

Wang, P., Gao, J., Zhao, Y., Zhang, M. & Zhou, S. 2021a Biodegradability of di-(2-ethylhexyl) phthalate by a newly isolated bacterium Achromobacter sp. RX. *Science of The Total Environment* **755**, 142476.

Wang, X., Wu, H., Wang, X., Wang, H., Zhao, K., Ma, B. & Lu, Z. 2021b Network-directed isolation of the cooperator Pseudomonas aeruginosa ZM03 enhanced the dibutyl phthalate degradation capacity of Arthrobacter nicotianae ZM05 under pH stress. *Journal of Hazardous Materials* **410**, 124667.

Weiss, J., Gustafsson, Å., Gerde, P., Bergman, Å., Lindh, C. & Krais, A. 2018 Daily intake of phthalates, MEHP, and DINCH by ingestion and inhalation. *Chemosphere* **208**, 40–49.

Wen, G., Ma, J., Liu, Z. & Zhao, L. 2011 Ozonation kinetics for the degradation of phthalate esters in water and the reduction of toxicity in the process of O_3/H_2O_2. *Journal of Hazardous Materials* **195**, 371–377.

Wen, G., Wang, S., Ma, J., Huang, T., Liu, Z., Zhao, L. & Xu, J. 2014 Oxidative degradation of organic pollutants in aqueous solution using zero valent copper under aerobic atmosphere condition. *Journal of Hazardous Materials* **275**, 193–199.

Xiao, J., Wang, C. & Liu, H. 2020 Fenton-like degradation of dimethyl phthalate enhanced by quinone species. *Journal of Hazardous Materials* **382**, 121007.

Xiong, Y.-H. & Pei, D.-S. 2021 A review on efficient removal of phthalic acid esters via biochars and transition metals-activated persulfate systems. *Chemosphere* **277**, 130256. https://doi.org/10.1016/j.chemosphere.2021.130256.

Xu, L., Chu, W. & Graham, N. 2013a A systematic study of the degradation of dimethyl phthalate using a high-frequency ultrasonic process. *Ultrasonics Sonochemistry* **20** (3), 892–899.

Xu, L., Chu, W. & Graham, N. 2013b Sonophotolytic degradation of dimethyl phthalate without catalyst: analysis of the synergistic effect and modeling. *Water Research* **47** (6), 1996–2004.

Xu, L., Chu, W. & Graham, N. 2014 Degradation of di-n-butyl phthalate by a homogeneous sono–photo–Fenton process with in situ generated hydrogen peroxide. *Chemical Engineering Journal* **240**, 541–547.

Xu, L., Chu, W. & Gan, L. 2015a Environmental application of graphene-based $CoFe_2O_4$ as an activator of peroxymonosulfate for the degradation of a plasticizer. *Chemical Engineering Journal* **263**, 435–443.

Xu, L., Chu, W. & Graham, N. 2015b Sonophotolytic degradation of phthalate acid esters in water and wastewater: influence of compound properties and degradation mechanisms. *Journal of Hazardous Materials* **288**, 43–50.

Xu, L., Qi, L., Sun, Y., Gong, H., Chen, Y., Pei, C. & Gan, L. 2020 Mechanistic studies on peroxymonosulfate activation by g-C_3N_4 under visible light for enhanced oxidation of light-inert dimethyl phthalate. *Chinese Journal of Catalysis* **41** (2), 322–332.

Yan, H., Lu, P., Pan, Z., Wang, X., Zhang, Q. & Li, L. 2013 Ce/SBA-15 as a heterogeneous ozonation catalyst for efficient mineralization of dimethyl phthalate. *Journal of Molecular Catalysis A: Chemical* **377**, 57–64.

Yang, Y. & Lin, Z. 2012 Study on the degradation of plasticizer Di (2-Ethylhexyl) phthalate by advanced oxidation process. *Advanced Materials Research* **529**, 463–467.

Yang, C., Wang, C. & Chen, C. 2013 Di-n-butyl phthalate removal using mixed cultures in batch reactors. *International Biodeterioration & Biodegradation* **85**, 587–591.

Yang, C., Wang, C. & Chen, C. 2014 Di-n-butyl phthalate removal by strain Deinococcus sp. R5 in batch reactors. *International Biodeterioration & Biodegradation* **95**, 55–60.

Yang, T., Ren, L., Jia, Y., Fan, S., Wang, J., Wang, J., Nahurira, R., Wang, H. & Yan, Y. 2018 Biodegradation of Di-(2-ethylhexyl) phthalate by Rhodococcus ruber YC-YT1 in contaminated water and soil. *International Journal of Environmental Research and Public Health* **15** (5), 964.

Yang, C., Ren, B., Wang, D. & Tang, Q. 2019 The synthesis of a nano-Fe@NdFeB/AC magnetic catalyst and its application in the degradation of dimethyl phthalate by a fenton-like process. *Desalination and Water Treatment* **144**, 257–262.

Yang, S., Song, Y., Chang, F. & Wang, K. 2020a Evaluation of chemistry and key reactor parameters for industrial water treatment applications of the UV/O_3 process. *Environmental Research* **188**, 109660.

Yang, Z., Chen, H., Wang, J., Yuan, R., Wang, F. & Zhou, B. 2020b Efficient degradation of diisobutyl phthalate in aqueous solution through electro-Fenton process with sacrificial anode. *Journal of Environmental Chemical Engineering* **8** (5), 104057.

Ye, Q., Wu, J., Wu, P., Rehman, S., Ahmed, Z. & Zhu, N. 2021 Enhancing peroxymonosulfate activation by Co-Fe layered double hydroxide catalysts via compositing with biochar. *Chemical Engineering Journal* **417**, 129111. https://doi.org/10.1016/j.cej.2021.129111.

Yi, K., Lei, M., Peng, L., Chen, A. & Luo, S. 2021 Sunlight-driven degradation of diethyl phthalate via magnetically modified biochar catalysts in water: internal electron transfer mechanism. *Chemosphere* **269**, 129366.

Yi, H., Wang, Y., Diao, L., Xin, Y., Chai, C., Cui, D. & Ma, D. 2022 Ultrasonic treatment enhances the formation of oxygen vacancies and trivalent manganese on α-MnO2 surfaces: mechanism and application. *Journal of Colloid and Interface Science* **626**, 629–638.

You, S., Hu, Y., Liu, X. & Wei, C. 2018 Synergetic removal of Pb(II) and dibutyl phthalate mixed pollutants on Bi_2O_3-TiO_2 composite photocatalyst under visible light. *Applied Catalysis B: Environmental* **232**, 288–298.

Yu, Y., An, H., Zhao, Y., Feng, J., Wei, T., Yu, S., Ren, Y. & Chen, Y. 2021 $MnFe_2O_4$ decorated graphene as a heterogeneous catalyst for efficient degradation of di-n-butyl phthalate using catalytic ozonation in water. *Separation and Purification Technology* **259**, 118097.

Zhang, L., Liu, J., Liu, H., Wan, G. & Zhang, S. 2015 The occurrence and ecological risk assessment of phthalate esters (PAEs) in urban aquatic environments of China. *Ecotoxicology* **24** (5), 967–984.

Zhang, X., Feng, M., Qu, R., Liu, H., Wang, L. & Wang, Z. 2016 Catalytic degradation of diethyl phthalate in aqueous solution by persulfate activated with nano-scaled magnetic $CuFe_2O_4$/MWCNTs. *Chemical Engineering Journal* **301**, 1–11.

Zhang, J., Zhang, C., Zhu, Y., Li, J. & Li, X. 2018a Biodegradation of seven phthalate esters by Bacillus mojavensis b1811. *International Biodeterioration & Biodegradation* **132**, 200–207.

Zhang, Z., Zhang, H., Zou, Y. & Yang, G. 2018b Distribution and ecotoxicological state of phthalate esters in the sea-surface microlayer, seawater and sediment of the Bohai Sea and the Yellow Sea. *Environmental Pollution* **240**, 235–247.

Zhang, L., Bi, X., Wang, Z., Ertürk, A., Elmaci, G., Zhao, H., Zhao, P. & Meng, X. 2022a Brønsted-acid sites promoted degradation of phthalate esters over MnO_2: mineralization enhancement and aquatic toxicity assessment. *Chemosphere* **291**, 132740.

Zhang, Y., Yu, H., Li, S., Wang, L., Huang, F., Guan, R., Li, J., Jiao, Y. & Sun, J. 2022b Rapidly degradation of di-(2-ethylhexyl) phthalate by Z-scheme Bi_2O_3/TiO_2@reduced graphene oxide driven by simulated solar radiation. *Chemosphere* **272**, 129631. https://doi.org/10.1016/j.chemosphere.2021.129631.

Zhao, H., Chen, Y., Peng, Q., Wang, Q. & Zhao, G. 2017 Catalytic activity of MOF(2Fe/Co)/carbon aerogel for improving H_2O_2 and OH generation in solar photo–electro–Fenton process. *Applied Catalysis B: Environmental* **203**, 127–137.

Zhao, H., Qian, L., Chen, Y., Wang, Q. & Zhao, G. 2018a Selective catalytic two-electron O_2 reduction for onsite efficient oxidation reaction in heterogeneous electro-Fenton process. *Chemical Engineering Journal* **332**, 486–498.

Zhao, M., Cheng, M., Zeng, G. & Zhang, C. 2018b Degradation of di (2-ethylhexyl) phthalate in sediment by a surfactant-enhanced fenton-like process. *Chemosphere* **198**, 327–333.

Zheng, X., Zhang, B. & Teng, Y. 2014 Distribution of phthalate acid esters in lakes of Beijing and its relationship with anthropogenic activities. *Science of The Total Environment* **476–477**, 107–113.

Zheng, D., Cao, J., Wang, P., Zhao, J., Zhao, Y., Zhang, T. & Li, C. 2020 Catalytic ozonation of dibutyl phthalate in the presence of Ag-doped $NiFe_2O_4$ and its mechanism. *Environmental Technology* **42** (28), 4528–4538.

Zhou, W., Yu, C., Fan, Q., Wei, L., Chen, J. & Yu, J. 2013 Ultrasonic fabrication of N-doped TiO_2 nanocrystals with mesoporous structure and enhanced visible light photocatalytic activity. *Chinese Journal of Catalysis* **34** (6), 1250–1255.

Zhou, P., Zhang, J., Zhang, G., Li, W., Liu, Y., Cheng, X., Huo, X., Liu, Y. & Zhang, Y. 2019 Degradation of dimethyl phthalate by activating peroxymonosulfate using nanoscale zero valent tungsten: mechanism and degradation pathway. *Chemical Engineering Journal* **359**, 138–148.

Zhou, P., Huo, X., Zhang, J., Liu, Y., Cheng, F., Cheng, X., Wang, Y. & Zhang, Y. 2020a Visible light induced acceleration of Fe(III)/Fe(II) cycles for enhancing phthalate degradation in C60 fullerenol modified Fe(III)/peroxymonosulfate process. *Chemical Engineering Journal* **387**, 124126.

Zhou, Y., Huang, M., Wang, X., Gao, J., Fang, G. & Zhou, D. 2020b Efficient transformation of diethyl phthalate using calcium peroxide activated by pyrite. *Chemosphere* **253**, 126662.

Zhu, T., Du, P., Zeng, L., Lü, H., Zhao, H., Li, Y., Mo, C. & Cai, Q. 2019 Variation in metabolism and degradation of di-n-butyl phthalate (DBP) by high- and low-DBP accumulating cultivars of rice (Oryza sativa L.) and crude enzyme extracts. *Science of The Total Environment* **668**, 1117–1127.

Ziembowicz, S., Kida, M., Koszelnik, P., Czarnota, J. & Miąsik, M. 2021 Fenton-like degradation of di-n-butyl phthalate in landfill leachate by endogenous catalysts or iron, copper and manganese loaded bottom sediments. *Environmental Nanotechnology, Monitoring & Management* **16**, 100551.

Zolfaghari, M., Drogui, P., Seyhi, B., Brar, S., Buelna, G. & Dubé, R. 2014 Occurrence, fate and effects of Di (2-ethylhexyl) phthalate in wastewater treatment plants: a review. *Environmental Pollution* **194**, 281–293.

First received 1 October 2022; accepted in revised form 7 April 2023. Available online 20 April 2023

doi: 10.2166/wst.2023.066

Floating immobilized TiO_2 catalyst for the solar photocatalytic treatment of micro-pollutants within the secondary effluent of wastewater treatment plants

Simon Mehling [a,*], Tobias Schnabel[b], Manuel Dutschke[c] and Jörg Londong[a]

[a] Department of Civil Engineering, Professorship of Urban Water Management, Bauhaus-Universität Weimar, Weimar, Germany
[b] FH Hof – University of Applied Science Hof, Hof, Germany
[c] MFPA Weimar – Materials Research and Testing Institute Weimar, Weimar, Germany
*Corresponding author. E-mail: simon.theodor.mehling@uni-weimar.de

SM, 0000-0002-9188-2074

ABSTRACT

Floating immobilized spherical titanium dioxide catalysts were used to degrade micro-pollutants by solar photocatalysis. The degradation of the micro-pollutants was performed in the secondary effluent of a wastewater treatment plant. During the experimental period, the continuous measurement of the solar ultraviolet (UV) radiation intensity was performed. The micro-pollutants were degraded to an average of 55% after 9 h of irradiation. A substance-specific degradation affinity was found, whereby degradation rates varied by a factor of up to 3.5. The substance-specific adsorption behavior was identified as a major limitation of the reaction performance. With an increasing influence of adsorption limitation, the degradation kinetics changed from the pseudo-first order to pseudo-zero order. A correlation between degradation rate and solar irradiance could only be found for substances with high degradation/adsorption affinity. For diclofenac, a 95% degradation rate could be achieved at a radiation dose of approximately 190 mWh/m^2. The investigated technology represents a promising possibility for a minimally invasive extension of wastewater treatment plants. Possibilities of implication were estimated and discussed within this work, whereby possibilities arise for large-scale as well as decentral treatment plants.

Key words: anthropogenic micro-pollutants, immobilized floating catalysts, photocatalysis, solar light, titanium dioxide, wastewater treatment

HIGHLIGHTS

- Novelty floatable TiO_2 catalysts for the micro-pollutant treatment in wastewater.
- Evaluation of reaction rates of micro-pollutants during solar photocatalysis.
- Correlation of individual reaction rate constants to measured UV-A radiation.
- Extrapolation of the cleaning performance of the micro-pollutant removal process.

GRAPHICAL ABSTRACT

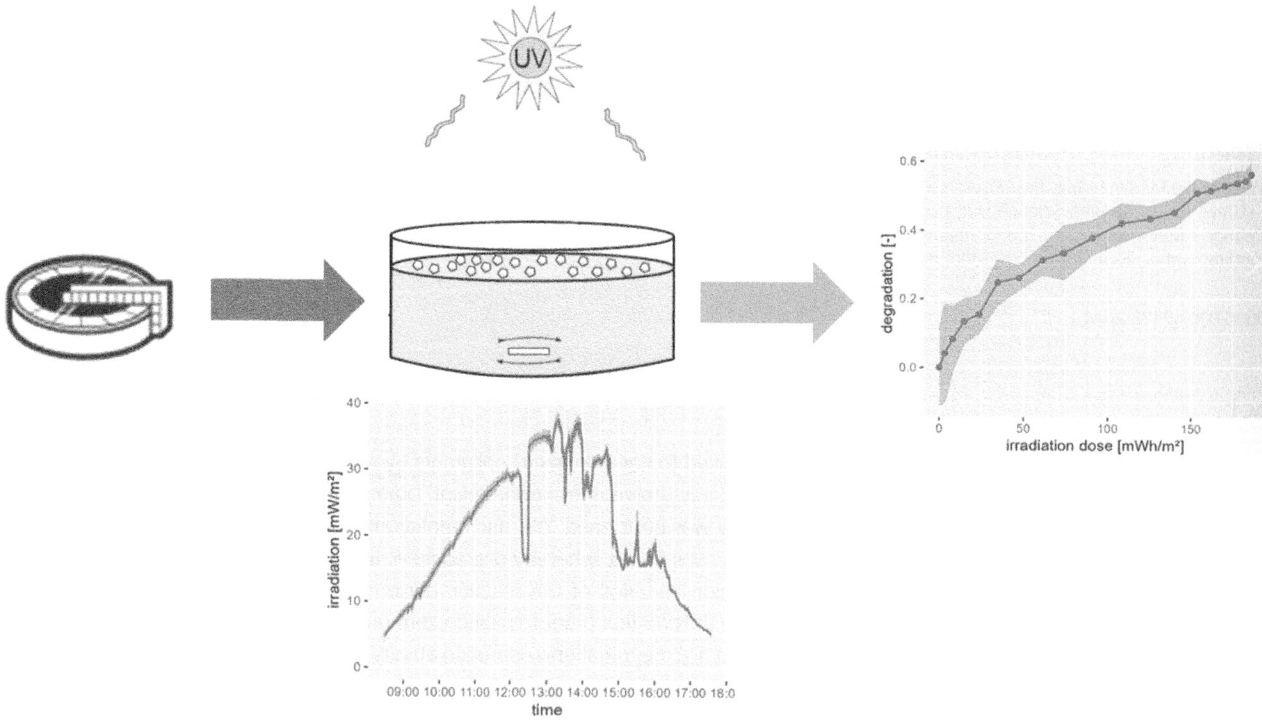

1. INTRODUCTION

Micro-pollutants released by anthropogenic activities represent one of the key challenges of future urban water infrastructure systems (Aemig *et al.* 2021). These compounds occur omnipresent in wastewater systems and are only partially removed by conventional treatment systems (Margot *et al.* 2015; Kanaujiya *et al.* 2019). Furthermore, the wastewater pathway represents one of the major entry pathways for these substance groups into the aquatic environment (Gallé *et al.* 2019). Due to this necessity, systems for the elimination or separation of these substances have been developed and implemented on an industrial scale. Ozonation and adsorption on granulated or powdered activated carbon, as well as various combinations of these processes, are established as state of the art (Bourgin *et al.* 2018; Guillossou *et al.* 2019, 2020; Teichgräber *et al.* 2021). Within research, various advanced oxidation processes (AOPs) continue to be discussed and investigated as possible systems to solve this challenge (Kanakaraju *et al.* 2018). Photocatalysis on titanium dioxide represents one possibility. During this process, hydroxyl radicals are generated by charge separation of photo semiconductor-material titanium dioxide (Pitre *et al.* 2017). Due to their high oxidation potential, these radicals enable the degradation of pollutants, theoretically up to CO_2 and H_2O (Kisch 2015; Kanakaraju *et al.* 2018). This complete mineralization prevents toxification, which is a major problem of ozone treatment (Gunten 2003; Rueda-Márquez *et al.* 2020a).

However, large-scale use of photocatalytic systems for water purification has only been implemented in a few pilot plants worldwide (Braham & Harris 2009; Abdel-Maksoud *et al.* 2016). On the one hand, the separation of frequently used particulate catalyst materials and treated wastewater prevents an efficient application of this technology (Abdel-Maksoud *et al.* 2016). Furthermore, the low energy efficiency of the processes often leads to high energy costs if artificial radiation sources are used or highly required surfaces for solar systems (Braham & Harris 2009). Solar systems enable energy-efficient treatment but are dependent on sunlight, so if permanent treatment is required, storage tanks, artificial radiation sources or a combination with other treatment technologies are necessary (Sundar & Kanmani 2020).

By (Rueda-Márquez *et al.* 2020b) using an immobilized TiO_2 catalyst placed in a turbulent tubular reactor at a cumulative UV-A and UV-B radiation dose of 33 Wh/m², the investigated micro-pollutants were degraded by over 60% after 6 h of treatment, with the exception of carbamazepine and atenolol. The reactor was filled with real wastewater treatment plant effluent. This relatively high radiation power was made possible by the 36° inclination of the reactor, a radiation-concentrating system and its geographical location in Spain. Diclofenac showed great affinity for photocatalytic treatment and was degraded by about 95%. By He *et al.* (2016), degradation rates of 75–100% were achieved in ultrapure water and secondary effluent water spiked with 5 mg/l of micro-pollutants after 96 h of photocatalytic treatment with TiO_2 immobilized on quartz sand (diclofenac: 100%, carbamazepine: 75%). The irradiation intensity in both experiments was 159 lux, generated with xenon high-intensity discharge lamps (approximately 0.2 W/m²), which emits radiation at 300–800 nm wavelength.

For solar photocatalytic treatment with particulate TiO_2, 50% carbamazepine degradation was achieved by Prieto-Rodriguez *et al.* (2012) after 6 h of treatment using actual wastewater at 250 W/m². Diclofenac was degraded by about 95% after this treatment period. By Awfa *et al.* (2020), first-order constants of 1.67×10^{-4} 1/s for the treatment of carbamazepine using particulate catalysts in wastewater are reported, which results in a 97% degradation after 6 h of treatment time. A 99.7% degradation rate of diclofenac after 180 min of treatment was achieved by Salaeh *et al.* (2016) with 12.5 W/m² radiation intensity. In Moreira *et al.* (2018), a 100% degradation rate of diclofenac and a 45% degradation rate of carbamazepine at a UV-radiation dose (300–400 nm) of 3.3 Wh/l are mentioned.

Immobilized catalyst systems seem to be limited in their reaction performance by available adsorption surfaces compared to particulate applications. The shape of the support material and the type of coating process thus have a significant influence on potential degradation rates. Furthermore, the UV-A absorption by overlaying water limits energy efficiency, especially for low-tech reactor designs (e.g. quartz sand catalysts). The comparison of existing research data on photocatalytic applications to carrier-based applications is hindered by the variability of surface structures, as well as the individual surface-to-volume ratio of each reactor. Furthermore, the variability of radiation sources and measurement methods used is a major obstacle.

In this work, TiO_2 catalysts immobilized on expanded glass granules are presented and their efficiency for the degradation of micro-pollutants in real wastewater matrix is investigated. As Nasir *et al.* (2020) described, there is previous development of floatable photocatalysts for wastewater treatment, but the database regarding micro-pollutant degradation is very limited. Only in Długosz *et al.* (2015), the degradation rate of sulfamethoxazole was investigated using floating catalysts, ultrapure water and artificial UV radiation sources. Simultaneous measurements of solar radiation power over the treatment period further provide initial data on the performance characteristics and optimization potentials of this type of catalyst when used in the secondary effluent of municipal wastewater treatment plants. The correlation between the reaction rate and the UV index measurement as methodology, which enables extrapolation based on existing large-scale datasets, is introduced. A key focus of the study concerns the analysis of the development of substance-specific reaction rates during treatment time as well as their influencing variables, which represents a research novelty. Based on the observed reaction rates, an initial assessment of the potential for integrating this technology into the existing wastewater infrastructure was carried out. Variants of application were defined and material recovery options for the developed catalysts were discussed.

2. METHODS

2.1. Chemicals and materials

Only chemicals with analytical purity grade were used. The calibration substances for LC–MS/MS analysis were purchased from Sigma-Aldrich. Bi-distilled water from a Milli Q system from Millipore was used for the eluents. The ammonium formiate for the eluents was purchased from Carl Roth. Titanium dioxide-coated expanded glass was used as the buoyant photocatalysts. The expanded glass granules used were produced by Liaver GmbH & Co KG and have an oval morphology with a grain size of 4–8 mm. Special appreciation lies in the excellent properties of this material in terms of buoyancy and porosity. Expanded glass granulate is produced from regenerated broken glass and can be recycled itself. Thus, it not only represents a highly resource-friendly substrate material for photocatalytic coating but also expands the range of applications for such coatings through the property of floatability. The total titanium dioxide content of the material used was 2.5 wt%. The titanium dioxide nanoparticles had a particle size of 14 nm and a Brunauer–Emmett–Teller (BET) surface area of 50 m²/g. The titanium dioxide particles were coated on the carrier material using an aqueous suspension including a small amount of an inorganic binder, a rotating plate and intermediate hot air drying at about 150 °C. The material and the coating process are based on the development of Lynatox GmbH. The scanning electron microscopy (SEM) images (Figure 1) show the uncoated and coated expanded glass spheres at different magnifications. Here, a very porous surface of the coated materials can be seen. The secondary effluent used in the experiments was obtained from the effluent of the Weimar-Tiefurt wastewater treatment plant immediately prior to the experiments.

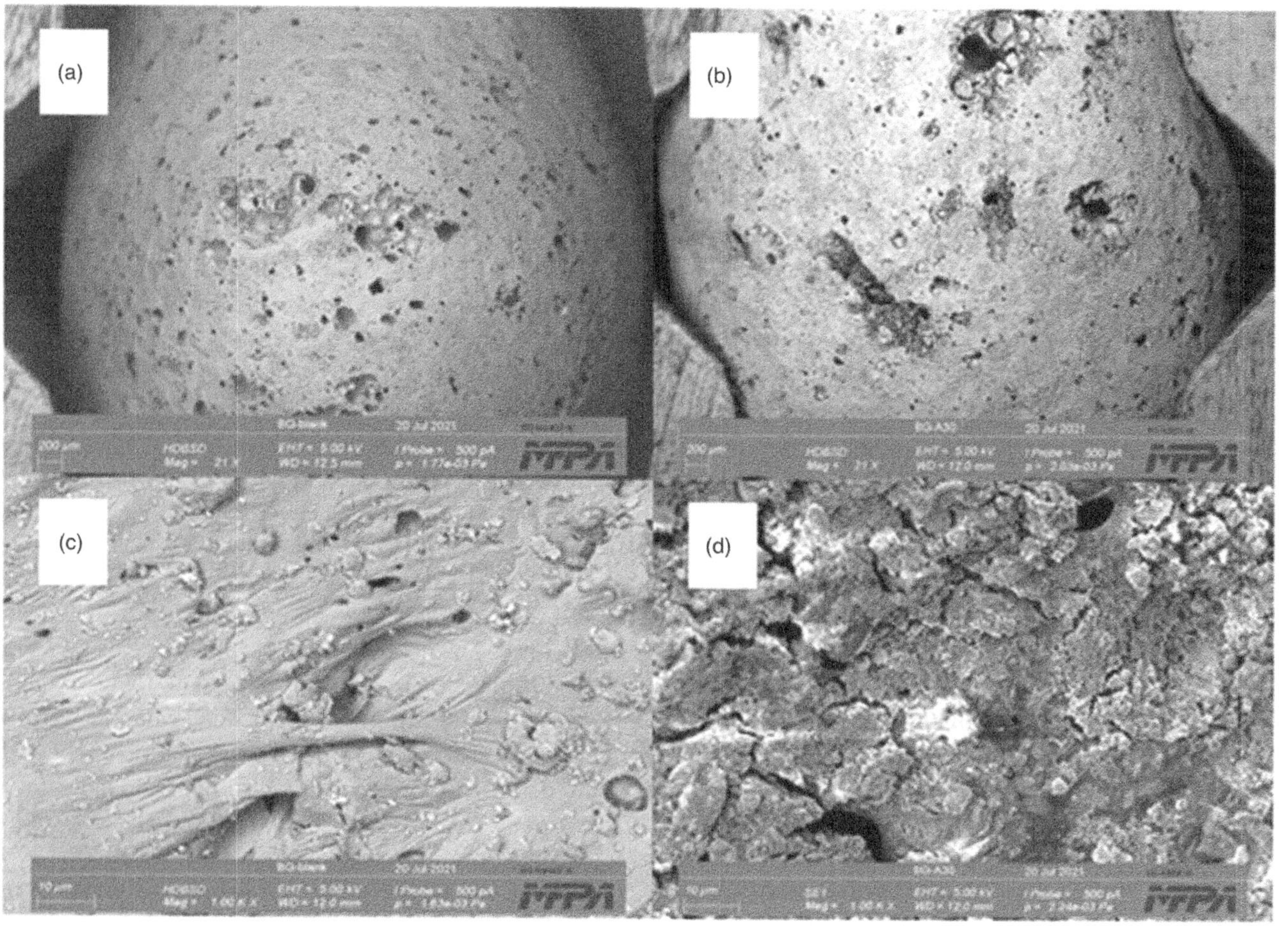

Figure 1 | Expanded glass uncoated ×21 magnification (a); expanded glass coated ×21 magnification (b); expanded glass uncoated ×1,000 magnification (c); and expanded glass coated ×1,000 magnification (d).

2.2. Experimental setup and procedure

The experiments on solar catalytic degradation were carried out in a self-designed batch reactor. The reactor consists of an acrylic glass cylinder with a diameter of 30 cm and a height of 15 cm. It has a hydraulic volume of 8 L and a solar-illuminated surface area of 706 cm^2, which was completely covered by one layer of the catalyst. The reactor liquid phase volume was completely mixed by a 5 cm magnetic stirrer at approximately 300 rpm during the whole test interval. The experimental unit is equipped with three UV sensors (Adafruit S12SD) with a precalibration for a wavelength from 240 to 370 nm and an output of radiation power in mW/m^2. The sensor has a photocurrent dependent on the wavelength of the abovementioned range. There is a linear increase in responsivity from 240 to 360 nm. From 360 nm, the responsivity decreases rapidly and reaches 0 at 370 nm. A calibration was performed according to the sensor datasheet, assuming a linear correlation to solar UV-A + B radiation power and UV index to determine the erythema-effective radiation. The type of radiation measurement as erythema-effective radiation was chosen to allow easy comparability of experimental conditions and the use of existing large data sets for extrapolation. For health protection reasons, for example, there are 33 permanent measuring stations in Germany whose measurement series go back as far as 1993 (Schmalwieser *et al.* 2017). The online measurement data were collected via a serial interface and Arduino Uno and finally transferred to a Raspberry pi for storage and distribution. The experiments were performed with natural sunlight (autumn sun in October at 50°5506N 11°1245E) and photocatalysts of identical composition. In the experiments, two reactors were operated simultaneously and 1 ml of sample was taken from the batch volume every 30 min. For every experiment, a new batch of photocatalysts was used. The samples were stored dark and cooled and were analyzed within 2 weeks. All measurement data are presented in the appendix. The schematic layout of the experimental setup is shown in Figure 2.

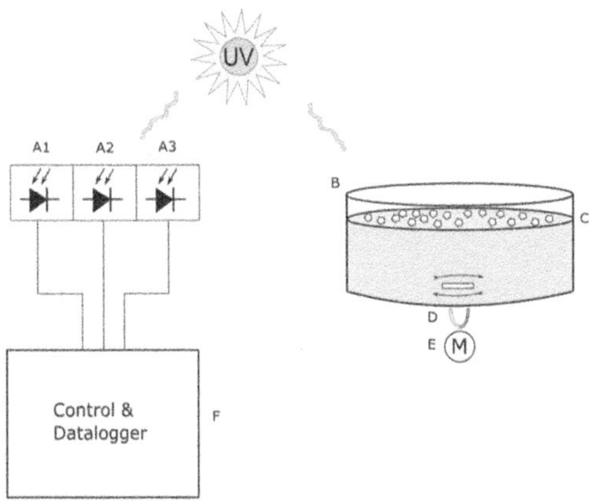

Figure 2 | Setup of the experiment: A1–A3: UV sensors; B: batch reactor; C: coated floatable glass catalysts; D and E: magnetic stirrer; F: Arduino data recording.

2.3. Analytical methods

The trace analytical determination of the micro-pollutants was performed by a triple quadrupole mass spectrometer (API 4000: Sciex) coupled with high-performance liquid chromatography (Dionex R3000) without enrichment directly from the sampled and filtered wastewater. The analytical separation column used was a Kinetex EVO C18 column from Phenomenex with a particle size of 2.5 μm, a length of 100 mm and a diameter of 2.1 mm. Acetonitrile and water with 20 mmol of ammonium formate as buffer were used as running medium. The gradient ran from 30% acetonitrile to 95% in 10 min. Each sample was analyzed twice. A control standard with a concentration of 1 μg/l was measured every 10 measurements to verify the results.

2.4. Kinetic and modeling

Photocatalytic reactions can be described according to formula 1 using the Langmuir–Hinshelwood (L–H) model (Lin *et al.* 2009; Asenjo *et al.* 2013). According to this, the reaction rate can be calculated as following when used in multicomponent mixtures:

$$r = -\frac{dC}{dt} = k_{LH}\frac{K_L C_{eq}}{1 + K_L C_{eq} + \sum_{i=1}^{n} K_i C_i (i = 1, n)} \tag{1}$$

where r is the reaction rate, k_{LH} is the specific rate constant, K_L is the Langmuir constant, and C_{eq} is the liquid concentration. K_i and C_i are the Langmuir constant and the concentration of each substance within the solution. However, in a real wastewater matrix, usually, only summary measurements of the concentrations C_i are possible (e.g. total organic carbon (TOC) and chemical oxygen demand (COD)). Furthermore, measured micro-pollutant concentrations are usually lower than measured matrix concentrations by a factor of 1,000. Therefore, in many papers, a reaction rate following pseudo-first order is applied (Gaya & Abdullah 2008; Nan Chong *et al.* 2010). Here, the term $K_L C_{eq}/1 + K_L C_{eq} + \sum_{i=1}^{n} K_i C_i (i = 1, n)$ is primarily defined by the wastewater matrix ($K_i C_i$), so that the reaction rate can be simplified as follows:

$$r = -\frac{dC}{dt} = k_{LH} K_L C_{eq} = k C_{eq} \tag{2}$$

If the substance of interest has a term $K_L C_{eq}$, which is significantly smaller than the influence of the wastewater matrix, the adsorption term approaches 0. A limitation of the available adsorption sites is given and the reaction can be simplified as zero order (formula 3). Both limiting cases further assume that no significant change in matrix composition occurs during the treatment, which is practically only partially given:

$$r = -\frac{dC}{dt} = k \tag{3}$$

In both cases, the rate constant k depends on the reactor configuration and the generated radical concentrations. Solar photocatalysis is thus significantly influenced by the variation of solar radiation during the day. This leads to variations in the concentration of radicals, which, in consequence, leads to a variation of k in the daily cycle. For quantum efficiency, a reciprocal quadratic relationship to the radiation power is known. Due to the relatively low intensity of usable radiation quantities in a solar non-concentrated application, the influence of the quantum efficiency can be neglected (Gerischer 1995). Following this, a linear correlation can be assumed:

$$k_{LH} \sim P_{solar} \tag{4}$$

The experimentally determined reaction rates, however, are influenced not only by the concentration of radical species but also by the diffusive and/or convective mass transport of the pollutant to the catalyst. The mass transport itself is influenced by substance properties as well as the hydraulic behavior of the reactor and is therefore assumed to be constant over the reaction period. In total, mass transport reduces the liquid concentration at the catalyst and overlays existing reaction mechanisms. This results in a substance-specific and reactor-specific reaction behavior whose order must be determined experimentally. In this work, this is demonstrated by correlating solar irradiation, as an essential variable operating parameter and the reaction constants of a zero- and first-order reaction.

For variable reactor geometries, the reaction rate is further determined by the available photocatalytic surface area. Similar to the simplification of the reaction order, a linear relationship between the volume-related reaction rate and the ratio of the surface area of the catalyst to the basin volume can be assumed as shown in the following equation:

$$\frac{r}{C_{eq}} = k \sim \frac{A_{catalyst}}{V_{reactor}} \tag{5}$$

Here, $A_{catalyst}$ is the available base surface area and $V_{reactor}$ the reactor volume. The analysis was performed using the open-source software R. First, the mean value and standard deviation were calculated from the individual measuring points of the concentrations and stored with the corresponding datetime values. The time profile of the reaction rates was calculated using the FDA function of the Doremi package, which is described in Chow *et al.* (2016). The order-dependent reaction constant k was calculated as follows:

$$r = C^n k \tag{6}$$

Here, r is the reaction rate, C is the measured micro-pollutant concentration, and n is the reaction order. The solar irradiance is included in the correlation as the mean value of the individual measurements over the sampling interval of the micro-pollutants. The following correlation was performed:

$$k_1 \sim A + B * P_{solar,mean} \tag{7}$$

The extrapolation of average purification potentials for the catalysts was calculated based on formula 5 as follows:

$$n_i = 1 - \exp\left(-k_i * HRT_i * \frac{H_{exp}}{H_i} * n_{solar}\right) \tag{8}$$

Here, k_i is the first-order reaction rate, HRT_i is the hydraulic retention time of the considered reactor, H_{exp} is the pool depth of the experimental reactor, H_i is the depth of the considered reactor, and n_{solar} is the fraction of time with solar irradiation. A plug-flow behavior of the reactor and a continuous and constant influx is assumed.

3. RESULTS AND DISCUSSION

3.1. Experimental results

The measured concentrations of micro-pollutants including error bars over the treatment period are shown in Figure 3. For amisulpride and diclofenac, the total degradation measured was approximately 90%. For citalopram, metoprolol, and venlafaxine, the degradation total was approximately 55%. The remaining micro-pollutants such as benzotriazole, candesartan, carbamazepine, and 1-methylbenzotriazole were reduced in concentration by 35–55%. The concentration course of carbamazepine, which initially appears to plateau at the beginning of the experiment, is notable. This delay in degradation could indicate slower diffusion of this substance due to lower adsorption capacities. A substance-specific affinity for photocatalytic degradation was found, whereby three groups can be formed based on the measurement results:

- high degradation affinity (amisulpride, diclofenac)
- medium degradation affinity (citalopram, metoprolol, venlafaxine)
- low degradation affinity (benzotriazole, candesartan, carbamazepine, 1-methylbenzotraizole)

For the experiment on photocatalytic treatment of wastewater treatment plant effluent, the time course of solar radiation during the day is shown in Figure 4(a). Low fluctuations of irradiation were measured around noon due to cloud coverage. The values expressed as UV index reach a maximum of about 1.5, which is a common value for this season and latitude. For Germany, UV indices of about 7–8 are achieved throughout the year (BFS 2022). Compared to previous works in Rueda-Márquez *et al.* (2020b), the radiation intensity was therefore relatively low.

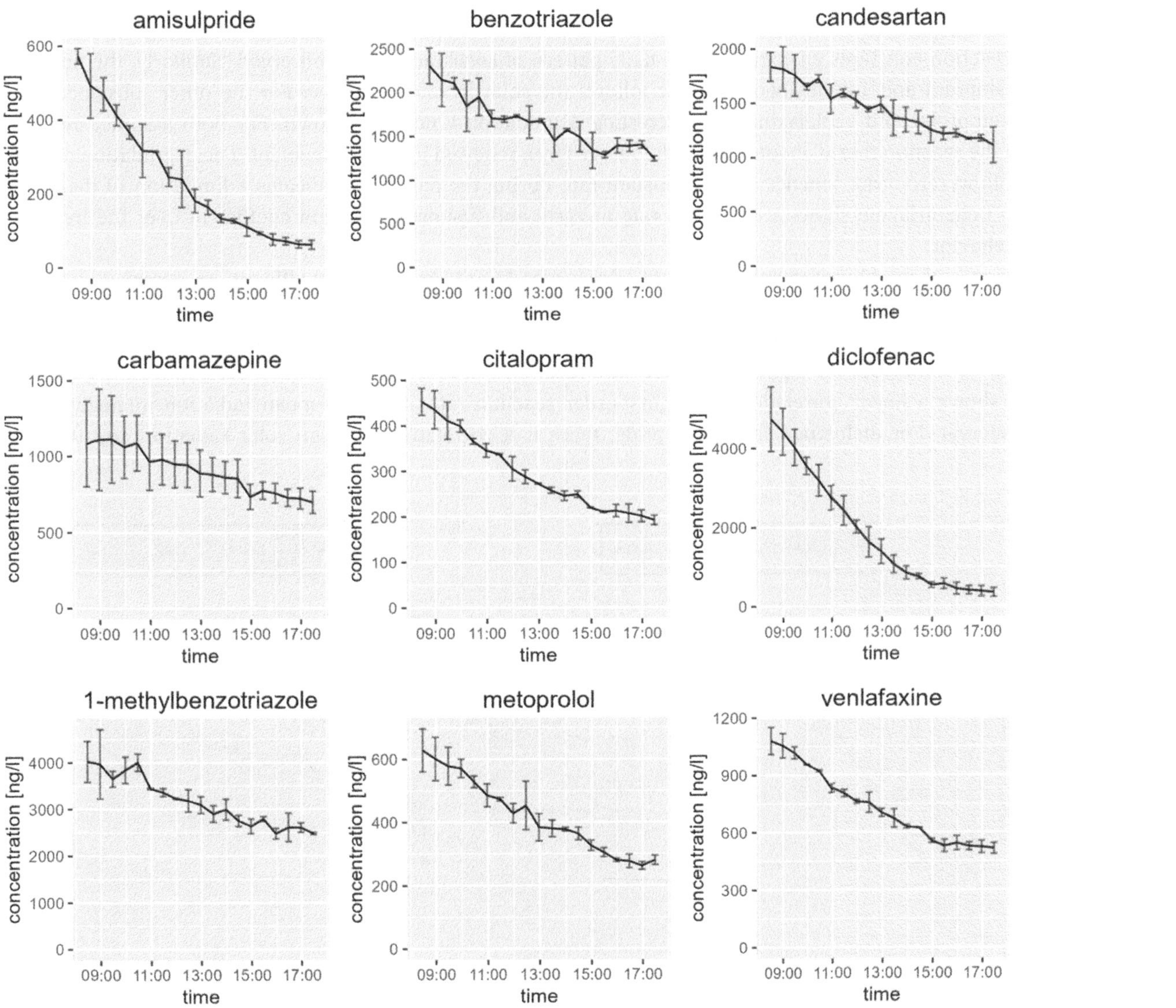

Figure 3 | Mean and standard deviation of measured micro-pollutant concentrations in relation to the time of day for solar photocatalytic treatment of secondary effluent.

The mean decrease in micro-pollutant concentration is shown in Figure 4(b) including the mean standard deviation. Furthermore, the decreased rates of diclofenac and carbamazepine are shown as the substance-specific best- and worst-case. Over the 9-h treatment period, a mean decrease of approximately 55% was achieved. There is a successive decrease in the slope of the mean degradation graph indicating predominant first-order reaction kinetics. For diclofenac, as an example of a substance with low inhibition of the photocatalytic reaction, a 75% degradation rate can be achieved at a radiation dose of 100 mWh/m². Carbamazepine showed the lowest degradation rate with a decrease rate of 20% at a dose of 100 mWh/m². Here, a linear zero-order reaction rate is indicated by the linear slope of the graph. This strong convergence to a zero-order kinetic rate indicates an adsorption limitation, which is mainly influenced by catalyst and substance charges and the pH value.

Compared to He *et al.* (2016), overall lower degradation rates were measured. However, the treatment time of 96 h was significantly higher. Using the reported first-order reaction constants, a degradation rate of 45% for diclofenac and 8.5% for carbamazepine was achieved after 6 h of treatment time. Compared to Rueda-Márquez *et al.* (2020b), similar degradation rates for diclofenac and carbamazepine were achieved after 6 h of treatment. The measured experimental erythema-effective radiation seems significantly lower than the UV-dose reported by Rueda-Márquez *et al.* (2020b); however, without the knowledge of the measured solar spectra, both experiments cannot be compared quantitatively.

The time profile of the kinetic reaction constants of the zero and first order of the micro-pollutants shows a substance-specific behavior analogous to the concentration profiles (Figure 5). For substances of affinity group 1, a pattern of the first-order reaction constants analogous to the daily pattern of solar radiation is observed. Similarly, the first-order reaction rate for citalopram and 1-methylbenzotriazole follows the solar irradiation curve. For the other substances of the affinity group 2 (metoprolol and venlafaxine), an interesting degradation kinetics is shown. For both assumed reaction orders, a decrease in the reaction rate at about 12 o'clock followed by a renewed increase is evident. The degradation of carbamazepine seems to follow zero-order kinetics with a constant rate during the day. This indicates a limitation of the reaction rate by adsorption. Benzotriazole shows a decreased rate in zero- and first-order reaction coefficient over the treatment period a deviating behavior.

Reaction rates of the photocatalytic treatment of micro-pollutants in secondary effluent are reported by He *et al.* (2016). Here, first-order constants of 2.83×10^{-5} to 4.17×10^{-6} 1/s were calculated for diclofenac and carbamazepine. Assuming a linear compensation of reactor depth (0.5–0.15 m), the reaction constants for diclofenac range from a 65% decrease to a maximum of 40% increase for higher irradiance. The reaction rates for carbamazepine were up to 145% higher.

Figure 6 shows a linear correlation of the first-order reaction rates with the mean radiation intensity of the pertinent sampling interval. For diclofenac and amisulpride, there is a correlation between solar radiation and the corresponding

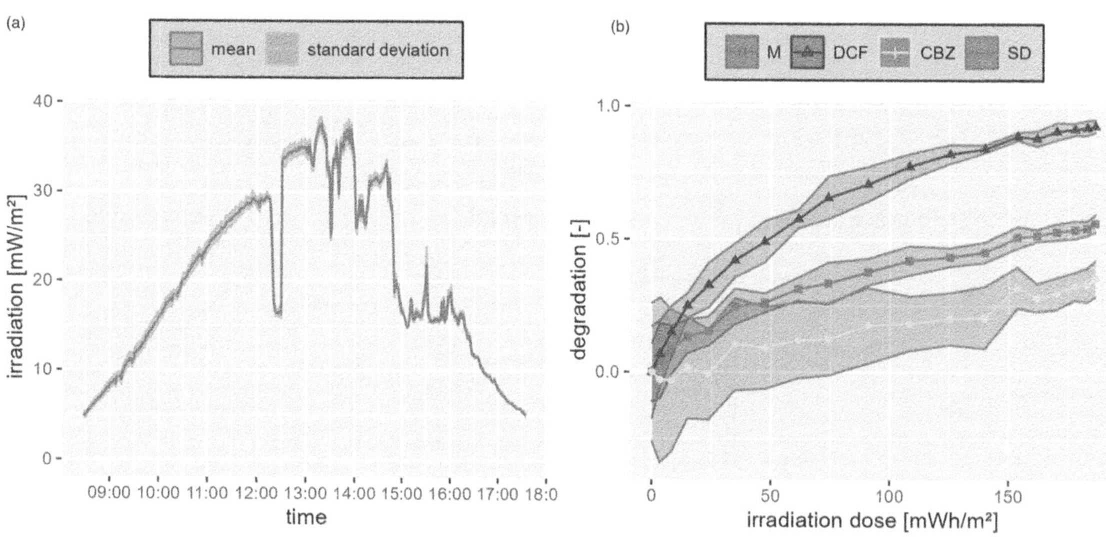

Figure 4 | Mean and standard deviation of erythema-effective solar UV irradiance during the treatment of secondary effluent (a) and micro-pollutant degradation rate in relation to irradiation doses (b) DCF: diclofenac; CBZ: carbamazepine; M: mean all substances; SD: specific standard deviation.

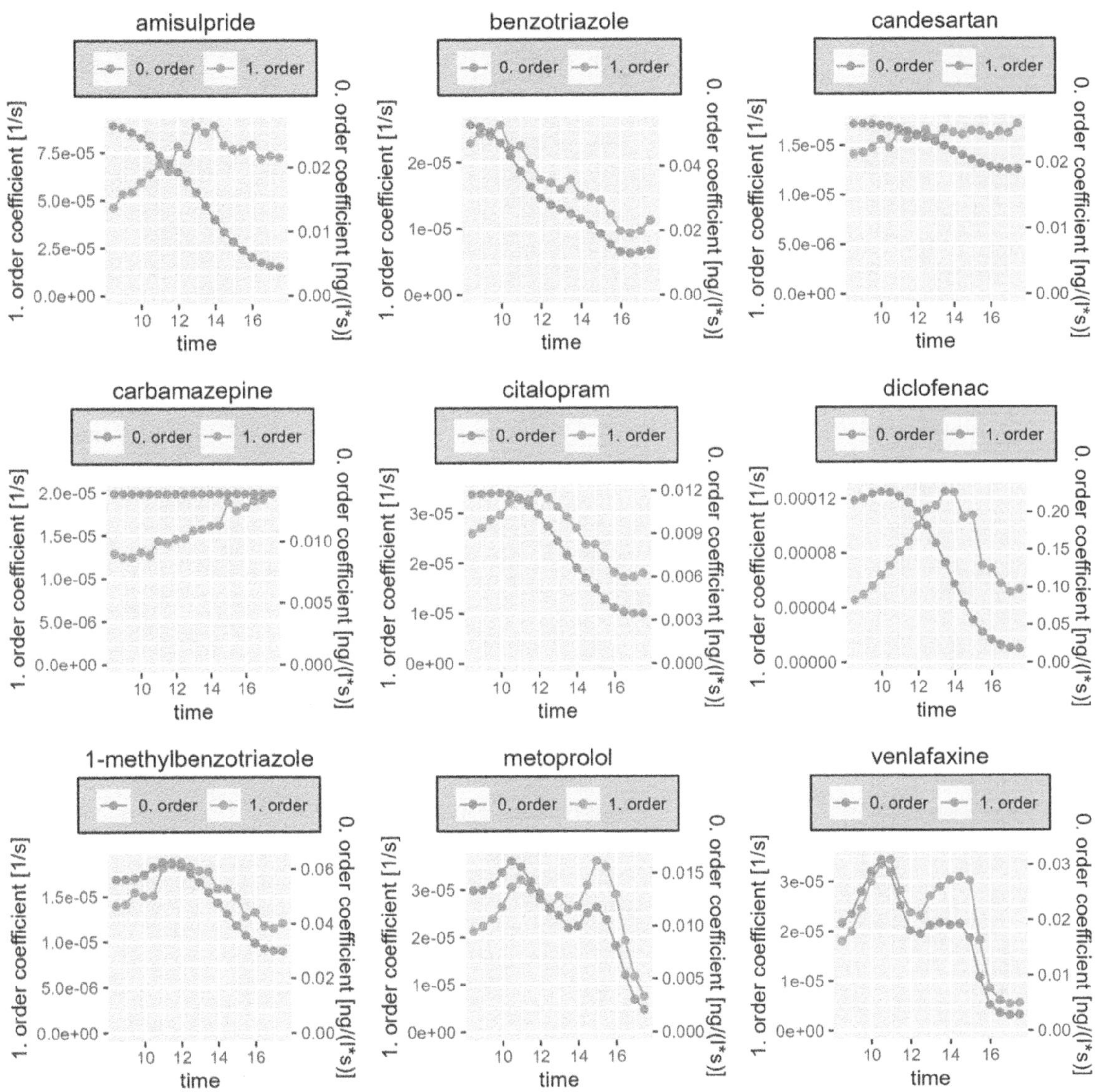

Figure 5 | Calculated zero- and first-order reaction rate coefficients in relation to the time of day for all measured micro-pollutants.

first-order reaction coefficient. Coefficients of determination of 0.73 and 0.48 are obtained. For amisulpride, this correlation appears to be superimposed by other influences, so that two linear curves are visible in the diagram. The substances of the moderate affinity group (citalopram, metoprolol, and venlafaxine) show a weak correlation between reaction rate and radiation intensity. A non-linear correlation is indicated, which also appears to be influenced by other factors. Coefficients of determination of 0.18–0.35 are achieved. Except for 1-methlybenzotriazole, the other micro-pollutants show no correlation between radiation intensity and degradation rate. The overall influence of measured UV radiation on the reaction rates seems moderate. Remarkably high reaction rates were observed, given the overall low UV radiation during the experiments. This behavior indicates a potential for solar photocatalysis for regions with overall lower solar radiation and a high proportion of diffusive radiation.

This substance-specific behavior, as well as the approximation of poorly degradable compounds to a zero-order reaction, can be explained by adsorption limitation using the L–H model. This also explains the poor correlations between the reaction rate and irradiance for micro-pollutants of moderate affinity. The two linear graphs that emerge are due to the uniformly distributed daily variation of solar irradiance. This results in similar irradiance values to which one reaction rate at high and one at low micro-pollutant concentration are attributed, respectively.

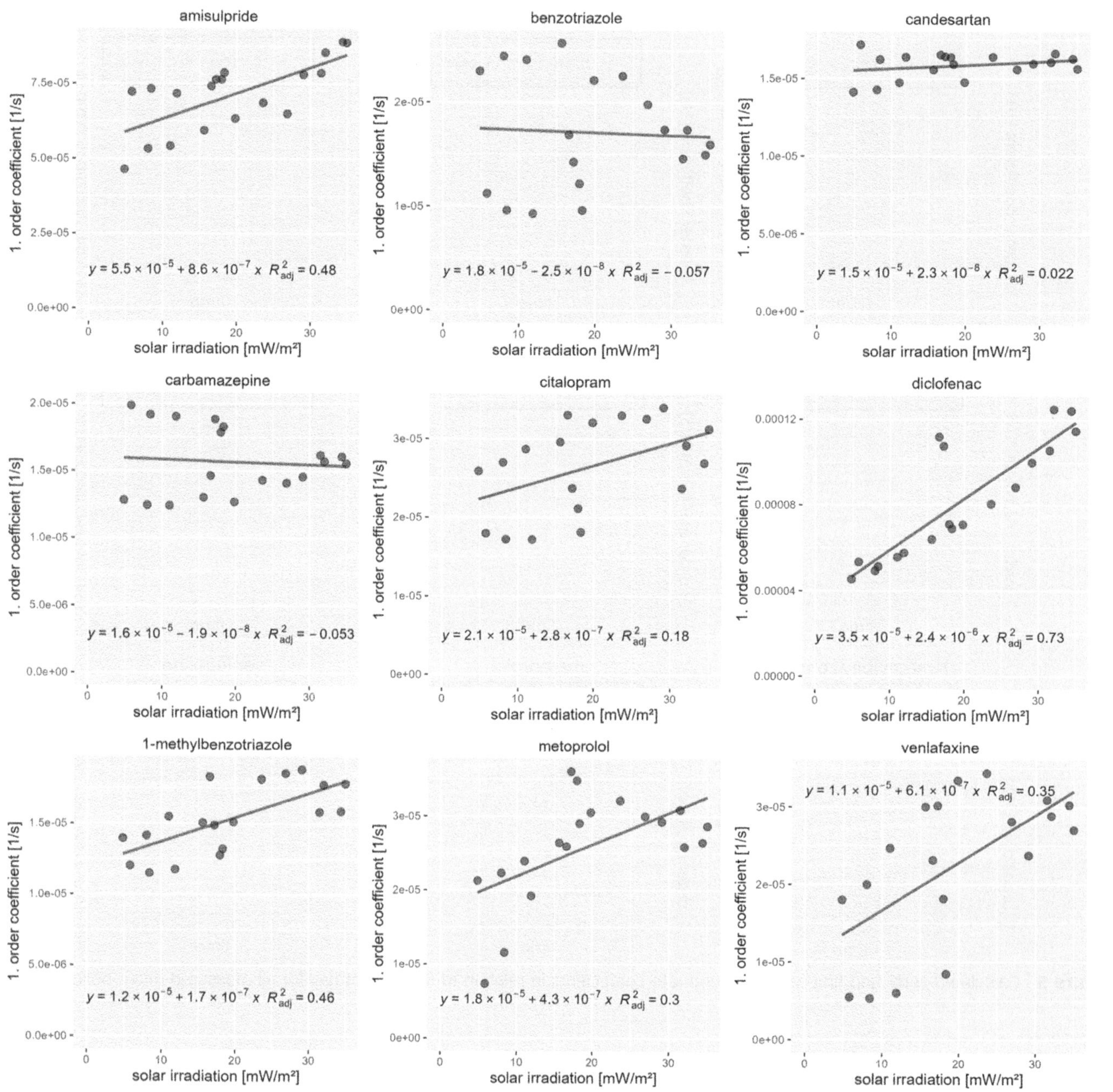

Figure 6 | Correlation of first-order reaction rate coefficient and solar irradiation for all investigated micro-pollutants.

Assuming a plug-flow behavior, which is a realistic assumption, especially for unmixed pond treatment systems, and similar degradation kinetics in all compartments, the achievable degradation rate is determined by the respective quotient of basin depth and the hydraulic retention time (here denoted vertical velocity) is shown in Figure 7. A proportion of annual hours with direct or diffusive irradiation of 0.5 was assumed. A mean 50% degradation cloud be achieved at a vertical velocity of 3×10^{-2} m/h (e.g. 33 h of hydraulic retention time with a 1.0 m reactor depth). When calculating the worst-case degradation with the lowest reaction rate constant (5×10^{-6} 1/s for venlafaxine at low radiation), an 50% degradation is achieved at a vertical velocity of 5×10^{-6} m/h (e.g. 20 days of hydraulic retention time with a 1.0 m reactor depth).

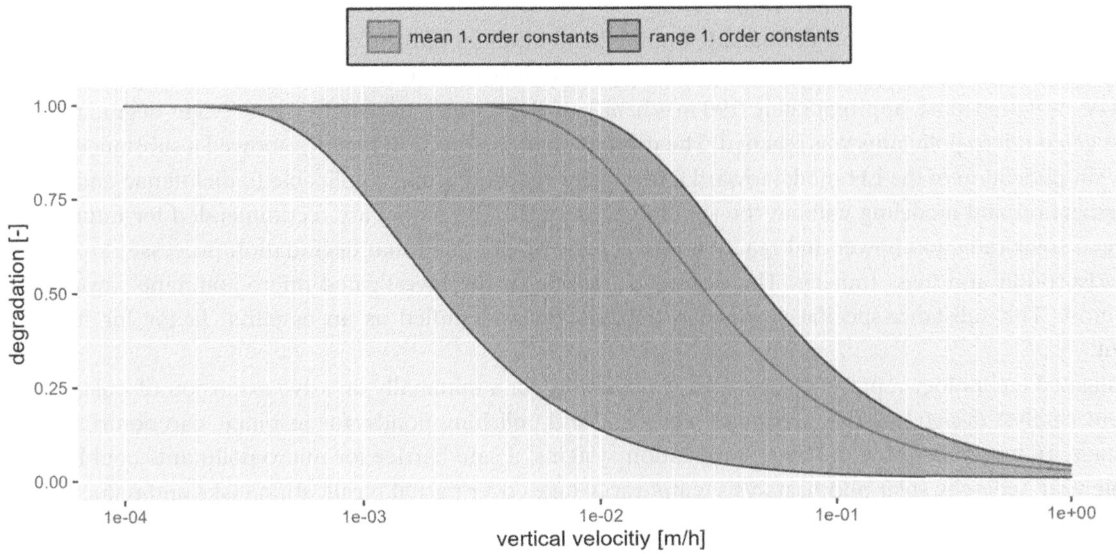

Figure 7 | Potential degradation following formula 8 as a function of the vertical velocity of the reactor and measured first-order reactions kinetics.

3.2. Discussion of application potentials

Following the results of the extrapolation in Figure 7, various applications in the field of wastewater treatment are plausible. The following discussion provides an initial assessment of the available measurement data in order to derive promising development potentials. Examples of integration could be small-scale, decentral wastewater treatment ponds or polishing ponds and secondary clarifiers at large-scale treatment plants. For unaerated wastewater treatment ponds, a capacity of 8–15 m^2/inhabitant and a pool depth of 1.0 m can be assumed. For this treatment method, a complete surface coverage is not possible due to the necessary oxygen production of algae and oxygen input via surface convection. A 50% coverage was assumed which would result in a mean vertical velocity of 3×10^{-3} m/h (DWA 2005). For polishing ponds, a hydraulic retention time of 48 h at 1.5 m pond depth and therefore a vertical velocity value of about 2×10^{-2} m/h are common (DWA 2005). The hydraulic retention time of the secondary clarifier of 9 h at 4.5 m depth (horizontal variant) can be estimated as a common value for big-scale applications (>100,000 inhabitants), following German normative regulations (DWA 2016). This results in a vertical velocity of about 5×10^{-1} m/h.

Concerning Figure 7, there is great potential for a use of this technology within wastewater treatment and polishing ponds. Existing treatment goals (in the German context) require an 80% degradation rate of specific indicator substances in relation to the raw wastewater influent concentrations (GSchV 2020). The degradation rate within the mechanical-biological stage of the wastewater treatment plant can be assumed to be 20% (Margot *et al.* 2015). Therefore, an 75% degradation rate within the photocatalytic treatment stage would be necessary to achieve compliance with existing regulations. For most examined reaction kinetics, an almost complete micro-pollutant degradation rate could be achieved with wastewater treatment ponds. Integration within a polishing pond could yield a 60% overall degradation rate for the micro-pollutants investigated. The use of secondary clarifiers could achieve a mean degradation rate of 2–3% with maximum values of up to 6%, which is an almost neglectable amount. When interpreting this extrapolation, it must be taken into account that the experiments were carried out at below-average UV radiation power. The exact behavior of this catalyst system at higher radiation powers is unknown, but an overall increase in reaction rates is likely. Consequently, the reaction rates achievable over the annual period are underestimated within this extrapolation. The reaction rates are also significantly influenced by matrix influences and the hydraulic behavior of the reactor. In summary, these influences lead to a reduction in the reaction rates. Especially for unaerated wastewater treatment ponds, a significant influence of partially degraded organic molecules within the reactor can be assumed. In summary, regarding the practical application of this catalyst system within wastewater treatment processes, there are positive and negative influences on expected whose quantification requires further experimental research.

4. CONCLUSIONS

Within this work, the suitability of novel floatable immobilized titanium dioxide catalysts for the decontamination of wastewater treatment plant effluents was demonstrated. A degradation was achieved for all nine investigated micro-pollutants. After 9 h (UV radiation dose approximately 190 mWh/m²) of wastewater treatment, an average degradation rate of 55% for all investigated micro-pollutants was reached. The evaluation of the reaction kinetics showed a substance-specific reaction behavior. A simplification of the LH model toward a first-order reaction is only applicable to diclofenac and only to a limited extent. Investigation and modeling without the simplification of the LH model are recommended for future work. A correlation between solar radiation power and reaction rate is present, provided that degradation processes are not significantly limited by adsorption and mass transfer. The degraded fraction of the investigated micro-pollutants varied by a factor of 3.5 at the most. The substance-specific adsorption behavior was identified as an essential factor for future application development.

The investigated technology represents a promising possibility for a minimally invasive extension of wastewater treatment. With efficient catalyst design, existing secondary clarifiers and polishing ponds, for instance, can contribute to the fourth treatment stage. In combination with active purification systems, a safe barrier for micro-pollutants could be implemented for the whole year, whereby solar photocatalysis temporarily takes over partial purification tasks and reduces total operating costs. Integration of these floating catalysts into existing wastewater treatment plants is primarily useful for components that have a free surface, a high surface-to-volume ratio and a low organic load of the wastewater. Possibilities of implication were estimated and discussed within this work, whereby possibilities arise for large-scale as well as decentral treatment plants.

Following this proof-of-concept further questions regarding the effects of photocatalytically produced radicals on biological treatment processes, the influence of dissolved organics and the mechanical stability and potential nano-particle emissions of developed coatings arise. There is initial data regarding the particle size of abraded coating. Sieve analyses of the abrasion of the coated granulate were carried out by IBU-tec advanced materials AG, which showed that the abraded particles occur on a microscale. Considering the widely known hazard potential of nanoscale titanium dioxide, this is of major importance.

DATA AVAILABILITY STATEMENT

All relevant data are included in the paper or its Supplementary Information.

CONFLICT OF INTEREST

The authors declare there is no conflict.

REFERENCES

Abdel-Maksoud, Y., Imam, E. & Ramadan, A. 2016 TiO_2 solar photocatalytic reactor systems: selection of reactor design for scale-up and commercialization – analytical review. *Catalysts* **6** (9), 138.

Aemig, Q., Hélias, A. & Patureau, D. 2021 Impact assessment of a large panel of organic and inorganic micropollutants released by wastewater treatment plants at the scale of France. *Water Research* **188**, 116524. doi:10.1016/j.watres.2020.116524.

Asenjo, N. G., Santamaría, R., Blanco, C., Granda, M., Álvarez, P. & Menéndez, R. 2013 Correct use of the Langmuir–Hinshelwood equation for proving the absence of a synergy effect in the photocatalytic degradation of phenol on a suspended mixture of titania and activated carbon. *Carbon* **55**, 62–69.

Awfa, D., Ateia, M., Fujii, M. & Yoshimura, C. 2020 Photocatalytic degradation of organic micropollutants: inhibition mechanisms by different fractions of natural organic matter. *Water Research* **174**, 115643. doi:10.1016/j.watres.2020.115643.

BFS 2022 *UV-Index Worldwide. German Federal Office for Radiation Protection (UV-Index weltweit. Bundesamt für Strahlenschutz).* Available from: https://www.bfs.de/DE/themen/opt/uv/uv-index/weltweit/weltweit.html; jsessionid=04E52C097C2754FFA69BDDDBA5B5F654.2_cid349 (last updated 2022, last checked on 01.05.2022).

Bourgin, M., Beck, B., Boehler, M., Borowska, E., Fleiner, J., Salhi, E. Teichler, R., von Gunten, U., Siegrist, H. & McArdell, C. 2018 Evaluation of a full-scale wastewater treatment plant upgraded with ozonation and biological post-treatments: abatement of micropollutants, formation of transformation products and oxidation by-products. *Water Research* **129**, 486–498. doi:10.1016/j.watres.2017.10.036.

Braham, R. J. & Harris, A. T. 2009 Review of major design and scale-up considerations for solar photocatalytic reactors. *Industrial and Engineering Chemistry Research* **48** (19), 8890–8905. doi:10.1021/ie900859z.

Chow, S.-M., Bendezú, J. J., Cole, P. M. & Ram, N. 2016 A comparison of two-stage approaches for fitting nonlinear ordinary differential equation models with mixed effects. *Multivariate Behavioral Research* **51** (2–3), 154–184. doi:10.1080/00273171.2015.1123138.

Długosz, M., Żmudzki, P., Kwiecień, A., Szczubiałka, K., Krzek, J. & Nowakowska, M. 2015 Photocatalytic degradation of sulfamethoxazole in aqueous solution using a floating TiO_2-expanded perlite photocatalyst. *Journal of Hazardous Materials* **298**, 146–153. doi:10.1016/j.jhazmat.2015.05.016.

DWA 2005 *Principles for the design, construction and operation of wastewater pond systems. (Grundsätze für Bemessung, Bau und Betrieb von Abwasserteichanlagen.)* August 2005. Hennef: German Association for Water Management Wastewater and Waste (DWA Worksheet, A 201).

DWA 2016 *Design of single-stage activated sludge plants. (Bemessung von einstufigen Belebungsanlagen.)* June 2016. Hennef: German Association for Water Management Wastewater and Waste (DWA Worksheet, A 131).

Gallé, T., Pittois, D., Bayerle, M. & Braun, C. 2019 An immission perspective of emerging micropollutant pressure in Luxembourgish surface waters: a simple evaluation scheme for wastewater impact assessment. *Environmental Pollution* **253**, 992–999. doi:10.1016/j.envpol.2019.07.080.

Gaya, U. I. & Abdullah, A. H. 2008 Heterogeneous photocatalytic degradation of organic contaminants over titanium dioxide: a review of fundamentals, progress and problems. *Journal of Photochemistry and Photobiology C: Photochemistry Reviews* **9** (1), 1–12. doi:10.1016/j.jphotochemrev.2007.12.003.

Gerischer, H. 1995 Photocatalysis in aqueous solution with small TiO_2 particles and the dependence of the quantum yield on particle size and light intensity. *Electrochimica Acta* **40** (10), 1277–1281. doi:10.1016/0013-4686(95)00058-M.

GSchV 2020 *Gewässerschutzverordnung (Water Protection Ordinance) Switzerland.* Available from: https://fedlex.data.admin.ch/filestore/fedlex.data.admin.ch/eli/cc/1998/2863_2863_2863/20210101/en/pdf-a/fedlex-data-admin-ch-eli-cc-1998-2863_2863_2863-20210101-en-pdf-a-1.pdf (accessed 1 November 2022).

Guillossou, R., Le Roux, J., Mailler, R., Vulliet, E., Morlay, C., Nauleau, F. Gaspéri, J. & Rocher, V. 2019 Organic micropollutants in a large wastewater treatment plant: what are the benefits of an advanced treatment by activated carbon adsorption in comparison to conventional treatment? *Chemosphere* **218**, 1050–1060. doi:10.1016/j.chemosphere.2018.11.182.

Guillossou, R., Le Roux, J., Brosillon, S., Mailler, R., Vulliet, E. & Morlay, C., Nauleau, F., Rocher, V. & Gaspéri, J. 2020 Benefits of ozonation before activated carbon adsorption for the removal of organic micropollutants from wastewater effluents. *Chemosphere* **245**, 125530. doi:10.1016/j.chemosphere.2019.125530.

Gunten, U. 2003 Ozonation of drinking water: part II. Disinfection and by-product formation in presence of bromide, iodide or chlorine. *Water Research* **37** (7), 1469–1487. doi:10.1016/S0043-1354(02)00458-X.

He, Y., Sutton, N. B., Rijnaarts, H. H. H. & Langenhoff, A. A. M. 2016 Degradation of pharmaceuticals in wastewater using immobilized TiO_2 photocatalysis under simulated solar irradiation. *Applied Catalysis B: Environmental* **182**, 132–141. doi:10.1016/j.apcatb.2015.09.015.

Kanakaraju, D., Glass, B. D. & Oelgemöller, M. 2018 Advanced oxidation process-mediated removal of pharmaceuticals from water: a review. *Journal of Environmental Management* **219**, 189–207.

Kanaujiya, D. K., Paul, T., Sinharoy, A. & Pakshirajan, K. 2019 Biological treatment processes for the removal of organic micropollutants from wastewater: a review. *Current Pollution Reports* **5** (3), 112–128. doi:10.1007/s40726-019-00110-x.

Kisch, H. 2015 *Semiconductor Photocatalysis – Principles and Applications.* Hoboken: Wiley.

Lin, Y., Ferronato, C., Deng, N., Wu, F. & Chovelon, J.-M. 2009 Photocatalytic degradation of methylparaben by TiO_2: multivariable experimental design and mechanism. *Applied Catalysis B: Environmental* **88** (1–2), 32–41.

Margot, J., Rossi, L., Barry, D. A. & Holliger, C. 2015 A review of the fate of micropollutants in wastewater treatment plants. *WIREs Water* **2** (5), 457–487.

Moreira, N. F. F., Narciso-da-Rocha, C., Polo-López, M. I., Pastrana-Martínez, L. M., Faria, J. L., Manaia, C. M. Fernández-Ibáñez P., Nunez O. C. & Silva A. M. T. 2018 Solar treatment (H_2O_2, TiO_2-P25 and GO-TiO_2 photocatalysis, photo-Fenton) of organic micropollutants, human pathogen indicators, antibiotic resistant bacteria and related genes in urban wastewater. *Water Research* **135**, 195–206. doi:10.1016/j.watres.2018.01.064.

Nan Chong, M., Jin, B., Chow, C. W. K. & Saint, C. 2010 Recent developments in photocatalytic water treatment technology: a review. *Water Research* **44** (44/2010), 2997–3027.

Nasir, A. M., Jaafar, J., Aziz, F., Yusof, N., Salleh, W. N. W., Ismail, A. F. & Aziz, M. 2020 A review on floating nanocomposite photocatalyst: fabrication and applications for wastewater treatment. *Journal of Water Process Engineering* **36**, 101300. doi:10.1016/j.jwpe.2020.101300.

Pitre, S. P., Yoon, T. P. & Scaiano, J. C. 2017 Titanium dioxide visible light photocatalysis: surface association enables photocatalysis with visible light irradiation. *Chemical Communications* **53** (31), 4335–4338. doi:10.1039/C7CC01952A.

Prieto-Rodriguez, L., Miralles-Cuevas, S., Oller, I., Agüera, A., Li Puma, G. & Malato, S. 2012 Treatment of emerging contaminants in wastewater treatment plants (WWTP) effluents by solar photocatalysis using low TiO_2 concentrations. *Journal of Hazardous Materials* **211–212**, 131–137. doi:10.1016/j.jhazmat.2011.09.008.

Rueda-Marquez, J. J., Levchuk, I., Fernández Ibañez, P. & Sillanpää, M. 2020a A critical review on application of photocatalysis for toxicity reduction of real wastewaters. *Journal of Cleaner Production* **258**, 120694. doi:10.1016/j.jclepro.2020.120694.

Rueda-Márquez, J. J., Palacios-Villarreal, C., Manzano, M., Blanco, E., Del Ramírez Solar, M. & Levchuk, I. 2020b Photocatalytic degradation of pharmaceutically active compounds (PhACs) in urban wastewater treatment plants effluents under controlled and natural solar irradiation using immobilized TiO_2. *Solar Energy* **208**, 480–492. doi:10.1016/j.solener.2020.08.028.

Salaeh, S., Juretic Perisic, D., Biosic, M., Kusic, H., Babic, S., Lavrencic, S. U. Dionysiou, D. D. & Bozic, A. L. 2016 Diclofenac removal by simulated solar assisted photocatalysis using TiO_2-based zeolite catalyst; mechanisms, pathways and environmental aspects. *Chemical Engineering Journal* **304**, 289–302. doi:10.1016/j.cej.2016.06.083.

Schmalwieser, A. W., Gröbner, J., Blumthaler, M., Klotz, B., Backer, H. d., Bolsée, D. Werner, R., Tomsic, D., Metelka, L., Eriksen, P., Jepsen, N., Aun, M., Heikkilä, A., Duprat, T., Sandmann, H., Weiss, T., Bais, A., Toth, Z., Siani, A.-M., Vaccaro, L., Diémoz, H., Grifoni, D., Zipoli, G., Lorenzetto, G., Petkov, B. H., Di Sarra, A. G., Massen, F., Yousif, C., Aculinin, A. A., Outer, P. den, Svendby, T., Dahlback, A., Johnsen, B., Biszczuk-Jakubowska, J., Krzyscin, J., Henriques, D., Chubarova, N., Kolarž, P., Mijatovic, Z., Groselj, D., Pribullova, A., Gonzales, J. R. M., Bilbao, J., Guerrero, J. M. V., Serrano, A., Andersson, S., Vuilleumier, L., Webb, A. & O'Hagan, J. 2017 UV index monitoring in Europe. *Photochemical & Photobiological Sciences* **16** (9), 1349–1370. doi:10.1039/C7PP00178A.

Sundar, K. P. & Kanmani, S. 2020 Progression of photocatalytic reactors and it's comparison: a review. *Chemical Engineering Research and Design* **154**, 135–150. doi:10.1016/j.cherd.2019.11.035.

Teichgräber, B., Jagemann, P., Hetschel, M., Bechtel, A. & Phan, L.-C. 2021 A module-based approach for elimination of organic micropollutants at wastewater treatment plants. *Water Science and Technology* **84** (2), 364–373. doi:10.2166/wst.2021.029.

First received 23 September 2022; accepted in revised form 19 February 2023. Available online 6 March 2023

doi: 10.2166/wst.2022.395

Extraction of iron from laterite soil and green synthesis of laterite nano iron catalyst (GLaNICs) for its application as Fenton's catalyst in the degradation of triclosan

Rashmishree K N*, Bhaskar S, Shrihari S and Arun Kumar Thalla

Department of Civil Engineering, National Institute of Technology Karnataka, Surathkal, P.O. Srinivasnagar, Mangalore 575025, India
*Corresponding author. E-mail: rashu.kn87@gmail.com

RK, 0000-0003-4557-7180

ABSTRACT

Laterite based nano iron particles were synthesized using natural laterite extract as a precursor and *Psidium guajava* plant extract for its application as Fenton's catalyst in the degradation of triclosan. Chemical digestion method was used for the extraction of iron from laterite soil. Synthesized nano iron catalyst was characterized using SEM-EDS, XRD and FTIR and evaluated for its catalytic application in the Fenton's oxidation of triclosan. Maximum triclosan degradation of 69.5% was observed with nano iron catalyst dosage of 0.1 g/L and hydrogen peroxide dosage of 200 mg/L at acidic pH of 3. Hydrogen peroxide influence on the process was observed with Fenton's oxidation. Role of iron in the process has been accessed by control experiment with no nano catalyst addition in which degradation is considerably low. Fenton's oxidation was compared with conventional Fenton's oxidation driven by a green nano iron catalyst. Study claims the usage of natural laterite iron as a replacement for commercial iron in Fenton's degradation of triclosan. Regeneration and reusability studies on catalyst were studied and synthesized catalyst was observed to be reusable in three consecutive cycles. Degradation of triclosan in Fenton's oxidation follows pseudo-second order reaction with linear fit.

Key words: Fenton's process, laterite, PCPs, *Psidium guajava*, triclosan

HIGHLIGHTS

- Iron extraction from laterite soil.
- Green synthesis of nanoparticles.
- Characterization of synthesized nanoparticles.
- Fenton's oxidation of triclosan.
- Pseudo-second order kinectics.

1. INTRODUCTION

Personal care products (PCP) are products used for the external application on body parts includes a wide variety of cosmetics and oral hygiene materials. PCPs differ from pharmaceutical products by their nature and production process. Some of the PCPs includes underarm deodorants, deodorant soaps, and dermatological and topical preparations for protection of skin (Bhargava & Leonard 1996). Current usage of personal care products has spiked up exponentially in past 50 years. Production of these PCPs includes many natural and synthetic chemicals among which triclosan has found its part exhibiting unique biocidal characteristics. Triclosan (2,4,4′-trichloro-2′-hydroxydiphenyl ether) is a synthetic, nonionic, broad spectrum antibiotic and antimicrobial agent that has been used extensively in personal care products, including underarm deodorants, deodorant soaps, livestock productivity and dermatological and topical preparations for protection of skin (Bhargava & Leonard 1996; Pragada & Thalla 2021). The mild nature of triclosan compared to alcohol, chlorhexidine gluconate, and para-chloro-meta-xylenol and greater sustained activity makes it one of the prime materials in the manufacturing of most of the PCPs. Despite of its usage, triclosan has shown a greater degree of toxicity with lethal dosage of 19–29 mg/kg for mice and rats. Clinical pathological studies have also shown reductions in red blood cell count, hemoglobin, and hematocrit in females and clotting time was increased in males (Bhargava & Leonard 1996). PCP industry wastewater contains many contaminants which are persistent to the degradation and resistant to biotransformation and accumulate in the environment if discharged untreated (Thompson *et al.* 2005; Adhikari *et al.* 2022; Bakare & Adeyinka 2022). Collection of pharmaceutical and PCP industries are sewerage systems.

Resistance to biotransformation made triclosan unsuitable for biological treatment. Lépesová *et al.* (2019), identified bacterial strain such as *Citrobacter freundii* and *Serratia spp.* known for its triclosan resistance in sewage sludge rendering resistance to biotransformation (Lépesová & Krahulcová 2019). Adsorption is known to remove triclosan effectively with its limitations in the maintenance and regeneration of adsorbent. Conventional treatment process like activated sludge process yields more satisfactory results in the triclosan removal than trickling filters (Heidler & Halden 2007; Nandikes *et al.* 2022). However, chemical treatment like advanced oxidation process is most advanced technologies in treating organic compounds. Despite its limitations, such as low mineralization with ozonation, formation of halogenated compounds with chlorination, and high energy consumption with electro Fenton's oxidation, it is proven to be highly efficient in triclosan removal (Luo *et al.* 2019). Divalent catalyst iron helps in the faster dissociation of hydrogen peroxide into hydroxyl radicals acting as a catalyst in the process. Investment on catalyst for its large-scale application and disposal of huge quantity of sludge generated marks the limitation of conventional Fenton's oxidation. To overcome the cost involved in the commercial iron purchase and to make the process a more cost-effective replacement for commercial iron has been studied by various researchers. Previous studies have claimed the application of iron extracted from natural laterite soil as a Fenton's catalyst in the degradation of organic compounds (Sangami & Manu 2019; Bhaskar *et al.* 2021). However, heterogeneous Fenton's oxidation driving by nano iron catalyst have proven efficient in the degradation of various organic contaminants (She 2010; Chen *et al.* 2017; Sangami & Manu 2018; Bhaskar *et al.* 2020).

Laterite soil by its characteristic has 30–40% of iron which makes at naturally available iron rich mineral. The investigation on usage of laterite iron as a catalyst in its natural form, iron leached from laterite soil by chemical method, bioleached laterite iron and nanoparticles synthesised from leached laterite iron confirms its role as a catalyst in the Fenton's oxidation (Sangami & Manu 2017; Bhaskar *et al.* 2021). *Psidium guajava,* a tropical plant extensively rich in phytochemicals such as alkaloids, flavonoids and polyphenols are used for the synthesis of nanoparticles. Polyphenols present in the plant extracts are proven to actively acts as capping agent around the particles reducing its size and increasing the surface area. Properties of nanoparticles like size, shape and surface area depends on the plant extract and the precursor used in the process of synthesis. Interaction of various organic functional groups with precursor used with the *Psidium gujava* makes it one of the prominent plants used in the green synthesis of wide variety of nanoparticles.

Potential application of nano catalyst synthesized using natural laterite iron extract as precursor are in need to be evaluated for its efficiency, cost-effectiveness and treatment time reduction. Nanoparticles synthesis using *Psidium guajava* leaf extract and laterite iron as percussor for its application in Fenton's degradation of triclosan has not been studied so far. In this view, the present study encompasses synthesis of green iron nano iron catalyst using laterite iron extract as a percussor and *Psidium guajava* extract for its application of in the degradation of triclosan.

2. MATERIALS AND METHODS

2.1. Materials

All the chemicals used for the synthesis of nanoparticles and triclosan removal were analytical grade reagents. Triclosan (99.5%) was procured from Sigma Aldrich. Sulphuric acid was procured from Merck India which is used to adjust the pH. Hydrochloric acid, hydrogen peroxide, sodium thiocyanate, potassium thiocyanate, titanium sulfate, 1–10 phenanthroline, ferrous ammonium sulfate, and sodium carbonate were purchased from Merck India.

The iron nanoparticles were synthesized from guava leaves. Instead of using iron salts, locally available laterite soil was used as a source of iron.

2.2. Extraction of iron from laterite soil

Iron laterite soil was extracted using chemical leaching method as described in Sangami & Manu (2017). One gram of laterite soil was weighed and mixed with 10 mL of concentrated HCl. The mixture was heated to dry and kept in an oven for 1 h to remove any organic matter if any. Cooled mixture at room temperature is again boiled on addition of 10 mL distilled water and 10 mL HCl for 1 min. Hot distilled water of about 20 mL was added when solution was boiling and cooled to room temperature. The solution was filtered using Whatman filter paper No. 42 and stored at 4 °C for further use. Iron content in the extracted solution was measured using UV-spectrophotometer (APHA Method 4500-F 1992).

2.3. Phytochemical synthesis of bioleached laterite iron nanoparticles

Chemically leached laterite iron solution was used as percussor for the preparation of nano iron catalyst using *Psidium guajava* leaf extract. Fresh leaves of *Psidium guajava* was collected, washed to remove any dirt particles with distilled water and dried in an oven at 40 °C for 2 days. Dried leaves were powdered and stored for the synthesis of nano catalyst. About 5 g of powdered *Psidium guajava* leaves was mixed with 100 mL distilled water and heated in a hot plate at 80 °C for 60 min. Solution was cooled and filtered using Whatman filtered paper 42 and stored for further use. Laterite iron and *Psidium guajava* leaf extract was mixed in a proportion of 1:1, filtered and stored in a moisture free container. Synthesized nanoparticles were characterized using SEM-EDS, XRD, BET and FTIR.

2.4. Fenton's oxidation of triclosan using green laterite nano iron catalyst (GLaNICs)

Fenton's oxidation of triclosan was carried out with initial of 2 mg/L (Sangami & Manu 2016, 2018). About 0.1 g of GLaNICs was added to solution containing triclosan taken in a conical flask. The solution was adjusted to pH 3 using 1 N H_2SO_4 and allowed for 10 min to ensure proper mixing. Experimentation was conducted with different dosages of H_2O_2 (100–400 mg/L). Samples were drawn at regular intervals for analysis. During sampling, each time 1 mL of sodium thiosulphate was added to arrest the reaction (Khan *et al.* 2009). Concentration of triclosan was measured using HPLC Agilent 1200 with a C18 reverse phase column (pore size 3.5 μm, 100 × 0.46 cm), using water and methanol (in the ratio of 58:42) as mobile phase injected with a flow rate of 1.0 mL/min employing diode array detector (DAD) (Anupama & Shrihari 2018; Bhaskar *et al.* 2019). Chemical oxygen demand (COD) was measured using calorimetric method (APHA Method 4500-F 1992). pH was measured using digital pH meter. Ferric iron was measured by potassium thiocyanate method using UV-Vis spectrophotometer (Mellon 1941). H_2O_2 was measured using UV-spectrophotometer (Eisenberg 1943). All the analyses were carried out in duplicates.

3. RESULTS AND DISCUSSION

3.1. Characterization of green laterite nano iron catalyst (GLaNICs)

Laterite iron extract was found to have an iron content of 5.02 g/L as total iron. Black color precipitation was observed on addition of laterite iron extract with *Psidium guajava* leaf extract at 1:1 ratio indicating the formation of nano iron particles. On morphological characterization formed iron nanoparticles were seen as agglomerated spherical particles arranged in group under scanning electron microscope (Figure 1). Table 1 represents the elemental composition of both raw laterite soil and formed nanoparticles. Raw laterite soil was found to have 32.3% of iron by weight. Aluminum, silica and oxygen accounts for other elements. Iron nanoparticles contains 27.6% iron by weight with potassium and chlorine being other elements in trace (Figure 2). Figure 3 dissipates XRD results with characteristics peaks of synthesised nanoparticles. Raw

Figure 1 | Scanning electron microscopic images of (a) raw laterite particles; (b) green laterite iron nano catalyst (GLaNICs) showing the morphological appearance.

Table 1 | Chemical composition of GLaNICs

Element	K	O	Al	Fe	Si	Cl
Raw laterite	–	55.0	8.9	32.3	3.8	–
GLaNICs	0.7	66.7	–	27.6	–	5.1

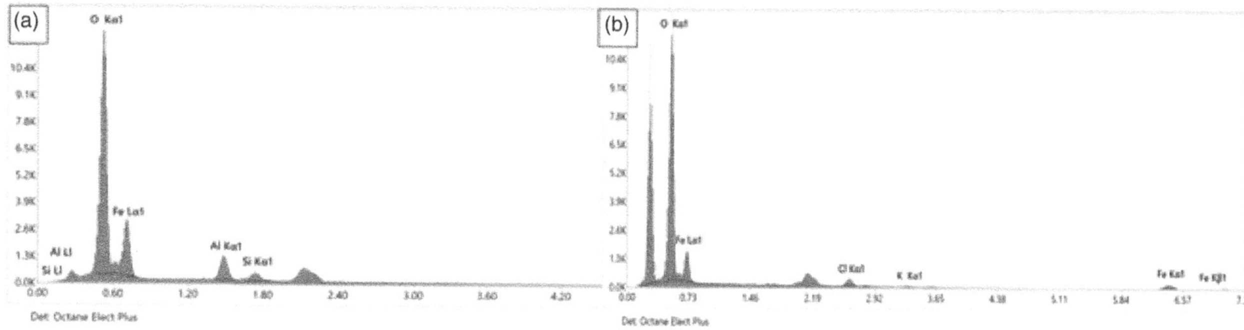

Figure 2 | EDS images of (a) raw laterite particles and (b) green laterite iron nano catalyst (GLINCs).

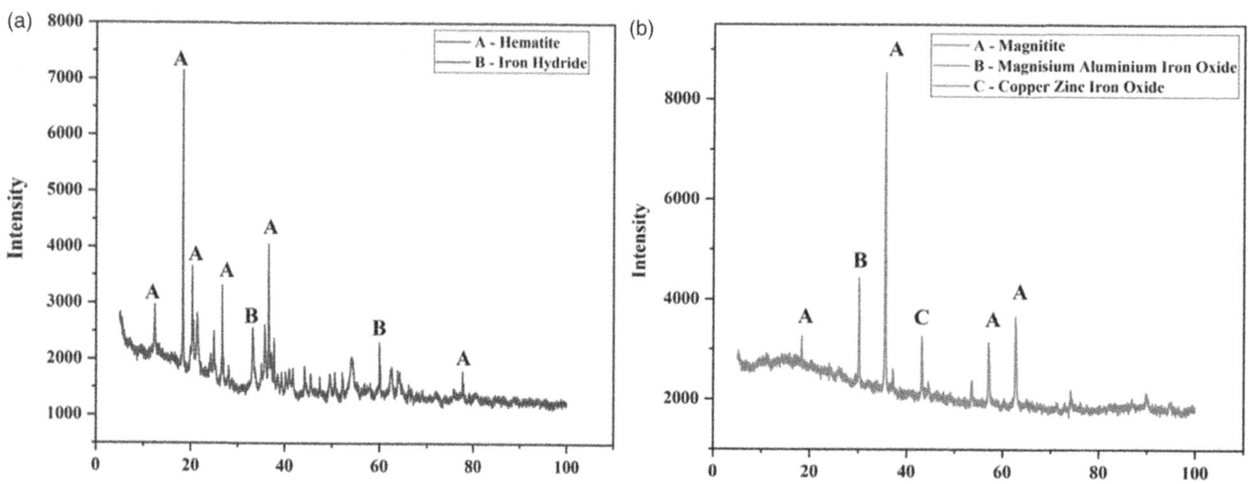

Figure 3 | XRD images of (a) raw laterite particles and (b) green laterite iron nano catalyst (GLaNICs) representing the corresponding peaks.

laterite soil exhibits six peaks at 2θ 12.401, 18.347, 21.448, 26.690, 35.7446 and 78.21 corresponds to iron oxide (PDF: 01-079-0007) and two peaks at 2θ 33.189 and 59.988 corresponds to iron hydride (PDF: 00-043-1321) while formed iron nanoparticles exhibits four peaks at 2θ 18.315, 35.487, 57.010 and 62. 624 corresponds to magnetite (PDF: 01-089-0691), one peak at 2θ 30.336 corresponds to magnesium aluminum iron oxide (PDF: 01-071-1234) and one peak at 2θ 43.116 corresponds to copper zinc iron oxide (PDF: 01-077-0012). Figure 4 shows FTIR spectrum of synthesized nanoparticles in which peaks at 3,323.22 corresponds to carbon to hydrogen stretching (C-H), 1,640.56 corresponds to carbon-to-carbon double bond stretching (C = C), 525.47 corresponds to iron to oxygen bonding (Fe-O) confirming the formation of iron oxide nanoparticles.

BET surface areas for BLFeNPs was found to be 11.707 m^2/g (Soon & Hameed 2011).

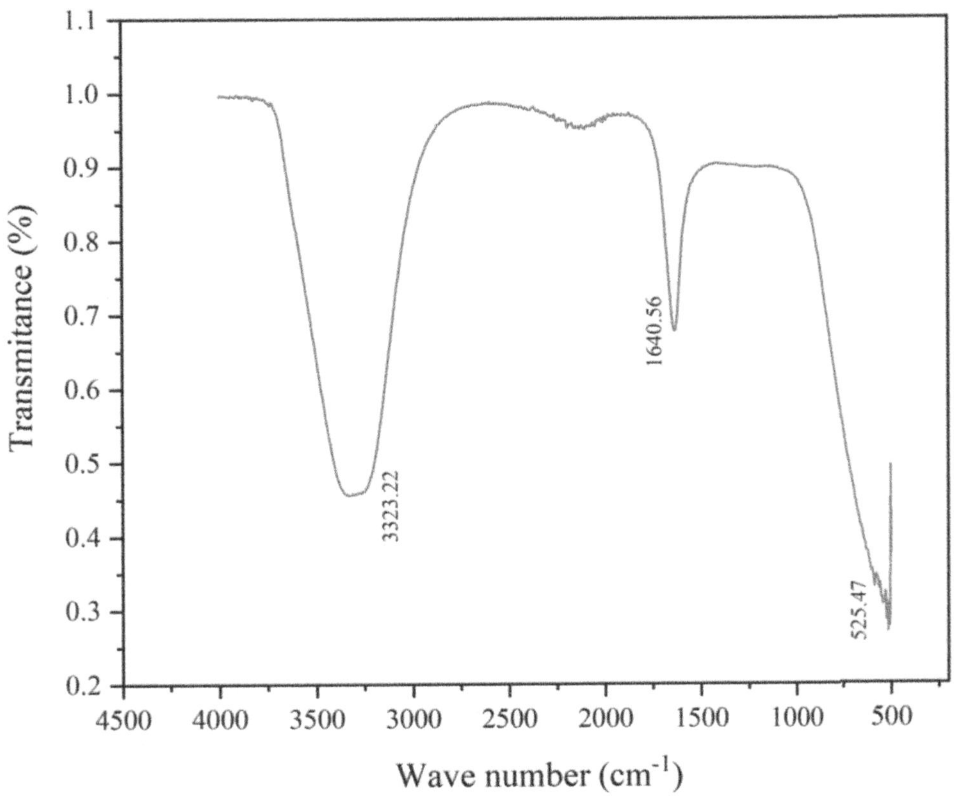

Figure 4 | FTIR spectrum of green laterite iron nano catalyst (GLaNICs).

3.2. Fenton's degradation of triclosan by green laterite iron nano catalyst (GLaNICs)

Degradation of triclosan was observed with the addition of catalyst varying different hydrogen peroxide dosage. High removal of triclosan was found to be 69.5% at catalyst to hydrogen peroxide ration 1:2 with the rate constant 0.0087/min. From the Figure 5(a) maximum degradation is observed within 20 min of treatment thereby the removal observed follows a constant rate. Increase in H_2O_2 dosage from 100 mg/L to 200 mg/L led to increase in removal efficiency by 29.5% with the rate constant doubled indicating the higher dissociation of H_2O_2, further increase to 300 mg/L, 400 mg/L led to drop in the rate of a reaction with a constant 0.0039/min indicating the scavenging effect. It is to be noted that treatment with nano catalyst time is reduced to half. Increase in catalyst load is more beneficial than increasing hydrogen peroxide since iron increases the speed of the reaction (Andrades *et al.* 2021). The rate of oxidation depends on the dissolution rate of ferrous ion that has leached out from the nanoparticles. Initial dissolution of ferrous ions is consumed and the rate of reaction will be more. It is the amount of catalyst addition indicates the adsorption part of heterogenous Fenton's oxidation that in turn contributes to maximum removal of triclosan in the process. Degradation of triclosan was observed within 60 minutes compare to Fenton's oxidation with commercial iron catalyst which indicates the reaction is heterogeneous involving adsorption of target compounds on the surface of nanoparticles. This is supported by previous studies (Chen *et al.* 2017; Bhaskar *et al.* 2019; Bhaskar *et al.* 2020). Maximum COD removal of 54.4% was observed with catalyst loading of 0.1 g/L and 200 mg/L H_2O_2 dosage while with H_2O_2 dosage of 100 mg/L, 300 mg/L and 400 mg/L COD removal observed was 38.99, 44.6 and 47.9%, respectively (Figure 5(b)). With the time increase in COD removal indicates better oxidation of triclosan and corresponding triclosan degradation stands in consistent with COD removal.

Sires *et al.* (2007) investigated suitability of electro-Fenton's oxidation in the degradation of both triclosan and triclocarban and identified oxalic acid, formic acid, maleic acid as degradation products with 2,4-dichlorophenol, 4-chlorocatechol and chlorohydroquinone as hydroxylated derivatives (Sir *et al.* 2007). Munoz *et al.* (2012) identified three major degradation compounds on Fenton's oxidation of triclosan and claims p-hydroquinone of triclosan forms when OH* radicals attack triclosan

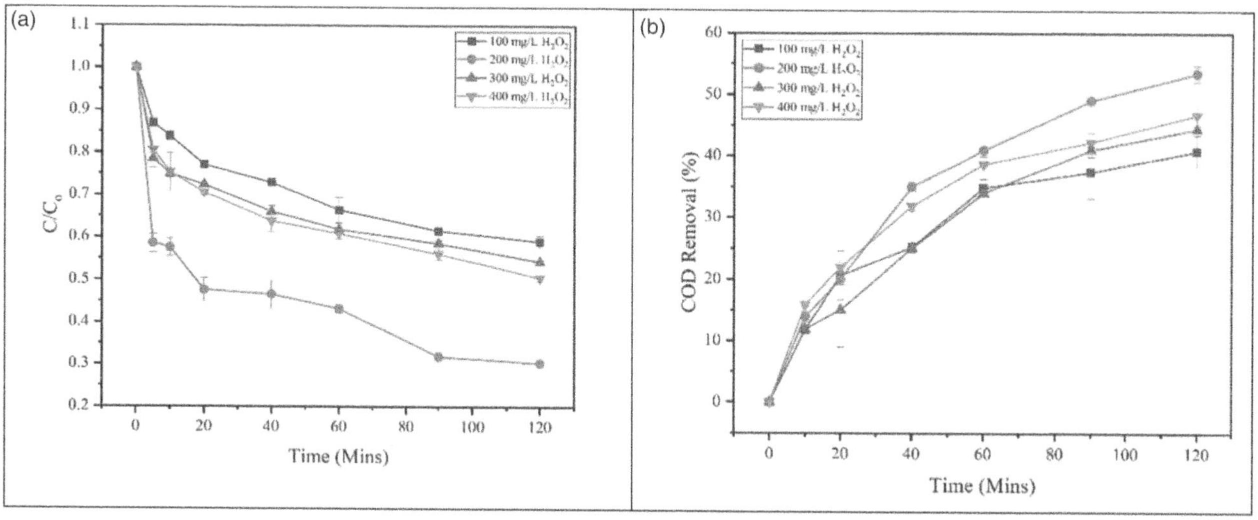

Figure 5 | Oxidative degradation of triclosan with 0.1 g/L GLaNICs catalyst loading and different H_2O_2 dosage: (a) Triclosan removal; (b) COD removal.

in para-position while 2,4-DCP and 4-CC forms when the OH* radical attacks to triclosan in the ortho-position leading to the opening of one of the aromatic rings (Munoz *et al.* 2012).

Dissociation of hydrogen peroxide during Fenton's oxidation is shown in Figure 6(a). It is observed that added hydrogen peroxide is decreased in concentration with time in all the initial concentration indicating continuous dissociation of hydrogen peroxide into hydroxyl radicals. This is inconsistent with the iron oxidation rate and is observed with Figure 6(b). Degradation of triclosan is observed to increase with increase in H_2O_2 concentration from 100 mg/L to 200 mg/L with increase in efficiency of 29.5% since more hydroxyl radicals liberated on higher H_2O_2 dosage. H_2O_2 dosage more than this does not lead to increase in triclosan degradation. This may be due to scavenging effect during the process.

Iron measured during the investigation shows maximum of 31 mg/L of iron leached out of nano catalyst participated in the degradation of triclosan. Initially ferrous form of iron was observed to be more with a concentration of 19.75 mg/L at 5 min later its ferric form of iron was observed to be more by 40 min. This change in iron form is in consistent with the degradation of triclosan where maximum degradation lasts for 40 min, thereby attaining a constant rate.

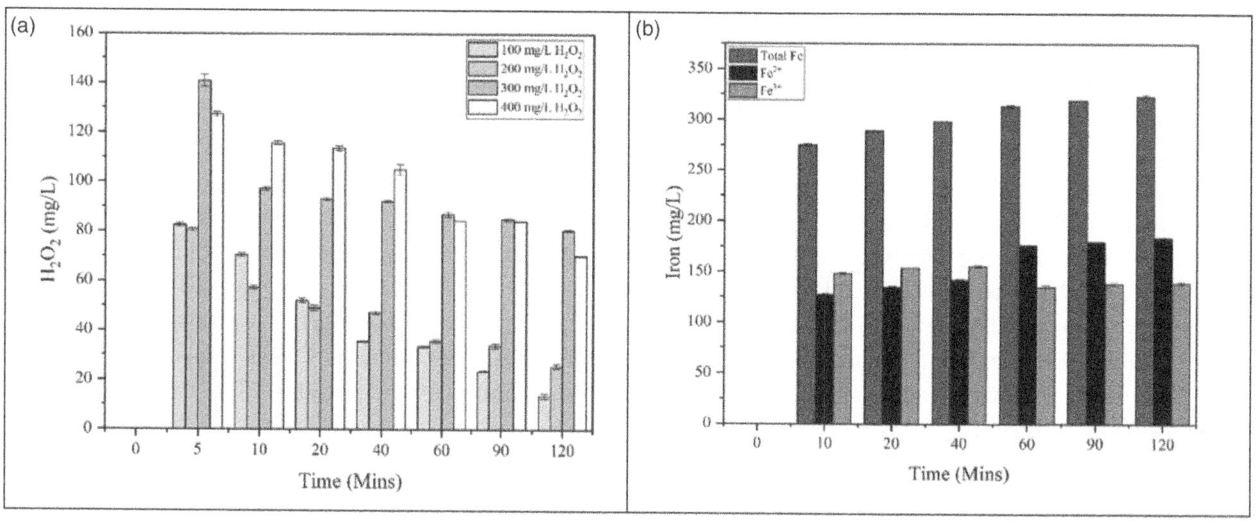

Figure 6 | Variation of (a) hydrogen peroxide and (b) iron on degradation of triclosan at 0.1 g/L GLaNICs catalyst loading and different initial H_2O_2 dosage.

Fenton's treatment of triclosan is shown to be active in acidic pH. During the Fenton's process hydrated ferrous ions get transformed into the colloidal ferric species forming the ferric hydroxyl complexes, thereby reducing efficiency at basic pH. During the study pH maintained was varied slightly from 3.0 to 4.4 which favoured the oxidation process forming hydroxyl radicals (Kang & Hwang 2000; Burbano *et al.* 2005).

Figure 7 shows the kinetic fit for triclosan degradation with GLaNICs as catalyst. Kinetic studies show that pseudo-second order rate kinetic model fits the process with rate constant 0.009/Min and R^2 0.9411 (Boussahel *et al.* 2007; Kang *et al.* 2016; Chen *et al.* 2017). Higher the degradation more the rate constant observed. Overview of Fenton's degradation of triclosan is tabulated in Table 2. Reusability studies conducted indicates that the catalyst can be reused for five consecutive cycles (Figure 8(a)). Degradation efficiency is observed to decrease by 46.5% for the fifth cycle of reusage. However, there is a decrease in degradation efficiency of triclosan by 9.5, 15.5 and 26% with rate constant of 0.0048/min, 0.0062/min and 0.0041/min for the first three cycles. Hence, the study claims the reuse and regeneration of spent catalyst for the first three cycles. Previous studies conducted on reusability of iron particles as catalyst in Fenton's oxidation are inconsistence with the present study (Bhaskar *et al.* 2022). Figure 8(b) presents the comparative study of classical Fenton's oxidation and laterite based – green nano iron particles driven Fenton's oxidation it is observed that both classical and nano iron particles-based Fenton's oxidation are efficient in triclosan degradation; however, laterite based-GLaNIC's Fenton's oxidation is eco-friendly and cost-effective treatment.

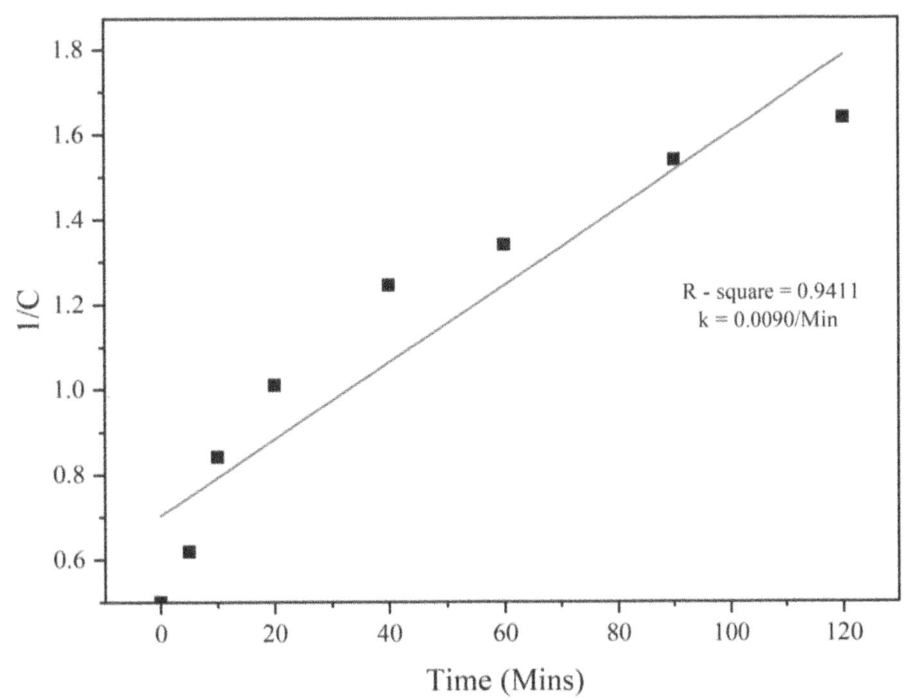

Figure 7 | Linear fit for Fenton's oxidation of triclosan at 0.1 g/L GLaNICs catalyst loading and 200 mg/L of H_2O_2 dosage.

Table 2 | Overview of Fenton's oxidation of triclosan

GLaNICs/H_2O_2	Triclosan degradation %	COD removal %	Rate constant (k)/Min
1:1	40.0	38.99	0.0037
1:2	69.5	54.4	0.0087
1:3	45.5	44.6	0.0039
1:4	49.0	47.9	0.0045

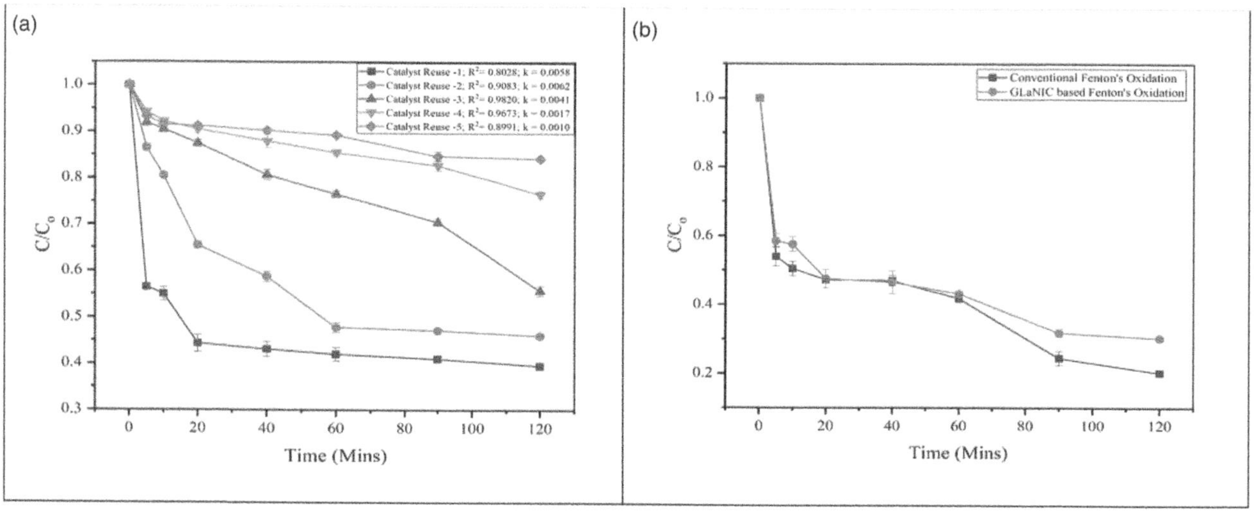

Figure 8 | (a) Reusability of GLaNIC catalyst and (b) comparision of conventional Fenton's oxidation and GLaNIC based Fenton's oxidation.

4. CONCLUSIONS

Laterite-based nano iron catalyst was successfully synthesized with extracted laterite iron as precursor using *Psidium guajava* plant extract by green synthesis. Synthesized nano iron catalyst was confirmed and evaluated for its application as a Fenton's catalyst in the degradation of triclosan. At acidic pH, triclosan is shown to be amenable for degradation with synthesized nano catalyst on Fenton's oxidation. Maximum degradation of 69.5% was observed at pH 3 with 0.1 g/L of catalyst and 200 mg/L of H_2O_2 at 1:2 ratio and corresponding COD removal observed was 54.4% within 120 min of treatment. Degradation of triclosan was observed to increase with increase in hydrogen peroxide dosage. Role of hydroxyl radicals in the degradation was evaluated by measuring hydrogen peroxide concentration and reported. Both classical Fenton's oxidation and GLaNIC-driven Fenton's oxidation are found to be effective for the oxidative degradation of triclosan. Reusability studies show that synthesized iron nano catalysts can be used for three more consecutive cycles. The study claims the use of natural laterite iron-based nanoparticles as a catalyst in Fenton's degradation of triclosan promising cost-effective treatment.

DATA AVAILABILITY STATEMENT

All relevant data are included in the paper or its Supplementary Information.

CONFLICT OF INTEREST

The authors declare there is no conflict.

REFERENCES

Adhikari, S., Kumar, R., Driver, E. M., Perleberg, T. D., Yanez, A., Johnston, B. & Halden, R. U. 2022 Mass trends of parabens, triclocarban and triclosan in Arizona wastewater collected after the 2017 FDA ban on antimicrobials and during the COVID-19 pandemic. *Water Research* **222**, 118894.

Andrades, J. A., Coello, M. D., Quiroga, J. M. & Lojo-l, M. 2021 Degradation of simazine by photolysis of hydrogen peroxide Fenton and photo-Fenton under darkness, sunlight and UV light. *Journal of Water Process Engineering* **42**. https://doi.org/10.1016/j.jwpe.2021.102115.

Anupama, S. S. & Shrihari, S. 2018 Triclosan removal from synthetic wastewater by TiO_2 /UV and O_3 /UV processes. *IOSR Journal of Environmental Science* **12** (3), 17–20. https://doi.org/10.9790/2402-1203021720.

APHA Method 4500-F 1992 *Standard Methods for the Examination of Water and Wastewater*, 18th edn. APHA, Washington, DC, USA, p. 552.

Bakare, B. F. & Adeyinka, G. C. 2022 Occurrence and fate of triclosan and triclocarban in selected wastewater systems across Durban Metropolis, KwaZulu-Natal, South Africa. *International Journal of Environmental Research and Public Health* **19** (11), 6769.

Bhargava, H. N. & Leonard, P. A. 1996 *ARTICLE Triclosan : Applications and Safety*. Association for Professionals in Infection Control and Epidemiology, Inc., Arlington, VA, USA.

Bhaskar, S., Manu, B. & Sreenivasac, M. Y. S. 2020 Bioleaching of iron from fly ash using a novel isolated Acidithiobacillus ferrooxidans strain and evaluation of catalytic role of leached iron in the Fenton's oxidation of cephelaxin. *Journal of the Indian Chemical Society* **97** (3), 360–367.

Bhaskar, S., Manu, B. & Sreenivasa, M. Y. 2019 Bacteriological synthesis of iron hydroxysulfate using an isolated Acidithiobacillus ferrooxidans strain and its application in ametryn degradation by Fenton's oxidation process. *Journal of Environmental Management* **232**. https://doi.org/10.1016/j.jenvman.2018.11.048.

Bhaskar, S., Manu, B. & Sreenivasa, M. Y. 2021 Bioleaching of iron from laterite soil using an isolated Acidithiobacillus ferrooxidans strain and application of leached laterite iron as Fenton's catalyst in selective herbicide degradation. *PLoS ONE* **16**. https://doi.org/10.1371/journal.pone.0243444.

Bhaskar, S., Rashmishree, K. N., Manu, B. & Sreenivasac, M. Y. 2022 Sustainable replacement of EDTA–Bíojarosite for commercial iron in the Fenton's and UV–Fenton's degradation of Rhowedamine B–a process optimization using Box–Behnken method. *Water Science & Technology* **86** (8), 2008–2019.

Boussahel, R., Harik, D., Mammar, M. & Lamara-mohamed, S. 2007 Degradation of obsolete DDT by Fenton oxidation with zero-valent iron. **206**, 369–372. https://doi.org/10.1016/j.desal.2006.04.059.

Burbano, A. A., Dionysiou, D. D., Suidan, M. T. & Richardson, T. L. 2005 Oxidation kinetics and effect of pH on the degradation of MTBE with Fenton reagent. *Water Research* **39**, 107–118. https://doi.org/10.1016/j.watres.2004.09.008.

Chen, F., Xie, S., Huang, X. & Qiu, X. 2017 Ionothermal synthesis of Fe_3O_4 magnetic nanoparticles as efficient heterogeneous Fenton-like catalysts for degradation of organic pollutants with H_2O_2. *Journal of Hazardous Materials* **322**, 152–162. https://doi.org/10.1016/j.jhazmat.2016.02.073.

Eisenberg, G. M. 1943 Colorimetric determination of hydrogen peroxide. *Industrial and Engineering Chemistry*, 327–328.

Heidler, J. & Halden, R. U. 2007 Mass balance assessment of triclosan removal during conventional sewage treatment. *Chemosphere* **66**, 362–369. https://doi.org/10.1016/j.chemosphere.2006.04.066.

Kang, Y. U. N. W. & Hwang, K. 2000 Effects of reaction conditions on the oxidation efficiency in the Fenton process. *Water Research* **34** (10), 2786–2790.

Kang, S., Liu, S., Wang, H. & Cai, W. 2016 Enhanced degradation performances of plate-like micro/nanostructured zero valent iron to DDT. *Journal of Hazardous Materials* **307**, 145–153. https://doi.org/10.1016/j.jhazmat.2015.12.063.

Khan, E., Wirojanagud, W. & Sermsai, N. 2009 Effects of iron type in Fenton reaction on mineralization and biodegradability enhancement of hazardous organic compounds. *Journal of Hazardous Materials* **161** (2–3), 1024–1034. https://doi.org/10.1016/j.jhazmat.2008.04.049.

Lépesová, K., Krahulcová, M., Mackuľak, T. & Bírošová, L. 2019 Sewage sludge as a source of triclosan-resistant bacteria. *Sciendo* **12** (1), 34–41. https://doi.org/10.2478/acs-2019-0006.

Luo, W., Zhu, L., Wang, N., Tang, H., Cao, M. & She, Y. 2010 Efficient removal of organic pollutants with magnetic nanoscaled BiFeO3 as a reusable heterogeneous Fenton-like catalyst. *Environmental Science & Technology* **44**(5), 1786–1791.

Luo, Z., He, Y., Zhi, D., Luo, L., Sun, Y., Khan, E., Wang, L., Peng, Y., Zhou, Y. & Tsang, D. C. W. 2019 Current progress in treatment techniques of triclosan from wastewater: a review. *Science of the Total Environment* **696**. https://doi.org/10.1016/j.scitotenv.2019.133990.

Mellon, J. T. W. & Woods, M. G. 1941 Thiocyanate method for iron. *Industrial & Engineering Chemistry Analytical Edition* **13** (8), 551–554.

Munoz, M., de Pedro, Z. M., Casas, J. A. & Rodriguez, J. J. 2012 Triclosan breakdown by Fenton-like oxidation. *Chemical Engineering Journal* **198–199**, 275–281. https://doi.org/10.1016/j.cej.2012.05.097.

Nandikes, G., Pathak, P., Razak, A. S., Narayanamurthy, V. & Singh, L. 2022 Occurrence, environmental risks and biological remediation mechanisms of triclosan in wastewaters: challenges and perspectives. *Journal of Water Process Engineering* **49**, 103078.

Pragada, S. C. & Thalla, A. K. 2021 Polymer-based immobilized Fe2O3–TiO2/PVP catalyst preparation method and the degradation of triclosan in treated greywater effluent by solar photocatalysis. *Journal of Environmental Management* **296**, 113305.

Sangami, S. & Manu, B. 2016 Fenton's treatment of actual agriculture runoff water. *Water Science and Technology*, 1–11. https://doi.org/10.2166/wst.2016.538.

Sangami, S. & Manu, B. 2017 Optimization of Fenton's oxidation of herbicide dicamba in water using response surface methodology. *Applied Water Science* **7** (8), 4269–4280. https://doi.org/10.1007/s13201-017-0559-8.

Sangami, S. & Manu, B. 2018 Catalytic efficiency of laterite based FeNPs for the mineralization of mixture of herbicides in water. *Environmental Technology*, 1–30. https://doi.org/10.1080/09593330.2018.1449899.

Sangami, S. & Manu, B. 2019 Catalytic efficiency of laterite-based FeNPs for the mineralization of mixture of herbicides in water. *Environmental Technology (United Kingdom)* **40** (20), 2671–2683. https://doi.org/10.1080/09593330.2018.1449899.

Sir, I., Oturan, N., Oturan, M. A., Mar, R. & Brillas, E. 2007 Electro-Fenton degradation of antimicrobials triclosan and triclocarban. **52**, 5493–5503. https://doi.org/10.1016/j.electacta.2007.03.011.

Sires, I., Oturan, N., Oturan, M. A., Rodríguez, R. M., Garrido, J. A., & Brillas, E. 2007 Electro-Fenton degradation of antimicrobials triclosan and triclocarban. *Electrochimica Acta* **52** (17), 5493–5503.

Soon, A. N. & Hameed, B. H. 2011 Heterogeneous catalytic treatment of synthetic dyes in aqueous media using Fenton and photo-assisted Fenton process. *Desalination* **269** (1–3), 1–16. https://doi.org/10.1016/j.desal.2010.11.002.

Thompson, A., Griffin, P., Stuetz, R. & Cartmell, E. 2005 The fate and removal of triclosan during wastewater treatment. *Water Environment Research* **77** (1), 63–67.

First received 15 September 2022; accepted in revised form 24 November 2022. Available online 2 December 2022

doi: 10.2166/wst.2022.325

Clindamycin removal from aqueous solution by non-thermal air plasma treatment: performance, degradation pathway and ensuing antimicrobial activity

Abbas Akbarzadeh[a], Hamid Reza Ghomi[b], Mohammad Rafiee[c,d], Omid Hosseini[e] and Mahsa Jahangiri-rad [f,*]

[a] Water and Wastewater Research Center (WWRC), Water Research Institute, Tehran, Iran
[b] Laser and Plasma Research Institute, Shahid Beheshti University, Evin, Tehran, Iran
[c] Department of Environmental Health Engineering, School of Public Health and Safety, Shahid Beheshti University of Medical Sciences, Tehran, Iran
[d] Air Quality and Climate Change Research Center, Shahid Beheshti University of Medical Sciences, Tehran, Iran
[e] Central Research Laboratories, Shahid Beheshti University of Medical Sciences, Tehran, Iran
[f] Water Purification Research Center, Tehran Medical Sciences, Islamic Azad University, Tehran, Iran
*Corresponding author. E-mail: m.jahangiri@iautmu.ac.ir

MJ-R, 0000-0002-1626-4014

ABSTRACT

The present study set out to investigate clindamycin (CLN) removal from aqueous solution using non-thermal plasma (NTP) under atmospheric air conditions and to address the effects of some variables including pH, initial concentration of CLN, and working voltage on CLN degradation. The result showed that the NTP system exhibited excellent degradation rate and mineralization efficiency on CLN in 15 min under neutral conditions, which exceeded 90 and 45%, respectively, demonstrating its conversion to other organic by-products. Furthermore, CLN degradation was largely dependent upon the initial pH of solution, applied voltage, and reaction time. Specifically, under acidic conditions (pH = 3), working voltage of 24 kV and after 15 min of reaction, almost 100% of CLN was degraded. NTP-initiated CLN degradation products through LC-MS/MS analysis, determined within 10 min of reaction, inferred that the complex structure of CLN has undergone deterioration by active radical species which subsequently generated small molecular organic compounds. Chemical processes involved in CLN degradation were found to be demethylation, desulfonylation, dechlorination, hydroxylation and deamination. Lastly, antimicrobial susceptibility tests revealed that the activity of CLN was reduced following NTP treatment, which is also in good agreement with the minimum inhibitory concentration (MIC) values obtained from microdilution analyses.

Key words: antibacterial susceptibility, clindamycin, cold plasma, degradation products, removal efficiency

HIGHLIGHTS

- Clindamycin degradation by non-thermal plasma was investigated.
- Almost 90% removal of CLN was achieved within 15 min of reaction.
- pH, initial CLN concentration and working voltage were the key factors influencing CLN removal.
- Intermediate products and possible CLN degradation pathway were proposed.

GRAPHICAL ABSTRACT

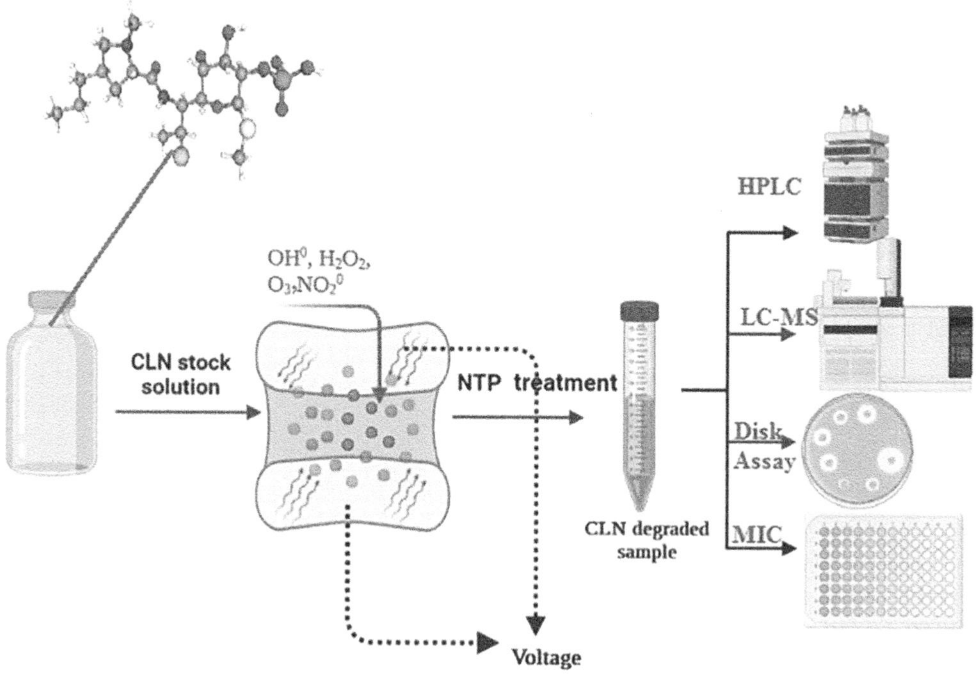

INTRODUCTION

The emergence of residual pharmaceutical compounds in environmental matrices has become a critical global public health issue. Owing to their low biodegradability and high chemical stability, a wide range of pharmaceutical compounds do not get completely degraded during typical wastewater treatment (Golchin *et al.* 2021; Majeed *et al.* 2021; Zamani *et al.* 2021). Owing to the incomplete removal of some antibiotics, their concentration has sharpened in the mixing and downstream sections of a local river receiving the effluent from conventional wastewater treatment plants (Dong *et al.* 2016). Clindamycin (CLN), as a class of lincosamide antibiotics, is primarily applied to treat anaerobic infections. A survey of the literature points out that roughly 10% of the administered CLN dose is excreted unchanged (Wynalda *et al.* 2003), while its main metabolized intermediates, namely clindamycin sulfoxide and N-desmethylclindamycin, are excreted in urine and feces. Whereas the former is bioactive, the latter is considered persistent. Indeed, a small portion of CLN has been reported to be removed by conventional barriers exploited in wastewater treatment plants (WWTPs) and thus it is one of the most frequently detected antibiotics in different sewage samples (Rodriguez-Mozaz *et al.* 2020). In contrast to the decline of most antibiotics in the treated effluent samples, an increase of CLN concentration in the final effluent in comparison to the influent of the sewage treatment plant has also been reported (Oertel *et al.* 2014), suggesting its persistence in biological treatment utilities. Hence, high concentrations of CLN and its residues have been detected in the effluent of WWTPs and also appears to be persistent in surface waters (Batt *et al.* 2006). Therefore, novel and advanced methods are required to remove the toxic and persistent CLN from contaminated effluents.

Advanced oxidation processes (AOPs) have gained considerable attention owing to their efficiency in the dissociation of persistent non-biodegradable compounds (Babu *et al.* 2019; Graumans *et al.* 2020). Traditional AOPs, however, suffer from serious potential limitations, including the consumption of oxidant reagents by oxidizable natural materials (they are nonselective), applicable to a narrow pH range, and little selectivity to oxidized substrates and thus low mineralization capacity (Wang & Zhuan 2020). Plasma deals with the fourth state of matter which consists of partial or fully ionized gas. Generally, non-thermal plasma (NTP) is generated by electricity, microwave, or laser discharges at low temperatures and low pressure. The most commonly used plasma systems are dielectric barrier discharges (DBD) defined as the electrical discharge between two electrodes separated by an insulating dielectric barrier (Jiang *et al.* 2020). NTP represents a non-thermal decontamination process that involves the generation of various types of reactive species containing high oxidation potential and thus results in the effective decomposition of complex organic pollutants (Li *et al.* 2016). Unlike thermal plasma, the NTP discharge does not show a localized thermodynamic equilibrium and can be obtained at atmospheric pressure using a low-power input (Chen *et al.* 2016). Preferably, this process involves the generation of oxidative species through the introduction of electrical, radiation, and/or chemical energy into the reaction zone, resulting in the conversion of complex chemical compounds into smaller organic acids (Coutinho *et al.* 2018; Lee *et al.* 2018). In a direct mode of NTP, the sample is exposed to the discharging zone or is in contact with the plasma transferred to the target site from the discharging zone. This mode of operation is highly effective for the decomposition of complex chemicals (Aggelopoulos & Tsakiroglou 2021; Hatzisymeon *et al.* 2021). Therefore, organic pollutants directly come into contact with the photons/UV radiation and subsequently the reactive chemical species (both short and long-life) and undergo degradation (Zhou *et al.* 2021). Chemical reactions occurring in a typical NTP process are quite sophisticated and involve a number of active species with lifetimes varying from a few seconds to hours. For instance, numerous reactive oxygen species (ROS) and reactive nitrogen species (RNS) are usually formed from air plasma (Ansari *et al.* 2021; Zhou *et al.* 2021).

NTP-based technology has been effectively considered for a range of applications in water treatment, food processing, agriculture, and pollution. Several researchers have evaluated the effectiveness of NTP to deal with pharmaceutical compounds in aqueous solutions such as ibuprofen (Aziz *et al.* 2017), ampicillin (Smith *et al.* 2017), sulfadiazine (Rong & Sun 2014), carbamazepine (Krause *et al.* 2009), paracetamol (Baloul *et al.* 2017), ciprofloxacin (Guo *et al.* 2019) and some other antibiotics (Magureanu *et al.* 2011; Xu *et al.* 2020b). The performance of the NTP process for the degradation of organic contaminants in an aqueous solution is based on the in-situ formation of $OH^{\bullet}$, H_2O_2, and O_3 (strong oxidizing agent). These oxidizing species are known to be responsible for the mineralization of organic pollutants into CO_2, H_2O, and inorganic acids (Sarangapani *et al.* 2019; Fahmy *et al.* 2020). Non–thermal plasma treatment has many advantages over other physical and chemical methods applied to persistent organics removal, in that (1) both chemical and heat treatments have inherent problems associated with the wastes in chemical processing and the huge energy losses in thermal processing which can be completely avoided using plasma. (2) The application of NTP for pollutant removal benefits from lower costs, higher processing and energy efficiency, reduced space volume and environmental friendliness, among others (Kale & Desai 2011). (3) An added benefit is the overall level of control over the treatment process. Plasma systems turn on and off instantly and settings are easily adjustable. With the development of atmospheric pressure, non-thermal plasma in recent decades, plasma technology offers an attractive means of online processing with wide applicability and low required input amounts of gas and electricity, which in turn make it energy efficient and environmentally friendly (Napartovich 2001).

As with traditional AOPs, the low mineralization capacity and thus the accumulation of intermediate by-products is a challenging issue in organics degradation by NTP. The literature already includes several studies focused on degradation products formed following treatment with plasma (Banaschik *et al.* 2018; Lee *et al.* 2018; Sang *et al.* 2019). Nonetheless, few reports have been published on the toxicity of final/intermediate products on specifically related organisms and elaborated on the possible pathway mechanisms of antibiotic degradation particularly when it comes to sophisticated organic molecules like CLN. The main objectives of this work were thus to study the effectiveness of atmospheric air plasma for degradation of CLN from aqueous solution and to elucidate the intermediate products. Additionally, the antimicrobial characteristics of treated CLN were evaluated by the loss of biological function of *Staphylococcus aureus* and *Clostridium perfringens* as aerobic and anaerobic microorganisms, respectively, using disk diffusion susceptibility assays. Furthermore, the toxicity of CLN degradation products was evaluated through comparative assessments of minimal inhibitory concentrations (MIC) before and after treatment with NTP. In addition, the mineralization and degradation mechanism was characterized and the possible degradation pathways of CLN were proposed based on the analytical results of LC-MS/MS.

MATERIALS AND METHODS

Chemicals

Clindamycin phosphate ($C_{18}H_{34}ClN_2O_8PS$) was provided from Pak Darou (Iran) and its structure is depicted in Figure 1. HPLC grade acetonitrile, phosphate buffer solution (PBS), hydrochloric acid (HCl), and sodium hydroxide (NaOH) were purchased from Merck company. Reinforced clostridial medium (RCM) was purchased from Q-lab (Canada). Gas pack (type-A) was purchased from Merck.

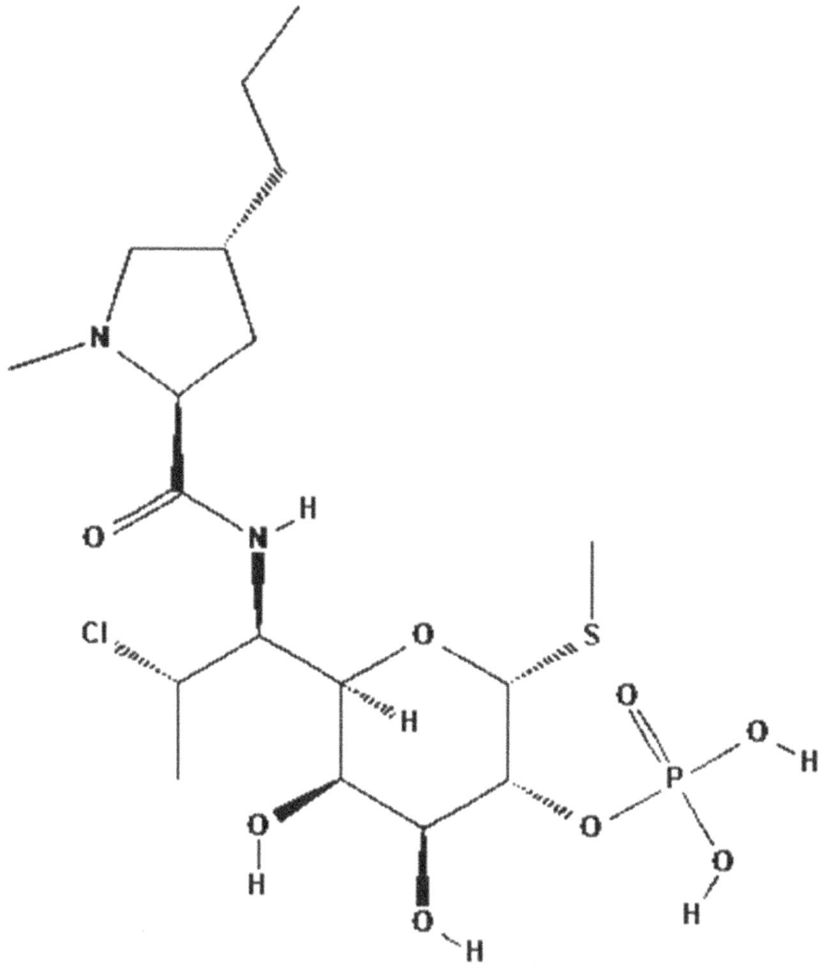

Figure 1 | Chemical structure of clindamycin phosphate (molecular weight = 505).

Experimental procedure

A series of experiments were carried out to explore the effect of solution pH (3,7,9), CLN concentration, reaction time, and initial working voltage on CLN removal efficiency and drug mineralization. CLN was dissolved in deionized water (DW) to obtain the solution of 10 and 20 mg L^{-1} (1,000 mL). These two concentrations (50 mL) were further used for each run of experiments. For each experiment, 50 mL of either 10 or 20 mg L^{-1} of CLN was subjected to NTP treatment (Enhanced-tech-15I). Atmospheric air plasma as the working gas at a variable voltage (15 and 24 kV), power of 50 and 80 W, and treatment duration (5, 10 and 15 min) was used. All the experiments were conducted in triplicate at room temperature (18–21 °C). The schematics of the plasma system are depicted in Figure 2.

Analytical procedures

The concentrations of CLN were analyzed using HPLC (Smartline from Knauer Company) equipped with a UV/Vis detector and a C18 column (250 mm × 4.6 mm, 5 μm). The mobile phase contained 80% phosphate buffered saline (PBS) and 20% acetonitrile. The flow rate was set at 1 mL min^{-1}. The detector wavelength was set at 220 nm. The detection of CLN was achieved after 11 min of injection. A standard calibration curve for CLN was established using a series of concentrations in deionized water ranging from 0.01 to 20 mg L^{-1}. The linear correlation coefficient (R^2) was 0.998.

Hydrogen peroxide (H_2O_2) concentration was determined according to the procedure described by Boehm *et al.* (Boehm *et al.* 2016). Summarily, H_2O_2 concentration was determined by the oxidation of potassium iodide (KI) to iodine and measured spectrophotometrically at 405 nm using spectrophotometer (Shimadzu RF-1501). Nitrate and nitrite concentrations were measured by ion chromatography (Metrohm 850). Dissolved ozone concentration was measured according to the procedure of Bader

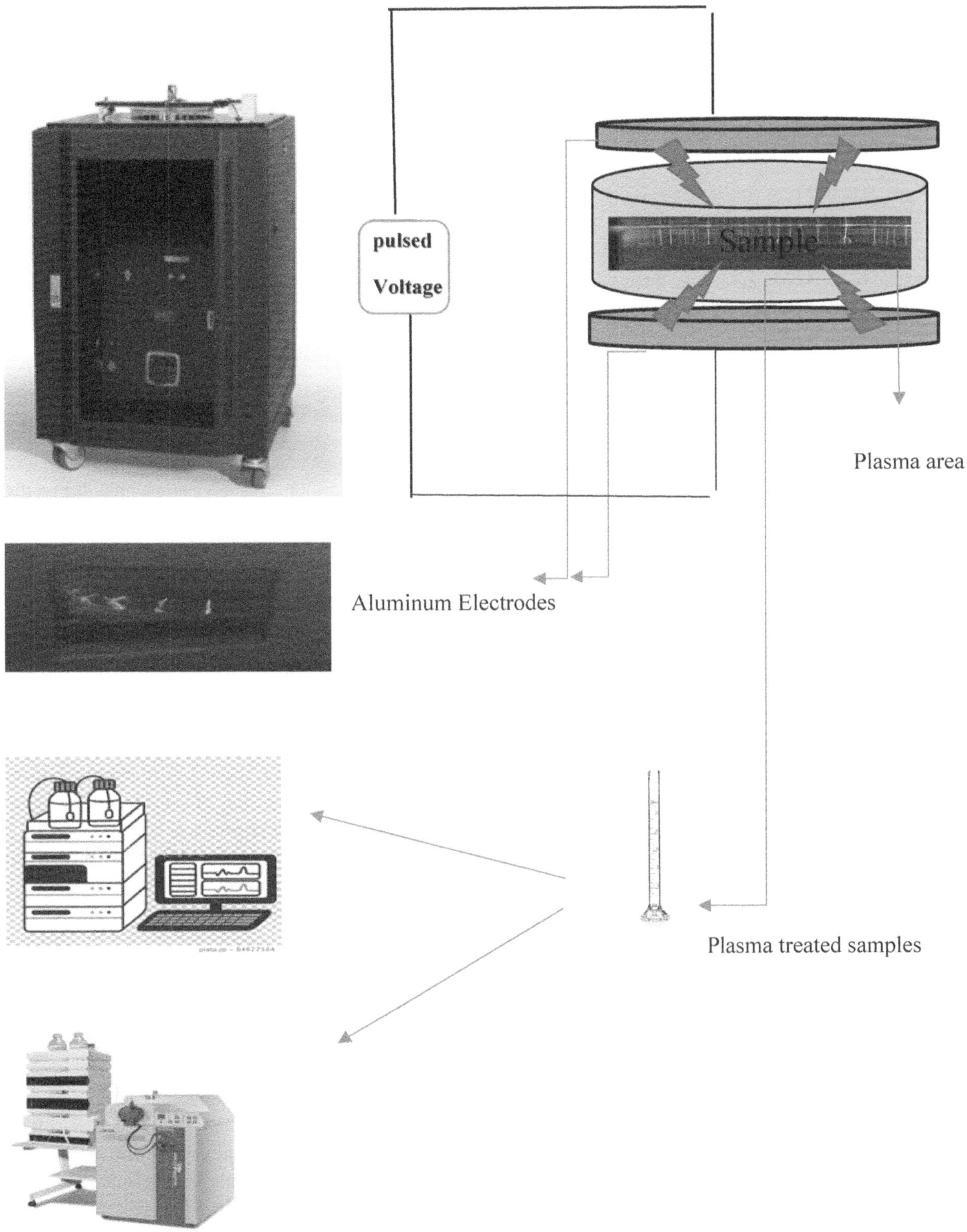

Figure 2 | The schematic of plasma systems and analytical analysis.

(1982). The total organic carbon (TOC) content of samples was determined using TOC analyzer (Shimadzu). The intermediates following treatment with the plasma discharge were analyzed by LC-MS/MS system using Waters Alliance 2695 HPLC – instrument consisting of UV/Vis detector coupled to a Micromass Quattro micro API Mass Spectrometer.

AC18 analytical column (2.1 × 100 mm, 3 μm) was employed for the separations at 35 °C, with a mobile phase (A: acetonitrile + 0.1% formic acid, and B: H_2O (60%) + 0.1% formic acid (40%)) at a flow of 0.2 mL/min. The injection volume was 5 μL.

Removal efficiency and degradation kinetics modeling

The removal efficiency of CLN was calculated based on the following formula:

$$RE\ \% = \frac{(C_0 - C_e)}{C_0} \times 100 \quad (1)$$

where *RE* represents removal efficiencies (%), C_o and C_e stand for the initial and remaining concentrations of CLN (mg L^{-1}) at time '*t*', respectively. As a general rule, it has been ascertained that the degradation of organic compounds following plasma treatment can be well explained by first-order kinetics (Magureanu *et al.* 2021), as:

$$C_e = C_0 \exp(-k_{obs}t) \quad (2)$$

where 'k_{obs}' is the degradation rate constant (min^{-1}) and '*t*' is the time of plasma treatment (min).

Energy yield calculation

The energy yield of NTP in the degradation of CLN, which is explained by the amount of energy consumed during the degradation process (Chandana & Subrahmanyam 2018), was calculated by the following equation:

$$Y = \frac{(C_0 - C_t) \times V}{t \times P} \quad (3)$$

where *Y* is the energy yield (g kWh^{-1}), C_0 and C_t are the initial and final concentration of CLN, respectively (g L^{-1}), *V* is the volume of solution (L), *P* is the discharge power (kW) and *t* is the discharge time (h).

Disk diffusion assay

In order to evaluate the antimicrobial activity of treated CLN, a standard disk diffusion assay was used involving two test species, namely *Closteridium perferigens* PTCC1765 and *Staphylococcus aureus* PTCC1826. Target microorganisms were obtained from the Iranian Research Organization for Science and Technology (IROST). *Staphylococcus aureus* was cultured in Muller Hinton agar (MHA) and incubated overnight at 37 °C; while *Closteridium perferigens* was cultured in RCM for 24 h under anaerobic conditions using anaerobic jar equipped with gas pack. Petri plates containing MHA were seeded uniformly over their entire surface with a standardized amount (McFarland 0.5) of each test organism. Both plasmas treated (10 and 20 min) and non-treated CLN solution was deposited on the solidified agar surface (diameter 6 mm).

Following standardized incubation for 24 h at 37 °C, the diameters of inhibition zones formed around the disks of CLN-treated solutions were compared with non-treated CLN.

Minimal inhibitory concentration

Minimal inhibitory concentration (MIC) of either treated (CLN treatment – 15 and 30 min) or pristine untreated antibiotic solutions for *Closteridium perferigens* and *Staphylococcus aureus* were determined using the broth dilution test (Balouiri *et al.* 2016). The MIC was determined by making a sequence of a total of 12 times decreasing concentrations (50–0.02%) of plasma-treated water in Muller Hinton broth (MHB), which was then inoculated with the test bacteria. Controls included each antibiotic concentration without bacterial cells (negative control), MHB and bacterial cells (positive control). Plates were incubated for 24 h at 37 °C. MICs of either treated or untreated antibiotics for two microorganisms were determined as the lowest concentration showing no visual turbidity as compared to corresponding blank controls.

RESULTS AND DISCUSSION

Effects of key factors on CLN degradation

Figure 3 shows the effect of initial CLN concentration on removal efficiency. CLN removal efficiency in the initial 5 min was more than 85%. The removal efficiency then increased with the reaction time linearly and reached above 90% after 15 min.

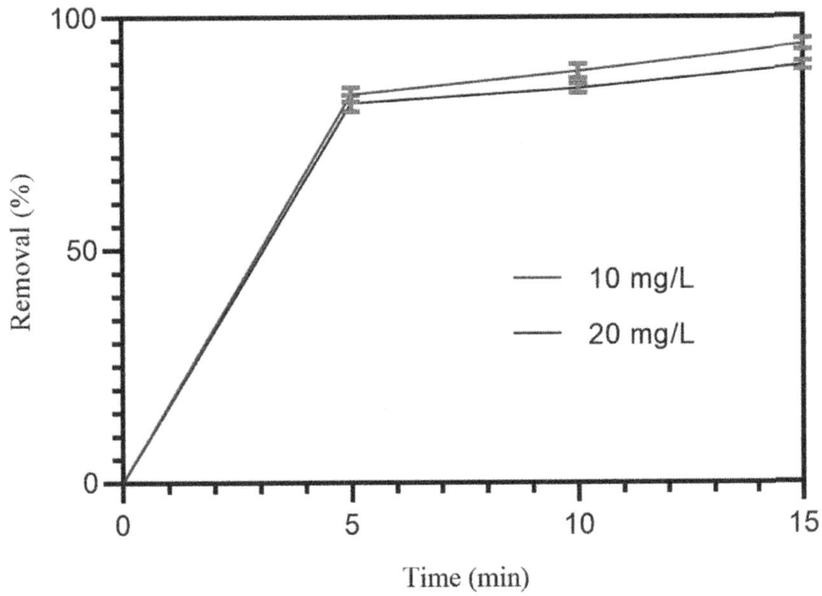

Figure 3 | Removal percentage of CLN solution at two initial concentrations (voltage = 24 kV, pH = 7.4).

The positive impact of treatment time on CLN removal following DBD system was clearly due to the increase in the reactive oxygen species over treatment time, so more CLN in aqueous solution could be consumed via reacting with reactive oxygen species over longer periods. Due to the normality of data obtained and equal variance among groups, One-Way ANOVA test was applied to evaluate the effect of reaction time in CLN degradation (Supplementary S1-a). The result revealed the significant role of treatment times in its removal ($p \leq 0.05$). The degradation efficiency of CLN gradually decreased from 92 to 85%, by the increasing initial concentration (Figure 3). Similarly, the reaction rate constant with 10 and 20 mg L^{-1} of CLN obtained 0.206 and 0.154 min^{-1}, respectively. This can be explained by the fact that the reactive oxidative species (ROS) produced during plasma treatment, target low concentration of CLN, subsequently generate a higher ratio of active species per CLN molecule which is consistent with other studies (Wang *et al.* 2018; Zhang *et al.* 2021). In accordance with the results obtained in our study, other researchers have also reported a decline in the removal efficiency of other organic compounds when the initial concentration increased (Lou *et al.* 2012; Aggelopoulos *et al.* 2018).

The effect of initial pH (3–9) on the CLN degradation in the NTP system was investigated under optimum conditions of concentration (10 mg L^{-1}) and voltage (24 kV). Figure 4 explores that acidic conditions favored the degradation process in which the degradation efficiency of CLN was increased with the declining initial pH of the solution, and reached almost 95% at pH of 3 during the first 5 min; whereas less than 80% removal was observed under more alkaline conditions (pH = 9) at the same time. This result could be attributed to the formation of more ROS under acidic conditions. These seem to indicate that the initial pH of water plays a pivotal role in the aqueous solution treatment with plasma, due in part to the alteration of the oxidation rate of ozonation and peroxone processes, as has been outlined before (Vanraes *et al.* 2017). To accurately evaluate the impact of changing pH solution on CLN removal Kruskal–Wallis test was applied due to the fact that analysis of variance, revealed differences among groups with various pHs (S1-b). The results of Kruskal–Wallis test also exhibited that the effects of varying pH on CLN degradation were also significant ($P \leq 0.05$).

In order to verify the above viewpoint, the pH of treated samples after the termination of the reaction was measured demonstrating that it reduced below 4 after 15 min of reaction (Figure 5). The decline in pH can be explained by the formation of H_3O^+ ions and organic acids (Murugesan *et al.* 2020; Xu *et al.* 2020a). Few studies have specifically dealt with the effects of pH in NTP process; however, it can be inferred that the role played by pH could be associated with the generation and activity of reactive species in the solution. For instance, with the reduced pH, the oxidation potential of O_3 – as the main product in NTP process – enhances. It is proved that the oxidizing potential of O_3 reduces from 2.08 V at acidic pH to 1.4 V in alkaline solutions. At high pH, O_3 in the aqueous solution undergoes decomposition into OH^- (Rahimpour *et al.* 2019). Furthermore, it has been found that H_2O_2 concentration decreases in alkaline conditions due to a higher concentration of OH ions, HCO_3^- and CO_3^{2-} (Jose & Philip 2019).

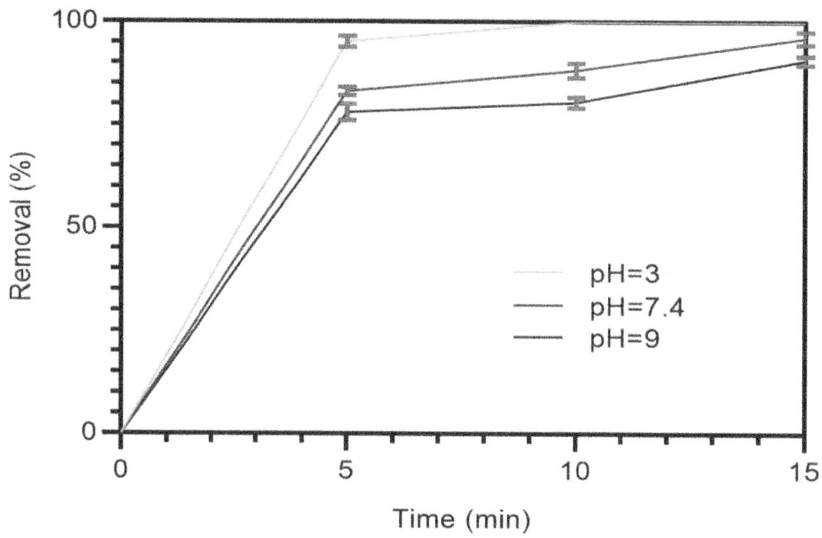

Figure 4 | Removal percentage of CLN at various initial pHs (voltage = 24 kV, initial concentration = 10 mg/L).

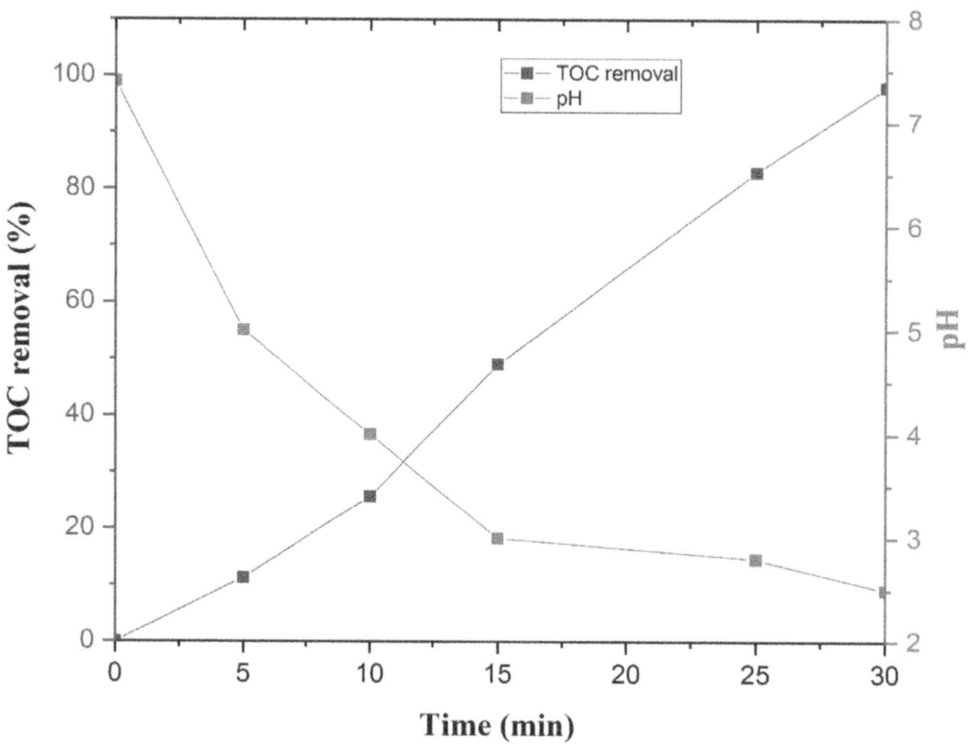

Figure 5 | Samples' variation of pH and TOC due to the CLN degradation by non-thermal plasma (CLN initial concentration = 20 mg/L, voltage = 24 kV).

The influence of applied voltage on CLN degradation was investigated and the results are shown in Figure 6. Two voltage values of 15 and 24 kV were tested and the corresponding discharge powers were 50 and 80 W, respectively. It was found that only 52% of CLN was removed in 5 min with the voltage of 15 kV, while roughly 85% removal was observed with the voltage of 24 kV in the same discharge time which demonstrates that the application of higher voltage led to the increase of active species, hence enhances the degradation of CLN. The results of the student's t-test between samples treated either by 15 kV or 24 kV were significant ($p \leq 0.05$), exhibiting the impact of higher voltage in efficient CLN degradation (S1-c).

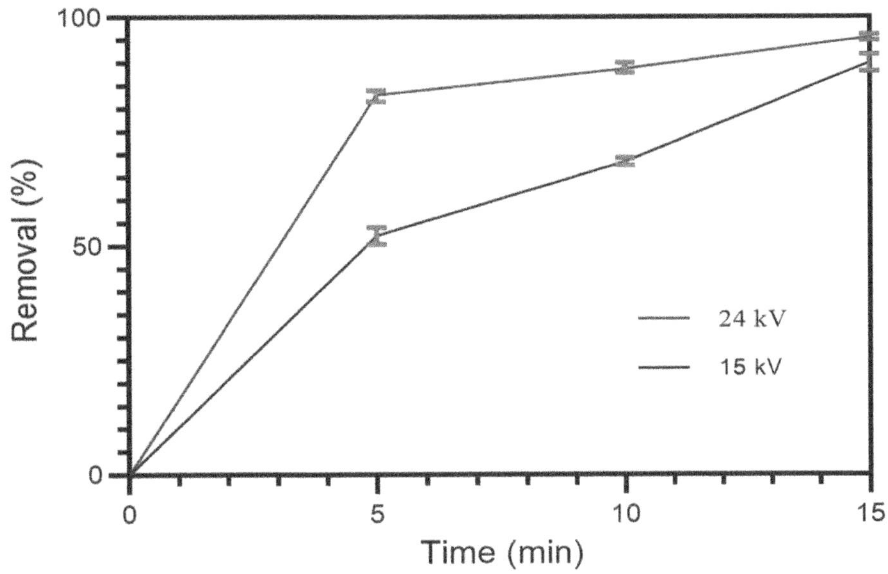

Figure 6 | CLN removal percentage at two working volatege (CLN initial concentration = 10 mg/L, pH = 7.4).

Similarly, other studies demonstrated the influential positive role of higher voltage in the degradation of organic compounds owing to the increased generation of active species (Sang *et al.* 2019; Zhang *et al.* 2021).

Analysis and roles of majors ROS

Chemical reactions taking place in a typical NTP process are quite complicated and encompass a great number of active species with a wide range of lifetimes from a few nanoseconds to hours. Numerous reactive species (ROS/RNS) are usually formed following air plasma application (Chandana *et al.* 2018). Ozone (O_3) and hydroxyl radicals ($OH^\bullet$) are the most common ROS formed, while nitric oxide ($NO^\bullet$) and nitrogen dioxide ($NO_2^\bullet$) radicals are important RNS. The recombination of short half-life reactive species can produce H_2O_2 and O_3 species via partial peroxone reactions which subsequently can improve the potency of contaminant removal. Hydrogen peroxide ($Eo = 1.77$ V/SHE) is one of the most reactive species formed in a gas–liquid plasma system. Hydrogen peroxide under ambient conditions is relatively stable in aqueous solutions with the reported stability of almost several weeks (Boehm *et al.* 2016). Although firm evidence is lacking, however, the generation of H_2O_2 can also be considered an efficient indicator of the $OH^\bullet$ radicals' generation in the DBD-containing plasma reactor. Accordingly, the efficacy of the DBD reactor in CLN removal was evaluated by the formation of H_2O_2 in antibiotic solutions (Figure 7). It can be observed that the concentration of H_2O_2 in the liquid phase generated by the DBD reactor is dependent on the treatment time. The concentration of H_2O_2 ranged from 48 to 890 mg L^{-1} and 44 to 470 mg L^{-1} for 10 and 20 mg L^{-1} CLN at 24 kV. This finding supports the literature that reports that the presence of pharmaceuticals or organic matter can enhance the formation of H_2O_2 (Li *et al.* 2020; Zhang *et al.* 2021). It was also approved that ozone is formed in considerable amounts in DBD reactors under oxygen-rich atmospheres. This strong oxidizer can disperse from the gas into the liquid and subsequently destroy organic contaminants (Magureanu *et al.* 2008). Hence, ozone can play a part in the present study as well, for the oxidative degradation of CLN. Figure 8 explores ozone concentration in the solution as a function of treatment time. As depicted, O_3 concentration increased rapidly in the first 10 min of reaction and reached an accumulated concentration of 0.6 mg L^{-1} and started to decrease fast afterward. These findings are consistent with those of Magureanu *et al.* (2011) who described O_3 changes for selected antibiotics treated by cold plasma. They found that O_3 concentration increased after around 5 min of reaction; however, it declined after 30–60 min of reaction depending on the type of antibiotics. They concluded that the decline of ozone might be ascribed to the formation of smaller degradation by-products that react also with ozone or its transfer to the effluent gas (Magureanu *et al.* 2011).

Similarly, the concentration of nitrate in the CLN-laden water solutions for 10 and 20 mg L^{-1} ranged from 69 to 233 mg L^{-1} and 23 to 208 mg L^{-1}, respectively (Figure 7). As depicted, an increase in both nitrate and H_2O_2 values were observed with increases in treatment time. The pH values of treated solutions in the current study plummeted drastically from 7.4 to 4.5

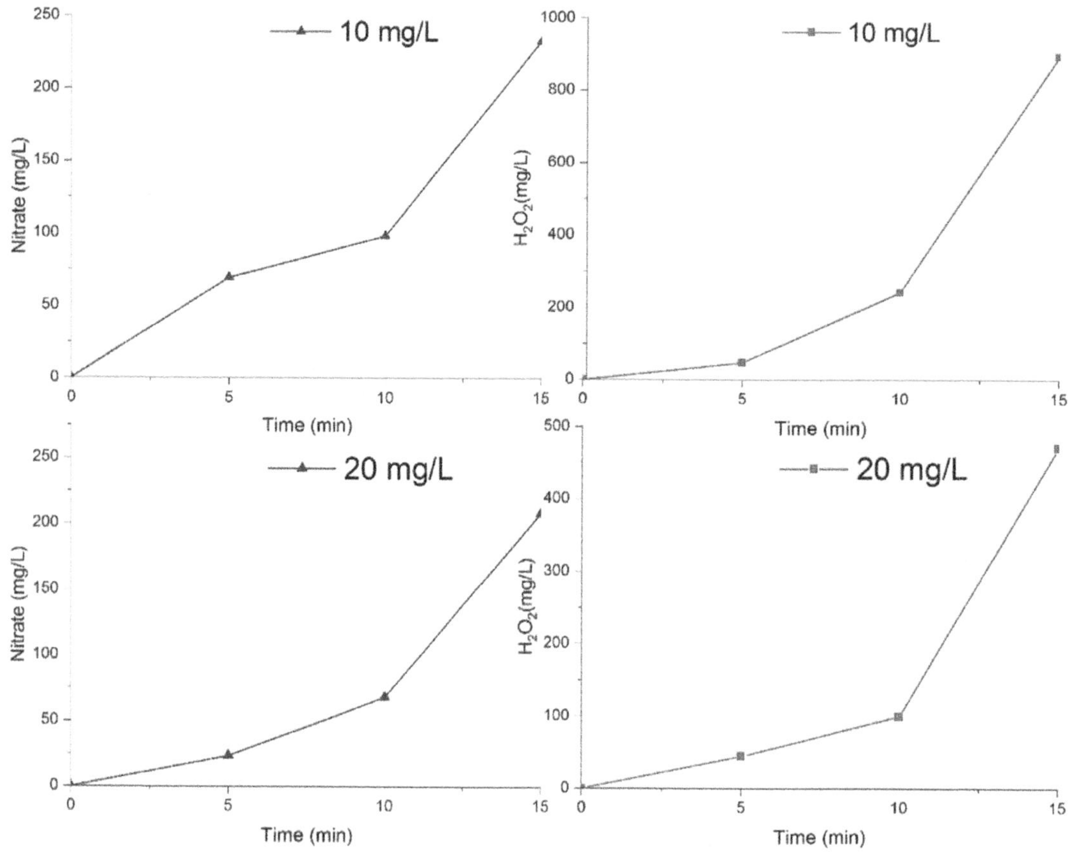

Figure 7 | Variation of NO_3 and H_2O_2 concentration in plasma treated CLN solutions (pH = 7.4, voltage = 24 kV).

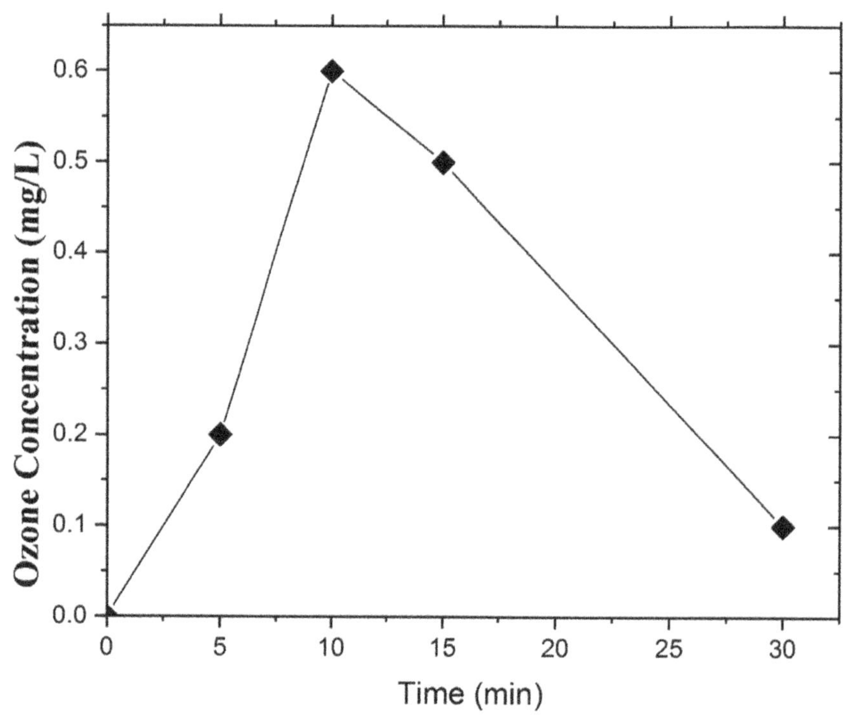

Figure 8 | Variations of concentrations of O_3 in plasma treated CLN solution (pH = 7.4, initial concentration = 20 mg/L and voltage = 24 kV).

after 5 min of reaction and reached 3 following 15 min (Figure 5). This result can be explained by the crucial role of nitrates in falling pH. Additionally, no nitrites were detected in the present study, which could be assigned to the easy conversion of nitrite to HNO_2 under acidic conditions. RNS is less capable than ROS due to the fact that they are formed by consuming the stronger oxidant $OH^\bullet$, which subsequently converts them to nitric acid (HNO_3) (Lukes *et al.* 2014).

Mineralization of CLN

In order to investigate the mineralization pathway, the changes in pH and TOC during the degradation time are important (Sang *et al.* 2019). As shown in Figure 5, during CLN degradation pH value decreased from 7.4 to 3 after 15 min of reaction which illustrates the substantial amount of acidic compounds that were generated. Based on detected compounds in the solution after plasma treatment aliphatic carboxylic acid compounds were detected which can reduce pH of the solution (Figure 9). Additionally, NH_4^+ was assumed to be one of the final products following degradation of CLN which drops pH of the solution during its conversion to NO_3 (Figure 9).

To obtain a broader picture, the mineralization of CLN by NTP was monitored within 30 min (Figure 5). It is evident that TOC removal steadily increased with the reaction time; however, the mineralization efficiency was not as high as CLN degradation. CLN degradation efficiency of the NTP system after 15 min was more than 95%, apparently higher than the mineralization degree in terms of TOC removal which reached about 50% at optimum conditions. It can be deduced that CLN was degraded to some intermediate by-products (small-molecule substances), and almost half of it underwent complete conversion to H_2O and CO_2. It can also be inferred that incomplete removal of organic carbon from the NTP process might be associated with insufficient photocatalytic oxidation efficiency or resistance of the intermediate products (Nasuhoglu *et al.* 2011). It is, however, worth mentioning here that CLN mineralization was continued by reaction time and almost complete mineralization of CLN (98%) occurred within 30 min of treatment, demonstrating that a large percentage of the drug molecules were degraded to mineral substances of CO_2 and H_2O. Other AOPs have also been studied for CLN removal from aqueous solution. As shown in Table 1, above 90% of CLN degradation was achieved using those methods. What stands out in our research is the complete removal of CLN in only 10 min of treatment without adding any materials to the final solution.

LC-MS/MS analysis and degradation pathway

According to the ROS analysis, the degradation of CLN is mainly attributed to the generation of oxidative radicals. Production of ozone and hydrogen peroxide in the solution can produce active radicals which in turn could improve the effectiveness of CLN degradation based on the following reactions (Lee *et al.* 2018):

$$H_2O_2 \leftrightarrow HO_2^- + H^+ \tag{4}$$

$$HO_2^- + O_3 \leftrightarrow HO_2^0 + O_3^{0-} \tag{5}$$

$$HO_2^0 \leftrightarrow O_2^0 + H^+ \tag{6}$$

$$O_3^{0-} + H^+ \rightarrow HO_3^0 \tag{7}$$

$$HO_3^0 \rightarrow O_2 + OH^0 \tag{8}$$

$$O_2^0 + O_3 \rightarrow O_2 + O_3^0 \tag{9}$$

Generally, species such as H_2O^+, H^0, O_3, HO_2^0 and OH^0 are considered the key elements responsible for the decontamination process in gas–liquid plasma treatment (Chauvin *et al.* 2017; Liao *et al.* 2017). To confirm and investigate the degradation pathways of CLN exposed to the plasma, the samples from the degradation process (pH = 7.4, t = 15 min, CLN = 20 mg L^{-1}) were collected and analyzed by LC-MS/MS. The corresponding LC spectrums of CLN treated with cold plasma are shown in Fig. S2. The characteristics of the identified intermediates, including the characteristic ions (m/z) and their structures, are shown in Figure 9 and the proposed pathway is exemplified. As depicted in Figure 9, three different pathways with the same final products were proposed. First path (blue color) initially contains a detachment of HPO_3^- from clindamycin phosphate molecule by hydroxyl radical attack and production of a stable compound with m/z = 424. This intermediate is attacked by hydroxyl radicals and then through the demethylation reaction, the methyl group attached to the atom loses its sulfide and the m/z = 410 intermediate is produced. Following this path and as a result of OH^0 attacks and through

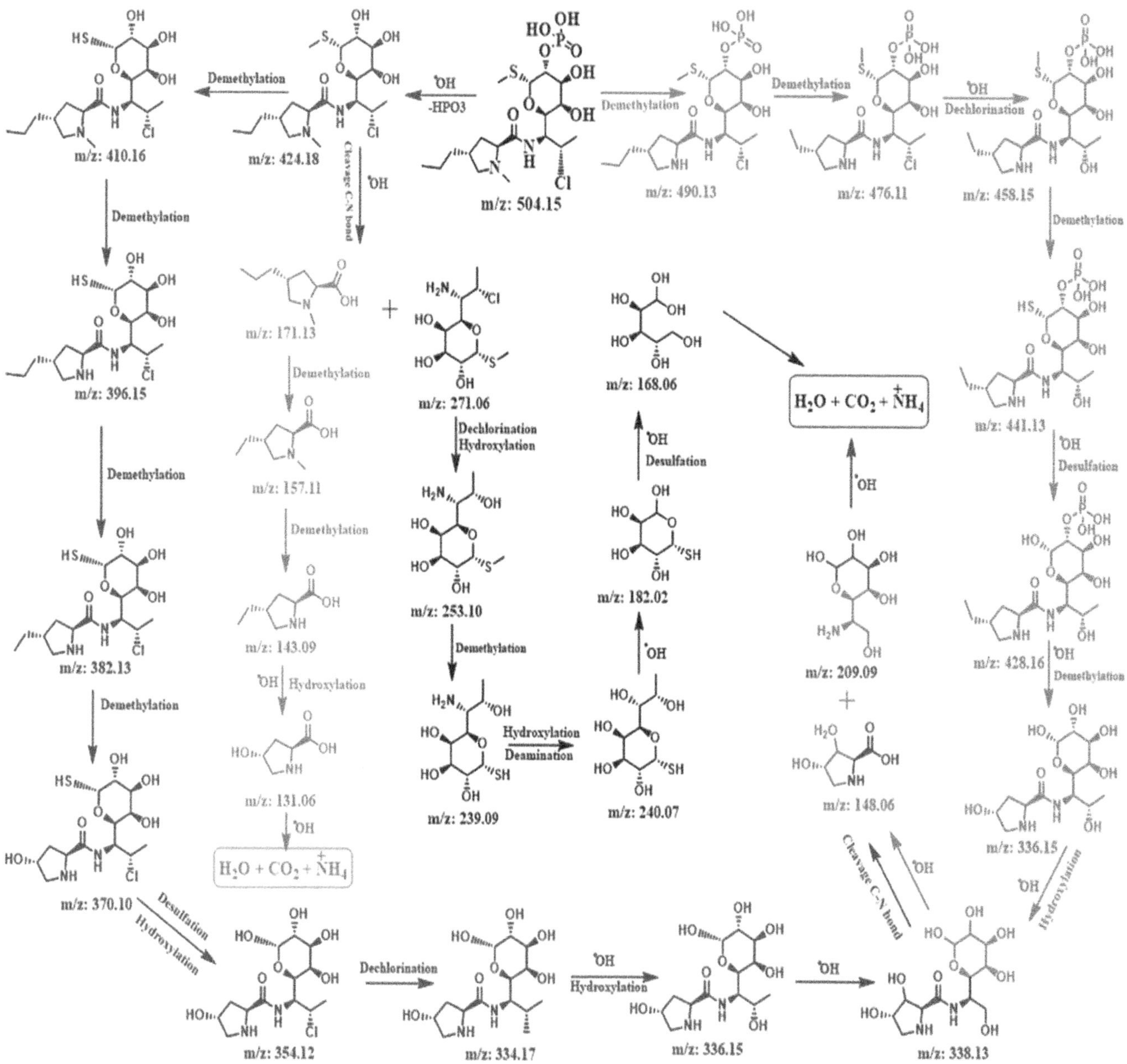

Figure 9 | Proposed degradation pathways for clindamycin phosphate.

the demethylation reaction, three other methyl groups are removed from different positions, resulting in the formation of intermediates with m/z of 396, 382 and 370. Hydroxylation reaction as well as desulfonation of m/z = 370 intermediate by the attack of hydroxyl radicals produces a new product with m/z = 354. In the following, the decolorization and hydroxylation reaction of the M/z = 354 intermediate by OH^0 produces new compounds with m/z = 334, m/z = 336 and m/z = 338, respectively.

The second route (red path) consists of two consecutive attacks of hydroxyl radicals on CLN molecule. Due to the elimination of two groups of methyl (demethylation reaction) from the primary clindamycin phosphate, two intermediates with masses m/z = 490 and m/z = 476 are formed, respectively. Hydroxylation with chlorination as well as desulfonylation and demethylation reactions, lead to the production of intermediates with m/z = 458, m/z = 441, m/z = 428 and m/z = 336. Both the first and second pathways jointly yield an intermediate with m/z = 338 resulting from sequential hydroxylation. Due to the subsequent attack of hydroxyl radicals to C-N bond, this molecule is broken down and forms products with masses

Table 1 | Summary of research studies on CLN degradation by other AOPs

Methods applied	Maximum removal	Condition	Reference
Nanoscale zero-valent iron/H_2O_2/US	90%	nZVI concentration (0.2 g/L), H_2O_2 concentration (to 180 mM), and ultrasound frequency (130 kHz/500 W).	(Gholami *et al.* 2016)
UV/H_2O_2	99.9%	medium-pressure UV lamp (MPUV) with hydrogen peroxide	(Keen & Linden 2013)
Electrochemical advanced oxidation processes (EAOPs)	100%	1 h action of the effluent wastewater process using the BDD anode	(Martínez-Pachón *et al.* 2022)
Sonochemical advanced oxidation processes	100%	Sono-Fenton, Sono-photo-Fenton, Sono-photo-Fenton/oxalic acid, 90 min of treatment	(Serna-Galvis *et al.* 2019)
Ozonation	>70%	A continuous system including 3 g O_3/m^3 next 5 g O_3/m^3 then 7 g O_3/m^3 after 20 min of reaction	(El-taliawy *et al.* 2017)
Non-thermal plasma	100%	Energy input 24 kV, reaction time = 10 min, initial pH = 3	Current study

m/z = 209 and m/z = 148. Finally, these two paths lead to complete degradation and mineralization of generated by-products by multiple attacks of hydroxyl radicals.

The third path (black and green path) includes breaking the C-N bond in the intermediate molecule) (m/z = 424), by hydroxyl radicals attack which can further produce two other compounds (m/z = 171 and m/z = 271). The m/z = 171 intermediate is converted into a series of acidic carboxylic aliphatic compounds through demethylation and hydroxylation reactions due to numerous attacks of hydroxyl radicals, and finally, these compounds are mineralized into water, carbon dioxide and ammonium ion. The product with m/z = 271 is also subjected to a hydroxylation reaction that causes the loss of chlorine ions in the molecular structure and yields m/z = 253 intermediate. With further reactions, due to numerous attacks of oxidants in the environment, including hydroxyl radicals and through hydroxyl reactions, demethylation, demining and desulfonation, the structure of the molecule becomes unstable, broken down and produces an aliphatic product with m/z = 168, which is ultimately mineralized by the attack of hydroxyl radicals like the other two routes to water and carbon dioxide. Amine removed from molecule structure with m/z = 239 is also likely to be released as ammonium ion. Thus it can be concluded that following the continuous hydroxyl attacking, the smaller molecular compounds are mineralized to CO_2 and H_2O. Likewise, the TOC of the CLN solution (Figure 5) gradually decreased with an increase in treatment time, with a 50% TOC removal obtained within 15 min treatment, suggesting that CLN can be oxidized and degraded into small inorganic compounds. With further oxidation/reactions and longer processing times, these molecules are expected to be decomposed or mineralized. As depicted after 30 min 98% TOC removal was obtained suggesting that almost all the remaining by-products have been converted to CO_2 and H_2O. In accordance with the results of the current study, Dobrin *et al.* (2013) reported that various short-chain acids like oxalic, maleic, malonic, succinic, formic, and acetic acids are formed following the degradation of diclofenac by plasma (Dobrin *et al.* 2013). Similarly, the main by-products from paracetamol degradation are organic acids and aromatic compounds (Taylor & Pirzada 1994).

Energy yield and economic analysis

Energy consumption is a deciding factor for evaluating the cost of the NTP system in practical applications (Zhang *et al.* 2012). The energy yield on CLN degradation as a function of the discharge power at different plasma exposure times was illustrated in Figure 10. As depicted after the first 5 min-treatment, the energy yield declined gradually with the increase of discharge power, and its peak was observed at 50 W within 5 min before declining. The energy yield subsequently decreased with an increase in discharge power as the treatment time extended to 10 and 15 min. Therefore, it can be concluded that the shorter the processing time the higher the energy yield will be. Balancing the energy yield and removal efficiency, the discharge power of 50 W was recommended for the following treatments. Based on the results of energy yield and economic analysis, the optimal energy efficiency of 2.34 g kWh^{-1} was obtained for CLN removal. Given the current average cost of electricity in Iran (700 Rilas per kWh^{-1}), it equals 150 Toman (150,000 Rials) per each gram of CLN degradation calculated by commercial power.

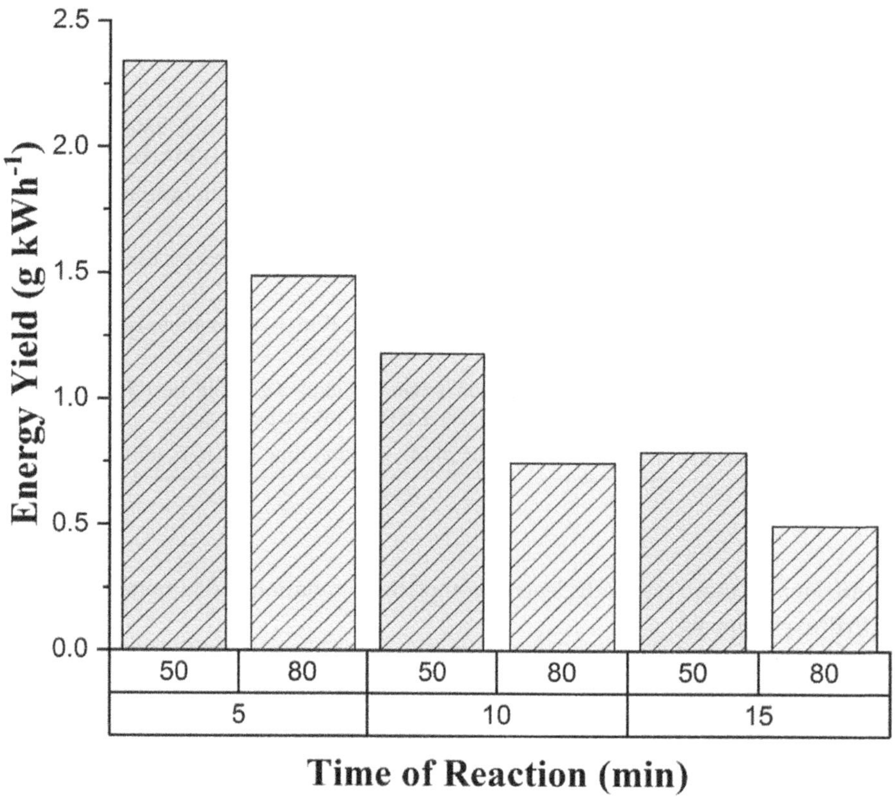

Figure 10 | The degradation energy yield of CLN using two voltages (50 and 80 W) and within 15 min of treatment (CLN concentration 10 mg/L and pH = 7.4).

Antibacterial activity

The degradation of antibiotics can be verified by the loss of their biological function. To examine this, the antibacterial activity of CLN dissolved in aqueous solution before and after NTP degradation was examined using the disk diffusion assay and measuring the inhibition zone diameter (mm) for *Staphylococcus aureus* under aerobic conditions and *Closteridium perferigens* under anaerobic environment. As shown in Figure 11(a) and 11(b), the inhibition zone diameter for non-treated CLN (20 mg L^{-1}) was obtained 27 and 25 mm for *Staphylococcus aureus* and *Closteridium perferigens*, respectively. However, after 5 and 15 min of treatment, the CLN molecules degraded to some extent which clearly resulted in the reduction of the diameters of inhibition zones formed around the disks suggesting the influence of reaction time on CLN degradation. What stands out in Figure 11 is that after 15 min of reaction the diameter of inhibition zones did not reach near zero which is indicative of remaining some degraded but still inhibitory or active parent or intermediate CLN by-products in the solution. This is also in accordance with the result obtained for CLN removal as in 15 min of reaction roughly 85% reduction in CLN was observed. Additionally, the results of the TOC and LC-MS/MS analysis show some other forms of organic structures still exist in the solution.

To elaborate more on the degradation efficacy of NTP process and/or the toxicity of the formed by-products towards selected microorganisms, the effect of the reaction time on the antimicrobial activity of the remaining CLN in samples was quantitatively assessed using the broth dilution method. This estimated the MIC values for the two selected microorganisms (Table 2). According to the MIC values for *Closteridium perferigens* and *Staphylococcus aureus*, the CLN samples exhibited lower antimicrobial activities (higher MICs) after NTP treatment as compared with the corresponding untreated control. It is abundantly clear that the reduction of the antimicrobial activity of target antibiotic in our study (CLN) is due to the plasma treatment, which is well in agreement with the result obtained from the disk diffusion assay. Moreover, it can be inferred that the degradation processes did not generate by-products with greater antimicrobial activity than the parent compound.

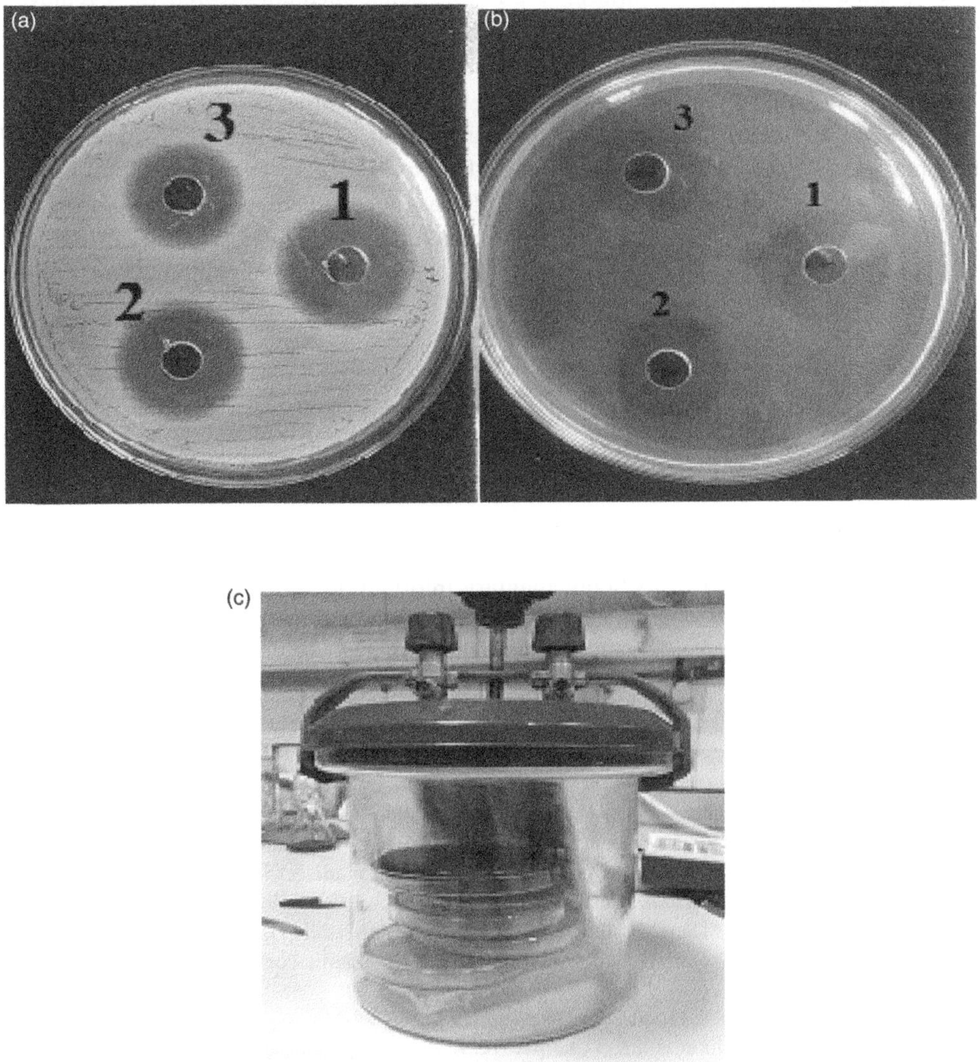

Figure 11 | Inhibition zone diameter for untreated CLN with the initial concentration of 20 mg/L (1), 5 min plasma-treated CLN (2), and 15 min plasma-treated CLN (3) against (a) *S-aureus* and (b) *C. perferingens*. Anaerobic condition for the growth of *C. perferingens* using gas pack (c).

Table 2 | Minimum inhibitory concentration of either treated (2: after 5 min, 3 = after 15 min) or untreated control (1: zero time) CLN for *Staphylococcus aureus* and *Closteridium perferigens*

Sample number / Organism	1	2	3
Staphylococcus aureus	6.25%(1/16)	12.5%(1/8)	25%(1/4)
Closteridium perferigens	6.25%(1/16)	3.1%(1/32)	6.25%(1/16)

CONCLUSION

The breakdown efficacy of CLN in water using NTP process was evaluated. Here, the effects of various parameters such as the initial pH of the solution, drug concentration, reaction time, and applied voltage on CLN degradation were worked out using atmospheric air plasma. Briefly, higher working voltage and acidic medium favored CLN removal. The results also revealed that the degradation rate was higher for a lower concentration of CLN. The reduced TOC values and analytical measures revealed that the target antibiotic was degraded to smaller intermediate molecules that eventually converted to NH_4^+, CO_2 and H_2O. Short-lived hydroxyl radicals have pivotal roles in this process. Based on the results of analytical techniques including LC-MS/MS, the possible degradation mechanism and pathways of CLN treated by DBD plasma were proposed. A relationship was found between the reduction of the diameter of inhibition zones in disk diffusion assays and reaction time of treatment demonstrating the effectiveness of plasma treatment on CLN degradation. The MIC values obtained indicated no toxic remaining by-products formed in the solution towards *Staphylococcus aureus* and *Closteridium perferigens*. NTP treatment showed its potency as an affordably promising method in complex-containing organic pollutants remediation.

ACKNOWLEDGEMENTS

This work was financially supported by the Water Research Institute of Iran (Grants No. 370/140011-06). We acknowledge support from the Water and Wastewater Research center and its laboratory operated by the Water Research Institute. The authors would like to thank Dr Parach at Water and Wastewater Laboratory for the help with the instrumental analysis.

DATA AVAILABILITY STATEMENT

All relevant data are included in the paper or its Supplementary Information.

CONFLICT OF INTEREST

The authors declare there is no conflict.

REFERENCES

Aggelopoulos, C. & Tsakiroglou, C. 2021 A new perspective towards in-situ cold plasma remediation of polluted sites: direct generation of micro-discharges within contaminated medium. *Chemosphere* **266**, 128969.

Aggelopoulos, C., Tataraki, D. & Rassias, G. 2018 Degradation of atrazine in soil by dielectric barrier discharge plasma–potential singlet oxygen mediation. *Chemical Engineering Journal* **347**, 682–694.

Ansari, M., Sharifian, M., Ehrampoush, M. H., Mahvi, A. H., Salmani, M. H. & Fallahzadeh, H. 2021 Dielectric barrier discharge plasma with photocatalysts as a hybrid emerging technology for degradation of synthetic organic compounds in aqueous environments: a critical review. *Chemosphere* **263**, 128065.

Aziz, K. H. H., Miessner, H., Mueller, S., Kalass, D., Moeller, D., Khorshid, I. & Rashid, M. A. M. 2017 Degradation of pharmaceutical diclofenac and ibuprofen in aqueous solution, a direct comparison of ozonation, photocatalysis, and non-thermal plasma. *Chemical Engineering Journal* **313**, 1033–1041.

Babu, D. S., Srivastava, V., Nidheesh, P. & Kumar, M. S. 2019 Detoxification of water and wastewater by advanced oxidation processes. *Science of the Total Environment* **696**, 133961.

Bader, H. 1982 Determination of ozone in water by the indigo method: a submitted standard method. *Ozone: Science & Engineering* **4** (8), 169–176.

Balouiri, M., Sadiki, M. & Ibnsouda, S. K. 2016 Methods for in vitro evaluating antimicrobial activity: a review. *Journal of Pharmaceutical Analysis* **6** (2), 71–79.

Baloul, Y., Aubry, O., Rabat, H., Colas, C., Maunit, B. & Hong, D. 2017 Paracetamol degradation in aqueous solution by non-thermal plasma. *The European Physical Journal Applied Physics* **79** (3), 30802.

Banaschik, R., Jablonowski, H., Bednarski, P. J. & Kolb, J. F. 2018 Degradation and intermediates of diclofenac as instructive example for decomposition of recalcitrant pharmaceuticals by hydroxyl radicals generated with pulsed corona plasma in water. *Journal of Hazardous Materials* **342**, 651–660.

Batt, A. L., Bruce, I. B. & Aga, D. S. 2006 Evaluating the vulnerability of surface waters to antibiotic contamination from varying wastewater treatment plant discharges. *Environmental Pollution* **142** (2), 295–302.

Boehm, D., Heslin, C., Cullen, P. J. & Bourke, P. 2016 Cytotoxic and mutagenic potential of solutions exposed to cold atmospheric plasma. *Scientific Reports* **6** (1), 1–14.

Chandana, L. & Subrahmanyam, C. 2018 Degradation and mineralization of aqueous phenol by an atmospheric pressure catalytic plasma reactor. *Journal of Environmental Chemical Engineering* **6** (3), 3780–3786.

Chandana, L., Sangeetha, C., Shashidhar, T. & Subrahmanyam, C. 2018 Non-thermal atmospheric pressure plasma jet for the bacterial inactivation in an aqueous medium. *Science of the Total Environment* **640**, 493–500.

Chauvin, J., Judée, F., Yousfi, M., Vicendo, P. & Merbahi, N. 2017 Analysis of reactive oxygen and nitrogen species generated in three liquid media by low temperature helium plasma jet. *Scientific Reports* **7** (1), 1–15.

Chen, H. H., Chang, H. C., Chen, Y. K., Hung, C. L., Lin, S. Y. & Chen, Y. S. 2016 An improved process for high nutrition of germinated brown rice production: low-pressure plasma. *Food Chemistry* **191**, 120–127.

Coutinho, N. M., Silveira, M. R., Rocha, R. S., Moraes, J., Ferreira, M. V. S., Pimentel, T. C., Freitas, M. Q., Silva, M. C., Raices, R. S., Ranadheera, C. S. & Borges, F. O. 2018 Cold plasma processing of milk and dairy products. *Trends in Food Science & Technology* **74**, 56–68.

Dobrin, D., Bradu, C., Magureanu, M., Mandache, N. & Parvulescu, V. 2013 Degradation of diclofenac in water using a pulsed corona discharge. *Chemical Engineering Journal* **234**, 389–396.

Dong, H., Yuan, X., Wang, W. & Qiang, Z. 2016 Occurrence and removal of antibiotics in ecological and conventional wastewater treatment processes: a field study. *Journal of Environmental Management* **178**, 11–19.

El-taliawy, H., Ekblad, M., Nilsson, F., Hagman, M., Paxeus, N., Jönsson, K., Cimbritz, M., la Cour Jansen, J. & Bester, K. 2017 Ozonation efficiency in removing organic micro pollutants from wastewater with respect to hydraulic loading rates and different wastewaters. *Chemical Engineering Journal* **325**, 310–321.

Fahmy, A., El-Zomrawy, A., Saeed, A. M., Sayed, A. Z., El-Arab, M. A. E., Shehata, H. & Friedrich, J. 2020 Degradation of organic dye using plasma discharge: optimization, pH and energy. *Plasma Research Express* **2** (1), 015009.

Gholami, M., Rahmani, K., Rahmani, A., Rahmani, H. & Esrafili, A. 2016 Oxidative degradation of clindamycin in aqueous solution using nanoscale zero-valent iron/H2O2/US. *Desalination and Water Treatment* **57** (30), 13878–13886.

Golchin, M., Khani, M., Sadani, M., Sadeghi, M. & Jahangiri-rad, M. 2021 Occurrence and fate of amoxicillin and penicillin G antibiotics in hospital wastewater treatment plants: a case study-Gonbad Kavous, Iran. *South African Journal of Chemistry* **75**, 98–105.

Graumans, M. H., Hoeben, W. F., Russel, F. G. & Scheepers, P. T. 2020 Oxidative degradation of cyclophosphamide using thermal plasma activation and UV/H2O2 treatment in tap water. *Environmental Research* **182**, 109046.

Guo, H., Jiang, N., Wang, H., Shang, K., Lu, N., Li, J. & Wu, Y. 2019 Pulsed discharge plasma induced WO3 catalysis for synergetic degradation of ciprofloxacin in water: synergetic mechanism and degradation pathway. *Chemosphere* **230**, 190–200.

Hatzisymeon, M., Tataraki, D., Tsakiroglou, C., Rassias, G. & Aggelopoulos, C. 2021 Highly energy-efficient degradation of antibiotics in soil: extensive cold plasma discharges generation in soil pores driven by high voltage nanopulses. *Science of The Total Environment* **786**, 147420.

Jiang, Y.-H., Cheng, J.-H. & Sun, D.-W. 2020 Effects of plasma chemistry on the interfacial performance of protein and polysaccharide in emulsion. *Trends in Food Science & Technology* **98**, 129–139.

Jose, J. & Philip, L. 2019 Degradation of chlorobenzene in aqueous solution by pulsed power plasma: mechanism and effect of operational parameters. *Journal of Environmental Chemical Engineering* **7** (6), 103476.

Kale, K. H. & Desai, A. 2011 Atmospheric pressure plasma treatment of textiles using non-polymerising gases. *Indian Journal of Fibre & Textile Research* **36**, 289–299.

Keen, O. S. & Linden, K. G. 2013 Degradation of antibiotic activity during UV/H2O2 advanced oxidation and photolysis in wastewater effluent. Environmental *Science & Technology* **47** (22), 13020–13030.

Krause, H., Schweiger, B., Schuhmacher, J., Scholl, S. & Steinfeld, U. 2009 Degradation of the endocrine disrupting chemicals (EDCs) carbamazepine, clofibric acid, and iopromide by corona discharge over water. *Chemosphere* **75** (2), 163–168.

Lee, D., Lee, J.-C., Nam, J.-Y. & Kim, H.-W. 2018 Degradation of sulfonamide antibiotics and their intermediates toxicity in an aeration-assisted non-thermal plasma while treating strong wastewater. *Chemosphere* **209**, 901–907.

Li, J., Liu, K., Yan, S., Li, Y. & Han, D. 2016 Application of thermal plasma technology for the treatment of solid wastes in China: an overview. *Waste Management* **58**, 260–269.

Li, H., Li, T., He, S., Zhou, J., Wang, T. & Zhu, L. 2020 Efficient degradation of antibiotics by non-thermal discharge plasma: highlight the impacts of molecular structures and degradation pathways. *Chemical Engineering Journal* **395**, 125091.

Liao, X., Liu, D., Xiang, Q., Ahn, J., Chen, S., Ye, X. & Ding, T. 2017 Inactivation mechanisms of non-thermal plasma on microbes: a review. *Food Control* **75**, 83–91.

Lou, J., Lu, N., Li, J., Wang, T. & Wu, Y. 2012 Remediation of chloramphenicol-contaminated soil by atmospheric pressure dielectric barrier discharge. *Chemical Engineering Journal* **180**, 99–105.

Lukes, P., Dolezalova, E., Sisrova, I. & Clupek, M. 2014 Aqueous-phase chemistry and bactericidal effects from an air discharge plasma in contact with water: evidence for the formation of peroxynitrite through a pseudo-second-order post-discharge reaction of H_2O_2 and HNO2. *Plasma Sources Science and Technology* **23** (1), 015019.

Magureanu, M., Piroi, D., Mandache, N. B. & Parvulescu, V. 2008 Decomposition of methylene blue in water using a dielectric barrier discharge: optimization of the operating parameters. *Journal of Applied Physics* **104** (10), 103306.

Magureanu, M., Piroi, D., Mandache, N., David, V., Medvedovici, A., Bradu, C. & Parvulescu, V. 2011 Degradation of antibiotics in water by non-thermal plasma treatment. *Water Research* **45** (11), 3407–3416.

Magureanu, M., Bilea, F., Bradu, C. & Hong, D. 2021 A review on non-thermal plasma treatment of water contaminated with antibiotics. *Journal of Hazardous Materials* **417**, 125481.

Majeed, H. J., Riquelme, M. V., Davis, B. C., Gupta, S., Angeles, L., Aga, D. S., Garner, E., Pruden, A. & Vikesland, P. J. 2021 Evaluation of metagenomic-enabled antibiotic resistance surveillance at a conventional wastewater treatment plant. *Frontiers in Microbiology* **12**, 657954.

Martínez-Pachón, D., Botero-Coy, A. M., Hernández, F., López, N. L., Torres-Palma, R. A. & Moncayo-Lasso, A. 2022 Elimination of contaminants of emerging concern and their environmental risk in world-real municipal wastewaters by electrochemical advanced oxidation processes. *Journal of Environmental Chemical Engineering* **10** (3), 107803.

Murugesan, P., Moses, J. & Anandharamakrishnan, C. 2020 Water decontamination using non-thermal plasma: concepts, applications, and prospects. *Journal of Environmental Chemical Engineering* **8** (5), 104377.

Napartovich, A. 2001 Overview of atmospheric pressure discharges producing nonthermal plasma. *Plasmas and Polymers* **6** (1), 1–14.

Nasuhoglu, D., Yargeau, V. & Berk, D. 2011 Photo-removal of sulfamethoxazole (SMX) by photolytic and photocatalytic processes in a batch reactor under UV-C radiation ($\lambda max = 254 nm$). *Journal of Hazardous Materials* **186**, 67–75.

Oertel, R., Schubert, S., Mühlbauer, V., Büttner, B., Marx, C. & Kirch, W. 2014 Determination of clindamycin and its metabolite clindamycin sulfoxide in diverse sewage samples. *Environmental Science and Pollution Research* **21** (20), 11764–11769.

Rahimpour, M., Taghvaei, H., Zafarnak, S., Rahimpour, M. R. & Raeissi, S. 2019 Post-discharge DBD plasma treatment for degradation of organic dye in water: a comparison with different plasma operation methods. *Journal of Environmental Chemical Engineering* **7** (4), 103220.

Rodriguez-Mozaz, S., Vaz-Moreira, I., Della Giustina, S. V., Llorca, M., Barceló, D., Schubert, S., Berendonk, T. U., Michael-Kordatou, I., Fatta-Kassinos, D., Martinez, J. L. & Elpers, C. 2020 Antibiotic residues in final effluents of European wastewater treatment plants and their impact on the aquatic environment. *Environment International* **140**, 105733.

Rong, S. & Sun, Y. 2014 Wetted-wall corona discharge induced degradation of sulfadiazine antibiotics in aqueous solution. *Journal of Chemical Technology & Biotechnology* **89** (9), 1351–1359.

Sang, W., Cui, J., Feng, Y., Mei, L., Zhang, Q., Li, D. & Zhang, W. 2019 Degradation of aniline in aqueous solution by dielectric barrier discharge plasma: mechanism and degradation pathways. *Chemosphere* **223**, 416–424.

Sarangapani, C., Ziuzina, D., Behan, P., Boehm, D., Gilmore, B. F., Cullen, P. J. & Bourke, P. 2019 Degradation kinetics of cold plasma-treated antibiotics and their antimicrobial activity. *Scientific Reports* **9** (1), 1–15.

Serna-Galvis, E. A., Botero-Coy, A. M., Martínez-Pachón, D., Moncayo-Lasso, A., Ibáñez, M., Hernández, F. & Torres-Palma, R. A. 2019 Degradation of seventeen contaminants of emerging concern in municipal wastewater effluents by sonochemical advanced oxidation processes. *Water Research* **154**, 349–360.

Smith, J., Adams, I. & Ji, H.-F. 2017 Mechanism of ampicillin degradation by non-thermal plasma treatment with FE-DBD. *Plasma* **1** (1), 1–11.

Taylor, P. R. & Pirzada, S. A. 1994 Thermal plasma processing of materials: a review. *Advanced Performance Materials* **1** (1), 35–50.

Vanraes, P., Ghodbane, H., Davister, D., Wardenier, N., Nikiforov, A., Verheust, Y. P., Van Hulle, S. W., Hamdaoui, O., Vandamme, J., Van Durme, J. & Surmont, P. 2017 Removal of several pesticides in a falling water film DBD reactor with activated carbon textile: energy efficiency. *Water Research* **116**, 1–12.

Wang, J. & Zhuan, R. 2020 Degradation of antibiotics by advanced oxidation processes: an overview. *Science of the Total Environment* **701**, 135023.

Wang, C., Qu, G., Wang, T., Deng, F. & Liang, D. 2018 Removal of tetracycline antibiotics from wastewater by pulsed corona discharge plasma coupled with natural soil particles. *Chemical Engineering Journal* **346**, 159–170.

Wynalda, M. A., Hutzler, J. M., Koets, M. D., Podoll, T. & Wienkers, L. C. 2003 In vitro metabolism of clindamycin in human liver and intestinal microsomes. *Drug Metabolism and Disposition* **31** (7), 878–887.

Xu, H., Ma, R., Zhu, Y., Du, M., Zhang, H. & Jiao, Z. 2020a A systematic study of the antimicrobial mechanisms of cold atmospheric-pressure plasma for water disinfection. *Science of the Total Environment* **703**, 134965.

Xu, Z., Xue, X., Hu, S., Li, Y., Shen, J., Lan, Y., Zhou, R., Yang, F. & Cheng, C. 2020b Degradation effect and mechanism of gas-liquid phase dielectric barrier discharge on norfloxacin combined with H_2O_2 or Fe^{2+}. *Separation and Purification Technology* **230**, 115862.

Zamani, L., Sadjadi, S., Ashouri, F. & Jahangiri-Rad, M. 2021 Carbamazepine removal from aqueous solution by synthesized reduced graphene oxide-nano zero valent iron (Fe0-rGO) composite: theory, process optimization, and coexisting drugs effects. *Water Science and Technology* **84** (9), 2557–2577.

Zhang, J., Liu, D., Bian, W. & Chen, X. 2012 Degradation of 2, 4-dichorophenol by pulsed high voltage discharge in water. *Desalination* **304**, 49–56.

Zhang, T., Zhou, R., Wang, P., Mai-Prochnow, A., McConchie, R., Li, W., Zhou, R., Thompson, E. W., Ostrikov, K. K. & Cullen, P. J. 2021 Degradation of cefixime antibiotic in water by atmospheric plasma bubbles: performance, degradation pathways and toxicity evaluation. *Chemical Engineering Journal* **421**, 127730.

Zhou, R., Zhang, T., Zhou, R., Mai-Prochnow, A., Ponraj, S. B., Fang, Z., Masood, H., Kananagh, J., McClure, D., Alam, D. & Ostrikov, K. K. 2021 Underwater microplasma bubbles for efficient and simultaneous degradation of mixed dye pollutants. *Science of The Total Environment* **750**, 142295.

First received 2 August 2022; accepted in revised form 4 October 2022. Available online 11 October 2022

doi: 10.2166/wst.2022.299

Sustainable replacement of EDTA–Biojarosite for commercial iron in the Fenton's and UV–Fenton's degradation of Rhowedamine B – a process optimization using Box–Behnken method

Bhaskar S [a,*], K. N. Rashmishree[a], B. Manu[a] and M. Y. Sreenivasa[b]

[a] Department of Civil Engineering, National Institute of Technology Karnataka, Surathkal, P.O. Srinivasnagar, Mangalore 575025, India
[b] Department of Studies in Microbiology, University of Mysore, Mysuru, Karnataka 570006, India
*Corresponding author. E-mail: baskarmalwanitk@gmail.com

BS, 0000-0002-0203-3629

ABSTRACT

Biojarosite as a replacement for commercial iron catalyst in the oxidative degradation of the dye Rhodamine B was confirmed and established. Investigations on the oxidative degradation by Fenton's oxidation and UV–Fenton's oxidation with EDTA at neutral pH were conducted and degradation of target compound was evaluated. UV–Fenton's oxidation was shown to be efficient over Fenton's oxidation in the degradation of Rhodamine B with removal efficiency of 90.0%. Design of Experiments was performed with Box–Behnken design. Investigation was conducted for the predicted values separately for both Fenton's oxidation and UV–Fenton's oxidation and the Rhodamine B removal was taken as response. Variable parameters biojarosite, H_2O_2 dosage and EDTA were optimized in the range of 0.1–1 g/L, 2.94–29.4 mM and 10–100 mM, respectively. A quadratic regression model is fitted for both Fenton's and UV–Fenton's oxidation. Analysis of variance (ANOVA) is performed and model fit is discussed.

Key words: Box–Behnken, Biojarosite, EDTA, Fenton's Process, Rhodamine B, UV–Fenton's process

HIGHLIGHTS

- Biojarosite – EDTA based Fenton's Oxidation.
- Biojarosite – EDTA based UV Fenton's Oxidation.
- Process optimization – Box Behnken method.
- Commercial iron replacement from Biojarosite.
- Comparison of Fenton's and UV–Fenton's oxidation for degradation of Rhodamine B.

GRAPHICAL ABSTRACT

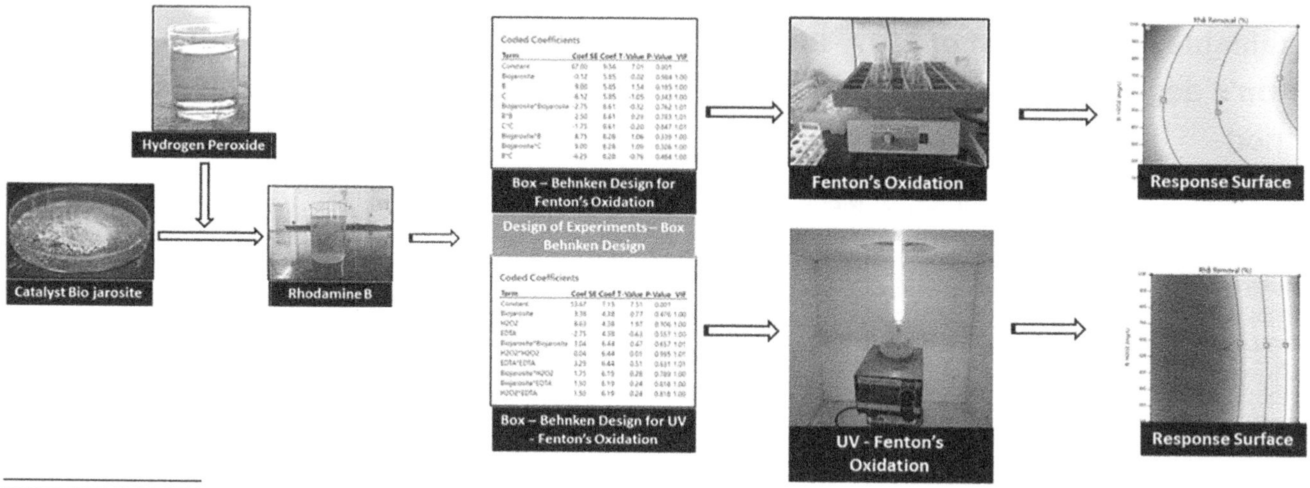

INTRODUCTION

Rapid development in the textile industries demands the usage of diverse chemical reagents, which leads to the generation of wastewater contaminated with environmental persistent chemicals. Dyes by the virtue of their nature are non-amenable to degradation on exposure to water and other chemicals (Dutta & Mukhopadhyay 2001). Xanthene dyes are widely used in textile industries and as fluorescent, among which Rhodamine B has a vast application in textile industries (Su *et al.* 2013). Dyes usually have benzene and naphthalene rings but may also contain aromatic or aliphatic groups. It is the side group attached to dye which imparts the color. It is the complex nature of dyes that make them more interesting in the field of chemical remediation. Discharge of wastewater containing dye to natural streams and rivers may harm the biodiversity posing toxicity to aquatic life (Islam & Mostafa 2018).

Many treatment methods like adsorption, biological degradation, filtration process and advanced oxidation have been tried over a decade by various researchers. Nanocomposites-based advanced oxidation process with metal oxides have been attempted in recent years claiming high removal efficiency with reduction in treatment time. Carbon iron–graphene oxide nanocomposite based sono–Fenton's oxidation and iron doped carbon gel based electro Fenton's oxidation are efficient in organic compound removal with 99% organic carbon removal (Hassani *et al.* 2017; Ahmadi *et al.* 2021; Karim *et al.* 2022). Xue *et al.* (2009), studied the effect of ferrous and ferric on sorption and Fenton's like degradation of Rhodamine B and claimed efficiency of Fenton's oxidation increases with increase in H_2O_2 dosage or with exposed surface area (Xue *et al.* 2009). AlHamedi *et al.* (2009) investigated degradation of Rhodamine B in presence of UV light and claimed its high efficiency in the dye degradation. The team derived low molecular weight aliphatic alcohols and acids as the degradation products and observed the degradation follows first order linear kinetics (AlHamedi *et al.* 2009). Ozonation, electrochemical method, and photochemical methods have been suggested for dye degradation, out of which photochemical method is proven to be effective because of its application (Dong & Tang 1993; Kim *et al.* 2004; Huang *et al.* 2008). Hydroxyl peroxide produced on the dissociation of hydrogen peroxides acts on the complex structure of organic dyes, breaking it and causing the degradation. OH radicals formed act as initiators for the degradation as well as scavengers (Zhou *et al.* 2016). High dosage of reactants and catalyst leads to increasing the treatment cost. It is thus necessary to optimize the oxidation process for the degradation of any organic compounds. Studies on the replacement of commercial iron with lateritic iron in Fenton's oxidation have been conducted, and it was reported to cut down the cost of catalyst (Karale *et al.* 2014; Sangami & Manu 2018; Bhaskar *et al.* 2021). Bhaskar *et al.* (2019), biologically synthesized jarosite using novel *Acidithiobacillus ferroxidans* strain and presented the application of synthesized biojarosite as a Fenton's catalyst in the degradation of Ametryn (Bhaskar *et al.* 2019). Apart from the catalyst and reactant dosage, a narrow range of acidic pH imposes another restriction on Fenton's oxidation.

Box–Behnken design (BBD) is a 3 interlocking 2^2 factorial design with points on the surface of a sphere surrounding the center of the design (Tripathi *et al.* 2009). BBD is used to optimize and investigate the influence of key process parameters as independent variables on the response as a dependent variable (Tak *et al.* 2015; Shokri *et al.* 2019). Being most suitable for chromatographic optimization of analysis, BBD indicates most significant factors and reveals many mutual interactions between the variable parameters (Ahmadi *et al.* 2017).

The present study encompasses novel replacement of commercial iron with biojarosite as a catalyst for the Fenton's and UV–Fenton's oxidation of Rhodamine B using EDTA as chelating agent. BBD is adopted for the optimization in which catalyst loading, H_2O_2 dosage and chelating agent concentration are taken as independent variables and rhodamine B removal efficiency is taken as response to fit the regression model.

MATERIALS AND METHODS

UV Fenton's and Fenton's oxidation of Rhodamine B

UV–Fenton's and Fenton's degradation of Rhodamine B was carried out with initial dye concentration of 5 mg/L for both the experiments (Zhao & Zhu 2006; Su *et al.* 2013). UV–Fenton's oxidation was carried out in a closed UV chamber with continuous stirring of solution in magnetic stirrer at an illumination of 254 nm with an intensity at the center of the solution being 5.6 mW/cm^2. The experimental set-up is shown in Figure 1. Synthesis and characterization of Biojarosite were discussed earlier (Bhaskar *et al.* 2019). Biojarosite was added as per design dose into Rhodamine B solution separately taken in a conical flask. The solution was adjusted to pH 7 using 1 N H_2SO_4 or 1 N NaOH and allowed for 50 minutes to ensure proper mixing and uniform distribution of Biojarosite powder in the solution before addition of H_2O_2. Initial investigation was conducted with 0.5 g/L of Biojarosite, 2.94 mM of H_2O_2 dosage and 50 mM of EDTA to ensure the degradation of Rhodamine B with Fenton's and UV–Fenton's oxidation before the run of design of experiments (DOE). Experimentation was as per design with a set of experimental conditions for both UV–Fenton's and Fenton's process separately. Samples were drawn at regular intervals for analysis. During sampling, each time 1 mL of sodium thiosulphate was added to arrest the reaction (Khan *et al.* 2009). Concentration of Rhodamine B was measured using double beam UV–Vis spectrophotometer. Chemical oxygen demand (COD) was measured using closed reflux calorimetric method (Lapara *et al.* 2000). pH was measured using digital pH meter. Ferric iron was measured by potassium thiocyanate method using UV-Vis spectrophotometer (Woods & Mellon 1941). H_2O_2 was measured using double beam UV-spectrophotometer (Eisenberg 1943).

Design of experiments – Box–Behnken design

Response surface methodology using BBD is adopted for the present study. BBD method was used in the present study. Response as the removal of Rhodamine B was analyzed via response surface methodology. Three levels (−1, 0, +1) were selected as independent variables and Rhodamine B removal was considered as dependent variable. 15 experimental runs were executed. The variable parameters were optimized in the range of 0.1 g/L to 1 g/L for Biojarosite, 2.94 mM to 29.4 mM for H_2O_2 dosage and 10–100 mM for EDTA concentration, and Rhodamine B is considered as response. Experimental data were fitted to regression model and 3D contour graphs were used to evaluate the optimized parameters which influence the responses (See Table 1).

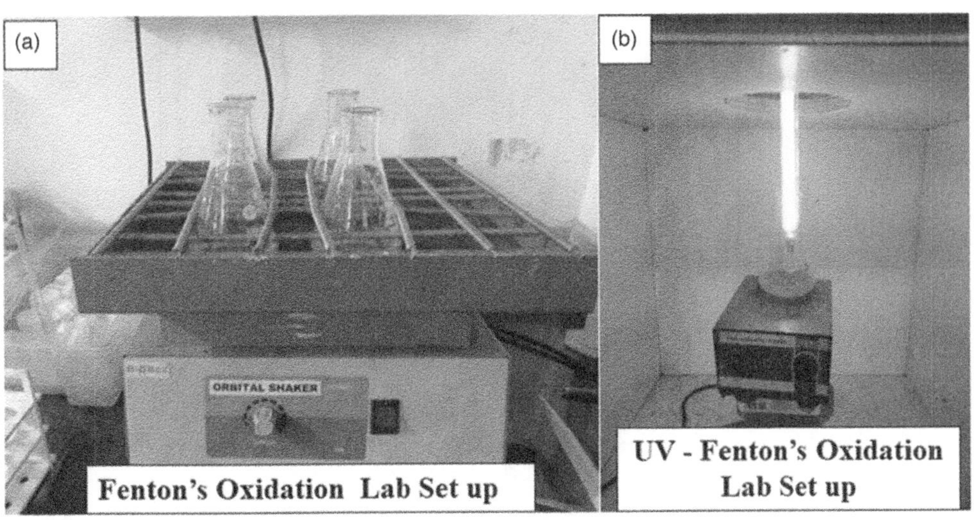

Figure 1 | Experimental Set-up for Fenton's and UV–Fenton's Oxidation of Rhodamine B.

Table 1 | Experimental levels of the variables

Dependent variables	Low Level (−1)	Mid-Level (0)	High Level (+1)
Biojarosite (g/L)	0.1	0.5	1.0
H_2O_2 (mM)	2.94	16.17	29.4
EDTA (mM)	10	50	100

RESULTS AND DISCUSSION

Fenton's oxidation and UV–Fenton's oxidation of Rhodamine B

UV–Fenton's oxidation is found to be most effective for the removal of Rhodamine B in the current study. High removal efficiency of 91.0% was observed with the Biojarosite:H_2O_2 ratio of 5:1 and 50 mM EDTA addition. For Fenton's oxidation, about 70.0% of degradation was observed within 120 minutes of treatment and almost complete degradation within 40 minutes with a rate constant 0.01895/min on 0.5 g/L of catalyst loading. Increase in catalyst load is more beneficial than increasing hydrogen peroxide since iron increases the speed of the reaction (Lojo-López *et al.* 2021). For UV–Fenton's oxidation, about 60.0% of degradation was observed within first 40 minutes of treatment and 27.6% of degradation was observed in another 40 minutes on a catalyst load of 0.5 g/L with a rate constant 0.0629/min. Degradation of rhodamine B via Fenton's oxidation and UV–Fenton's oxidation is presented in Figure 2.

About 120 mg/L of leached iron out of Biojarosite has participated in the Fenton's and UV–Fenton's oxidation. There is an increase in the ferric form of iron at initial stage as leached iron gets oxidized to ferric form and later there is a drop in the concentration of ferric form of iron in the solution. The rate of oxidation depends on the dissolution rate of ferrous ion that has leached out from the jarosite. Initial dissolution of ferrous ions is consumed and the rate of reaction will be more. It is the amount of catalyst addition which indicates the adsorption part of this heterogenous Fenton's and UV–Fenton's oxidation that in turn contributes to maximum removal of target compounds in the process. Observed results are consistent with previous works (Bhaskar *et al.* 2019, 2021; Bhaskar & Basavaraju Manu 2020). On UV–Fenton's oxidation, high removal efficiency is correlated to photo reduction of formed ferric iron species (Arslan & Tuhkanen 1999; Tekbas *et al.* 2008). From Figure 2, it is observed that maximum removal is observed within 20 mins of the treatment with UV-Fenton's oxidation, which is attributed to photo reduction on the surface of Biojarosite. Bel *et al.* (2014) demonstrated non-catalytic UV based phenol degradation and Fe clay based UV based catalytic degradation of organic compounds (Bel *et al.* 2014). Iron leached

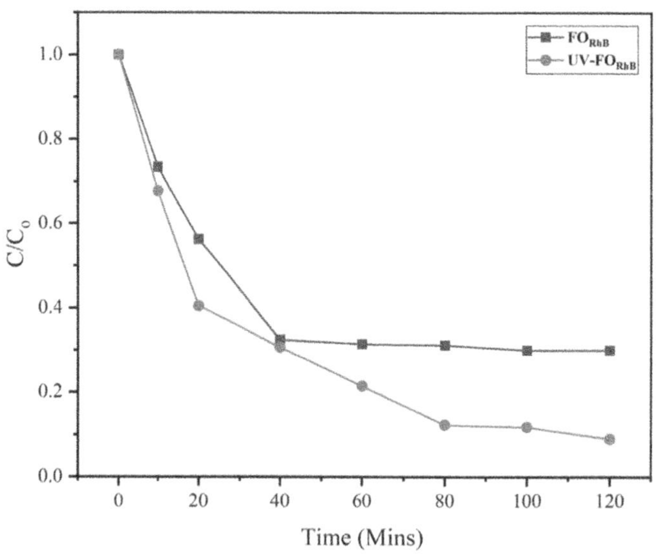

Figure 2 | Degradation of Rhodamine B on Fenton's and UV–Fenton's oxidation with 0.5 g/L of catalyst loading, 100 mg/L of H_2O_2 dosage and 500 mM of EDTA addition.

out of Biojarosite actively participated in Fenton's oxidation by the oxidation from ferrous to ferric form. Ferric iron reduced back to ferric in presence of UV light, which increases the dissociation of hydrogen peroxide to form hydroxyl radicals. OH radicals thus formed attacks the xanthene ring of rhodamine, initiating degradation.

In both Fenton's oxidation system and UV–Fenton's oxidation system, on addition of EDTA, an organic complex is formed at neutral conditions and increases the solubility of ferric iron in solution, limiting the ferric precipitation. This mechanism was observed by the previous works (Huang *et al.* 2008; Zhou *et al.* 2014; Tang *et al.* 2019). Regeneration of Fe-II EDTA complex accelerates the generation of OH radicals, promoting degradation.

Organic carbon removal was determined by the COD measurements in the present study. Maximum reduction in COD observed was 57.0% and 69.0% for the optimum removal of Rhodamine B via Fenton's oxidation and UV–Fenton's oxidation, respectively. Reduction in COD indicates better oxidation of target compound via both processes (Figure 3). The reduction in COD is also attributed to the adsorption process other than Fenton's oxidation (Masomboon *et al.* 2009).

Many researchers have discussed the degradation mechanism of Rhodamine B via Fenton's oxidation. Rhodamine B is susceptible to oxidation by the decomposition of the conjugated xanthene ring (Zhao & Zhu 2006). Degradation of Rhodamine B follows N-de-ethylation, chromophore cleavage, ring opening and mineralization stages. Zhou *et al.* (2016) identified three intermediate products due to de-ethylating steps of N,N'-diethylammonium groups such as N,N,N'-triethyl rhodamine, N,N'-diethyl rhodamine and rhodamine (Zhou *et al.* 2016). Hou *et al.* (2011) claim oxalic acid, formic acid and acetic acid as chromophore cleavage and ring opening intermediates. Tertiary amines formed during the degradation can be oxidized to amine oxides, which are further oxidized to aldehyde intermediates and then to carboxylic acids (Hou *et al.* 2011).

Catalytic efficiency of Biojarosite on reuse is tabulated in Table 2. Degradation on first cycle of reuse was 68.2% and 86.0% with rate constants 0.009/min and 0.0141/min for Fenton's oxidation and UV–Fenton's oxidation, respectively, while on second cycle of reuse the efficiency was 67.6% and 82.8% with rate constants 0.010/min and 0.0135/min, respectively. For the third cycle of reuse degradation efficiency was 65.6% and 82.4% with rate constants 0.009/min and 0.0127/min, respectively.

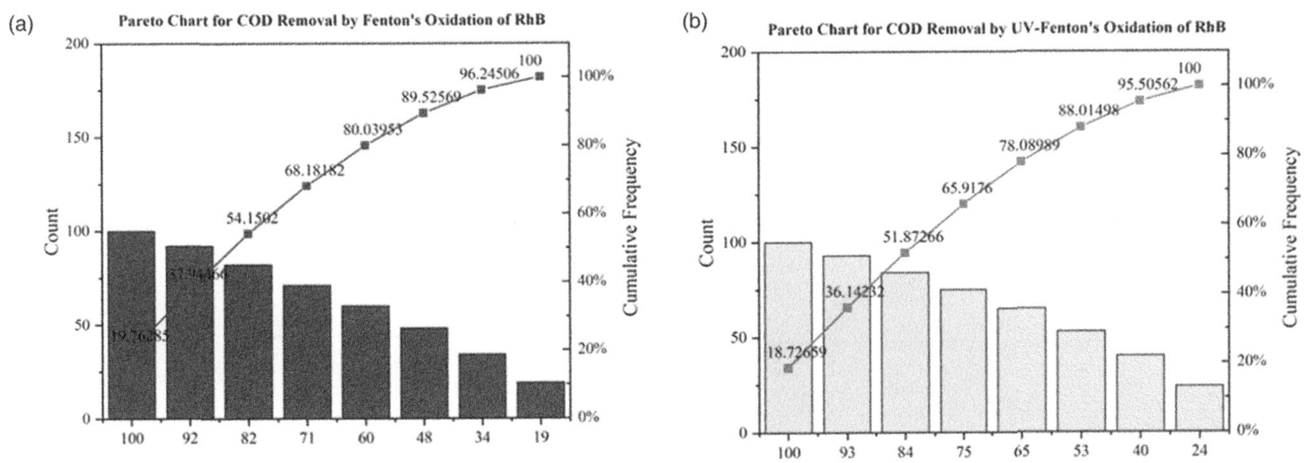

Figure 3 | Pareto chart for COD Removal on degradation of Rhodamine B. (a) Fenton's (b) UV-Fenton's Oxidation with 0.5 g/L of catalyst loading, 100 mg/L of H_2O_2 dosage and 500 mM of EDTA addition.

Table 2 | Reusability of catalyst on Fenton's and UV–Fenton's oxidation of Rhodamine B

	Fenton's oxidation		UV- Fenton's oxidation	
Cycle	**k/min**	**R^2**	**k/min**	**R^2**
1	0.0090	0.9701	0.0141	0.97653
2	0.0099	0.97816	0.0135	0.95436
3	0.0091	0.9052	0.0127	0.96835
4	0.0092	0.92416	0.0129	0.95889

Degradation on fourth cycle was 65.2% and 82.6% with rate constants 0.009/min and 0.0129/min, respectively (Figure 4). The degradation efficiency was found to be constant after third consecutive cycle of the catalyst, marking the limit for reuse.

Fenton's oxidation and UV–Fenton's oxidation of Rhodamine B – design optimization

Model F-value 89.80 indicates that the model is significant for the Rhodamine B removal by Fenton's oxidation. There is only 0.01% chance that F-value could occur due to noise. Probability value of <0.0001 indicates that the model is significant. A regression co-efficient $R^2 = 0.9939$ is a measure of goodness of fit of the model, indicating there is a high degree of correlation between observed value and predicted value. The value suggests that more than 99.39% of variance is attributed to variables showing high significance for Fenton's oxidation of Rhodamine B model. Only 0.61% of total variance cannot be explained by the model. The analysis of variance (ANOVA) quadratic regression model demonstrated that the model was highly significant for Fenton's oxidation of Rhodamine B with a low probability ($p < 0.0001$) in the F-test and an insignificant contribution from lack of fit of the model. The F – value of 0.75 for lack of fit implies that lack of fit does not contribute significantly to the total error. There is a 61.48% chance that a lack of fit F-value this large could occur due to noise. The predicted R^2 value of 0.9414 is in reasonable agreement with adjusted R^2 value 0.9828, i.e, the difference is less than 0.2.

For UV–Fenton's oxidation, the model F-value 69.21 indicates that the model is significant for the Rhodamine B removal by UV–Fenton's oxidation. There is only a 0.01% chance that this F- value could occur due to noise. Probability value of <0.0001 indicates that the model is significant. A regression co-efficient $R^2 = 0.9920$ is a measure of goodness of fit of the model, indicating that there is a high degree of correlation between observed value and predicted value. The value suggests that more than 99.20% of variance is attributed to variables showing high significance for UV–Fenton's oxidation of Rhodamine B model. Only 0.80% of total variance cannot be explained by the model.

The ANOVA quadratic regression model demonstrated that the model was highly significant for UV–Fenton's oxidation of Rhodamine B with a low probability value ($p < 0.0001$) in the F-test and insignificant contribution from lack of fit of the model. A lack of fit F-value of 7.75 implies that lack of fit is not significant in relation to pure error. There is a 11.64% chance that a lack of fit F-value this large could occur due to noise. The predicted R^2 value of 0.8813 is in reasonable agreement with adjusted R^2 value 0.9777, i.e., the difference is less than 0.2.

Co-efficient of variance (CV) obtained in the model for Fenton's oxidation is 2.47%; hence, the model can be considered to be reproducible. The CV value indicates high precision and reliability of the experiments. The *p* values of regression co-efficient suggest that among the independent test variables linear, quadratic and interaction effect the Fenton's oxidation of Rhodamine B and are highly significant. For UV–Fenton's oxidation, co-efficient of variance obtained in the model is 3.41%; hence, the model can be considered to be reproducible. The CV value indicates high precision and reliability of

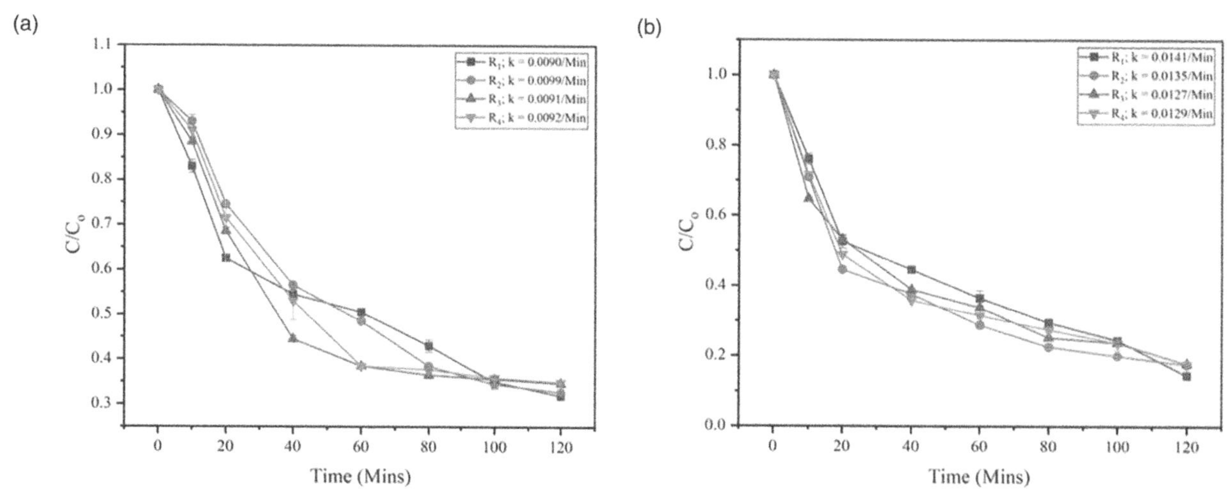

Figure 4 | Reusability of catalyst for the degradation of Rhodamine B. (a) Fenton's Oxidation (b) UV–Fenton's Oxidation with 0.5 g/L of catalyst loading, 2.94 mM of H_2O_2 dosage and 500 mM of EDTA addition.

the experiments. The p values of regression co-efficient suggest that among the independent test variables, there are linear, quadratic and interaction effects on the Fenton's oxidation of Rhodamine B, which are highly significant.

In the present study, A, C, AB and B^2 are significant model terms for the Fenton's oxidation of Rhodamine B. Thus, statistical analysis of all the experimental data showed that Biojarosite loading, H_2O_2 dosage and EDTA concentration had a significant effect on Fenton's oxidation. It is observed that Biojarosite loading and EDTA dosage exerted more linear influence on Fenton's oxidation, which indicates that Rhodamine B removal by the process was positively influenced by Biojarosite loading, H_2O_2 dosage and addition of EDTA. Higher positive linear effect of iron dosage over H_2O_2 in Fenton's oxidation had been reported by the researchers in previous work. Again, Biojarosite loading has the highest positive linear

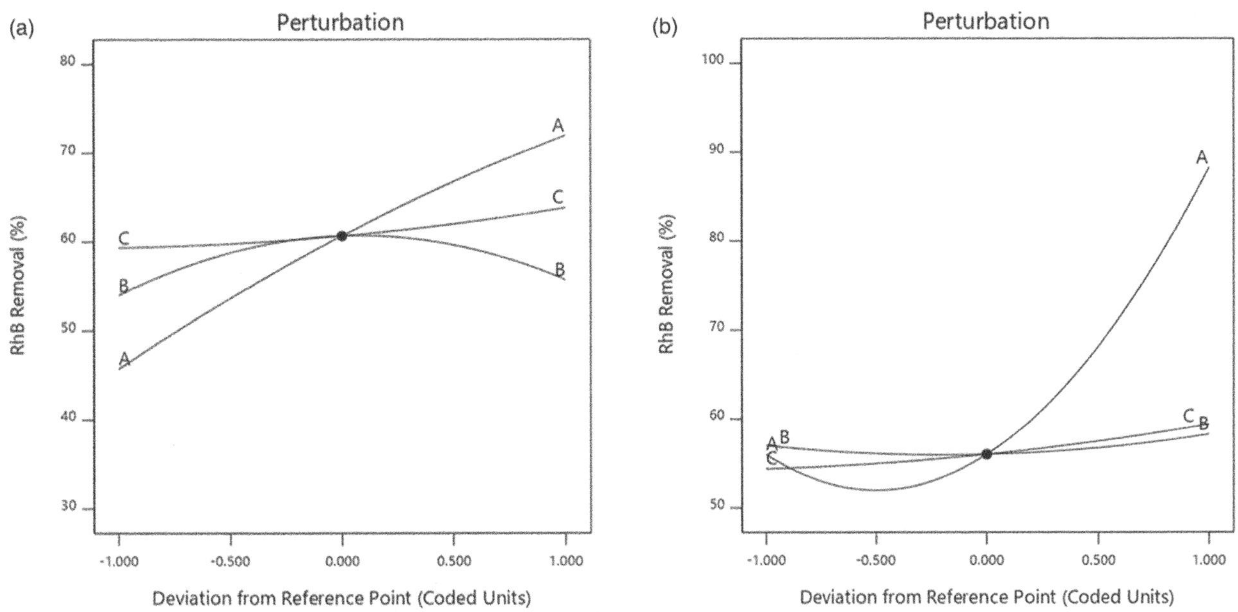

Figure 5 | Perturbation plot for the degradation of Rhodamine B. (a) Fenton's Oxidation (b) UV–Fenton's Oxidation.

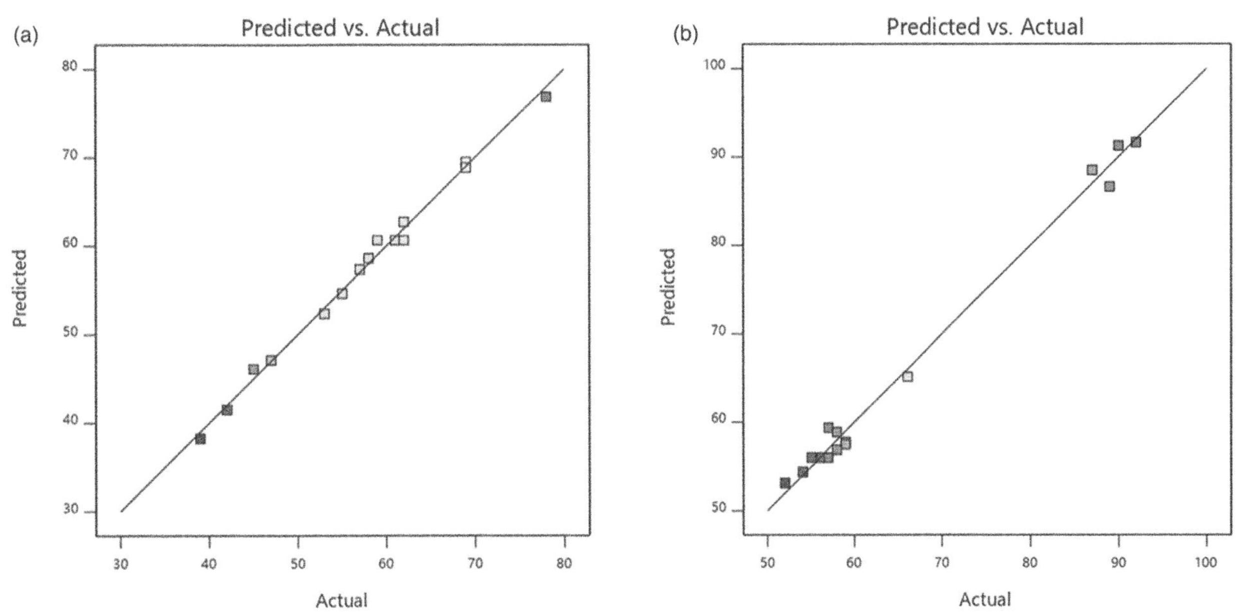

Figure 6 | Prediction and actual response for the degradation of Rhodamine B. (a) Fenton's Oxidation (b) UV–Fenton's Oxidation.

influence (high co-efficient) than influence of EDTA. The quadratic effect of the independent variables on Fenton's oxidation of Rhodamine B was negative.

The linear regression quadratic equation of the model fit is explained by

$$Y = 60.67 + 13.12A + 0.87B + 2.25C + 2.5AB + 1.75AC - 0.25BC - 1.83A^2 - 5.83B^2 + 0.916C^2.$$

A, C, BC and A^2 are significant model terms for the UV–Fenton's oxidation of Rhodamine B. Thus, statistical analysis of all the experimental data showed that Biojarosite loading and EDTA addition had a significant effect on UV–Fenton's oxidation. It is observed that Biojarosite loading and EDTA dosage exerted more linear influence on UV–Fenton's oxidation, which indicates that Rhodamine B removal by the UV–Fenton's oxidation was positively influenced by Biojarosite loading and addition

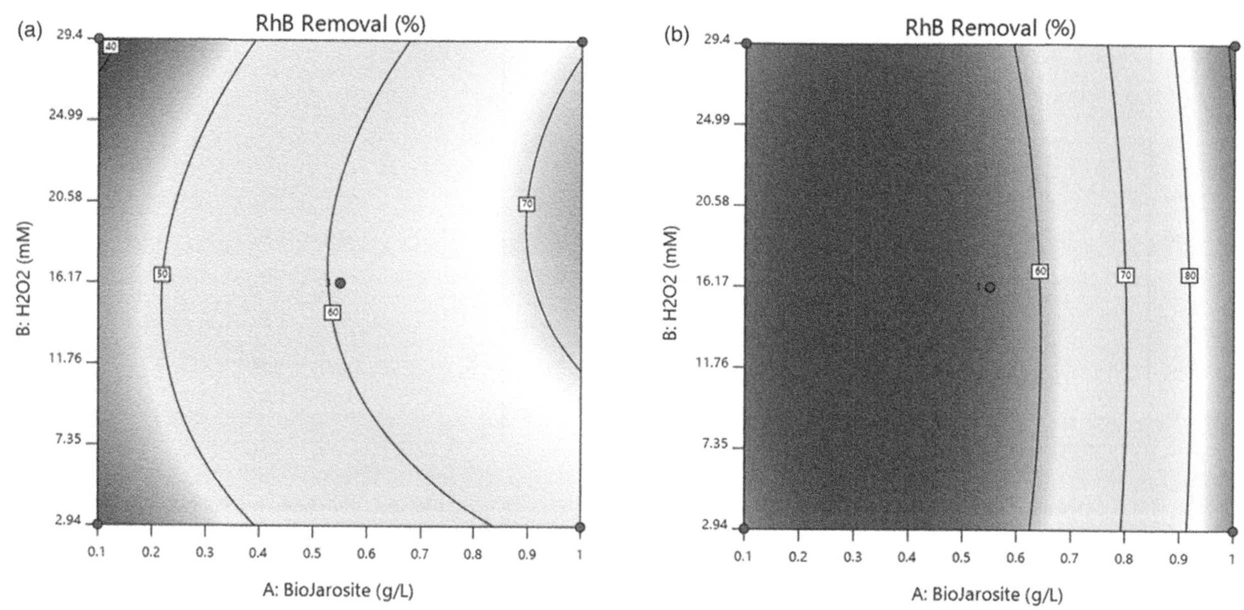

Figure 7 | Contour plot for the degradation of Rhodamine B. (a) Fenton's Oxidation (b) UV–Fenton's Oxidation.

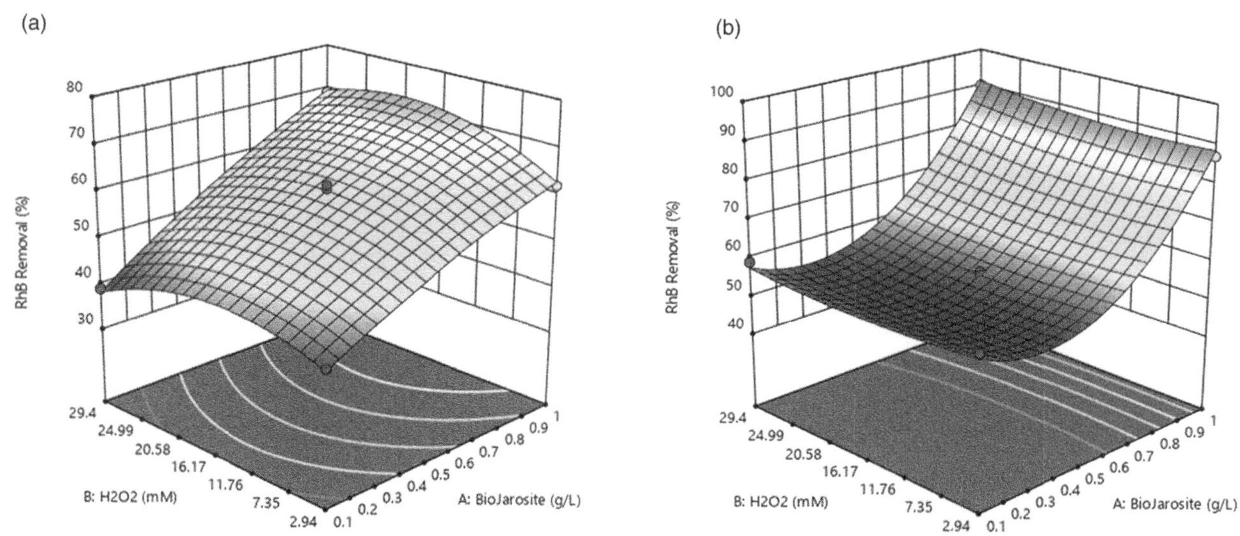

Figure 8 | Response surface plot for the degradation of Rhodamine B. (a) Fenton's Oxidation (b) UV–Fenton's Oxidation.

of EDTA. Again, Biojarosite loading has a higher positive linear influence (16.12) than influence of EDTA concentration (2.50). The quadratic effect of the independent variables on Fenton's oxidation of Rhodamine B was negative.

The linear regression quadratic equation of the model fit is explained by

$$Y = 56.00 + 16.12A + 0.625B + 2.5C + 0.75AB + 3.5BC + 16.13A^2 + 1.63B^2 + 0.875C^2.$$

In comparison with Fenton's oxidation, UV–Fenton's oxidation has all the co-efficients like interaction, linear and quadratic positive, while in the Fenton's oxidation all the interaction and linear co-efficients are positive and the three quadratic

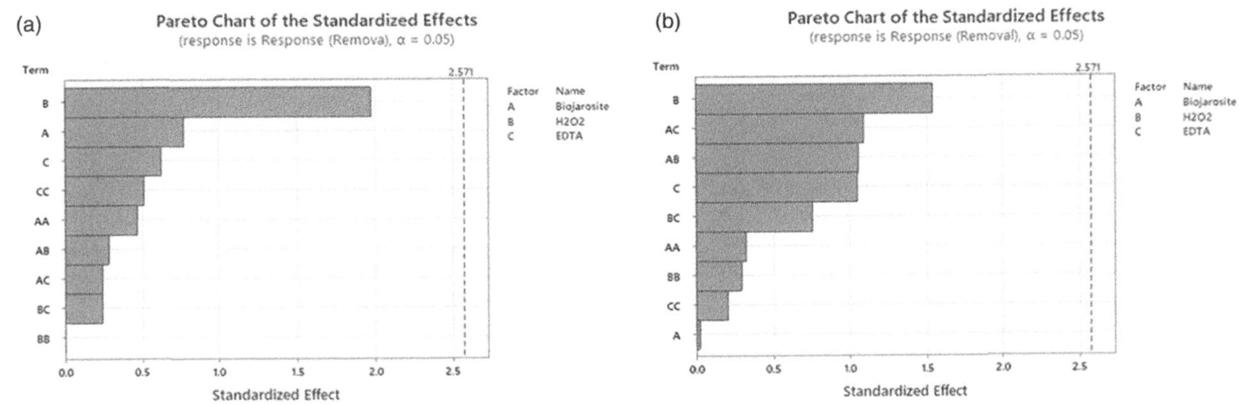

Figure 9 | Pareto chart for response on degradation of Rhodamine B. (a) Fenton's Oxidation (b) UV-Fenton's Oxidation.

Table 3 | ANOVA for the quadratic response surface model fitting for Fenton's oxidation of Rhodamine B

	SS	df	MS	F-value	*p*-value	Remarks
Residual Model	1,603.02	9	178.11	89.80	<0.0001	Significant
Lack of fit	5.25	3	1.75	0.75	0.6148	Not significant
Pure error	4.67	2	2.33			
Total correlation	1,612.93	14				

$R^2 = 0.9939$.
Adjusted $R^2 = 0.9828$.
Predicted $R^2 = 0.9414$.
Adequate Precision = 33.5905.

Table 4 | Co-efficient of the model for Fenton's oxidation of Rhodamine B

Factor	Co-efficient estimate	Standard error	F-value	*p*-value	Remarks
X	60.67	0..81	89.80	<0.0001	Significant
A	13.12	0.49	694.85	<0.0001	Significant
B	0.875	0.49	3.09	0.1392	Not significant
C	2.25	0.49	20.42	0.0063	Significant
AB	2.50	0.70	12.61	0.016	Significant
AC	1.75	0.70	6.18	0.055	Significant
BC	−0.25	0.70	0.12	0.737	Not significant
A^2	−1.83	0.73	6.26	0.054	Significant
B^2	−5.83	0.73	63.35	0.005	Significant
C^2	0.9167	0.73	1.56	0.266	Not significant

co-efficients are negative. Figure 5 shows a perturbation plot in which response is plotted over rhodamine removal over Biojarosite loading, H_2O_2 dosage and EDTA concentration 0.55, 550 and 55 as actual factors. Actual and predicted response for both Fenton's oxidation and UV–Fenton's oxidation are highly relevant (Figure 6). From Figure 7 it is observed that Fenton's oxidation for Rhodamine B removal works in a narrow range of Biojarosite and H_2O_2 dosage, while UV–Fenton's oxidation has a comparatively broad range. Figure 8 shows 3D response surface plot for Fenton's oxidation and UV–Fenton's oxidation.

Influence of independent variables on dependent variables was analyzed using Pareto chart (Figure 9). In case of Fenton's oxidation, B (H_2O_2 dosage), A (Biojarosite dosage) and C (EDTA concentration) have the significant effects, while for UV–Fenton's oxidation, B (H_2O_2 dosage), AC (Combinational effect of Biojarosite and EDTA), AB (Combinational effect of Biojarosite and H_2O_2), C (EDTA concentration) and BC (Combinational effect of H_2O_2 and EDTA concentration) have significant effects. In both the cases, H_2O_2 dosage was shown to have highest effect (See Tables 3–7).

Table 5 | ANOVA for the quadratic response surface model fitting for UV–Fenton's oxidation of Rhodamine B

	SS	df	MS	F-value	*p*-value	Remarks
Residual Model	3,145.68	9	349.52	69.21	0.0001	Significant
Lack of fit	23.25	3	7.75	7.75	0.1164	Not significant
Pure error	2.00	2	1.00			
Total correlation	3,170.93	14				

$R^2 = 0.9920$.
Adjusted $R^2 = 0.9777$.
Predicted $R^2 = 0.8813$.
Adequate Precision = 20.9827.

Table 6 | Co-efficient of the model for UV–Fenton's oxidation of Rhodamine B

Factor	Co-efficient estimate	Standard error	F-value	*p*-value	Remarks
X	56	1.30	69.21	0.0001	Significant
A	16.12	0.7945	411.91	<0.0001	Significant
B	0.6250	0.7945	0.6188	0.4671	Not significant
C	2.50	0.7945	9.90	0.0255	Significant
AB	0.75	1.12	0.4455	0.5340	Not significant
AC	0.00	1.12	0	1.00	Not significant
BC	3.5	1.12	9.70	0.0264	Significant
A^2	16.13	1.17	190.11	<0.0001	Significant
B^2	1.63	1.17	1.93	0.2234	Not significant
C^2	0.8750	1.17	0.5598	0.4880	Not significant

Table 7 | Overview of Fenton's and UV–Fenton's oxidation of Rhodamine B optimized with Box–Behnen design

Method	Biojarosite:H_2O_2	EDTA (mM)	RhB removal (%)
UV–Fenton's Oxidation	1:1	55	90.0
Fenton's Oxidation	1:0.55	100	78.0

CONCLUSIONS

Fenton's oxidation and UV–Fenton's oxidation for the degradation of Rhodamine B using synthesized Biojarosite as a catalyst and EDTA as chelating agent at neutral pH were optimized with model fitting using Box–Behnken design. Initial investigation with Biojarosite and EDTA reveals that UV–Fenton's oxidation is more efficient in the degradation of Rhodamine B. Actual response over predicted response on Box–Behnken method was analyzed and both the model is fitted the actual response with a high degree of precision. Influencing parameters in the model for both methods are discussed and a quadratic equation is derived. The study claims that both Fenton's oxidation and UV–Fenton's oxidation work effectively for the degradation of dye Rhodamine B, and Biojarosite can be used as a replacement for the commercial iron catalyst at neutral pH using EDTA.

DATA AVAILABILITY STATEMENT

All relevant data are included in the paper or its Supplementary Information.

CONFLICT OF INTEREST

The authors declare there is no conflict.

REFERENCES

Ahmadi, M., Rahmani, K., Rahmani, A. & Rahmani, H. 2017 Removal of benzotriazole by Photo-Fenton like process using nano zero-valent iron: response surface methodology with a Box-Behnken design. *Polish Journal of Chemical Technology* **19** (1), 104–112.

Ahmadi, A., Zarei, M., Hassani, A., Ebratkhahan, M. & Olad, A. 2021 Facile synthesis of iron (II) doped carbonaceous aerogel as a three-dimensional cathode and its excellent performance in electro-Fenton degradation of ceftazidime from water solution. *Separation and Purification Technology* **278**, 119559. https://doi.org/10.1016/j.seppur.2021.119559.

AlHamedi, F. H., Rauf, M. A. & Ashraf, S. S. 2009 Degradation studies of Rhodamine B in the presence of UV/H_2O_2. *Desalination* **239** (1–3), 159–166. https://doi.org/10.1016/j.desal.2008.03.016.

Arslan, I. & Tuhkanen, T. 1999 Oxidative treatment of simulated dyehouse effuent by UV and near UV light assisted Fenton's reagent. *Chemosphere* **39** (15), 2767–2783.

Bel, H., Da, P., Beaunier, P., Elena, M. & Ben, M. 2014 Fe-clay-plate as a heterogeneous catalyst in photo-Fenton oxidation of phenol as probe molecule for water treatment. *Applied Clay Science* **91–92**, 46–54. https://doi.org/10.1016/j.clay.2014.01.020.

Bhaskar, S. & Basavaraju Manu, M. Y. S. 2020 Green synthesis of Bioleached Flyash Iron Nanoparticles (GBFFeNP) using *Azadirachta indica* leaves and its application as Fenton's catalyst in the degradation of dicamba. In: *Recent Trends in Civil Engineering* Springer, pp. 365–371.

Bhaskar, S., Manu, B. & Sreenivasa, M. Y. 2019 Bacteriological synthesis of iron hydroxysulfate using an isolated *Acidithiobacillus ferrooxidans* strain and its application in ametryn degradation by Fenton's oxidation process. *Journal of Environmental Management* **232**, 236–242. https://doi.org/10.1016/j.jenvman.2018.11.048.

Bhaskar, S., Manu, B. & Sreenivasa, M. Y. 2021 Bioleaching of iron from laterite soil using an isolated *Acidithiobacillus ferrooxidans* strain and application of leached laterite iron as Fenton's catalyst in selective herbicide degradation. *PLoS ONE* **16**. https://doi.org/10.1371/journal.pone.0243444.

Dong, C. P. H. C. & Tang, Z. 1993 Advanced chemical oxidation: its present role and potential future in hazardous waste treatment. *Waste Management* **13**, 361–377.

Dutta, K. & Mukhopadhyay, S. 2001 Chemical oxidation of methylene blue using a Fenton-like reaction. *Journal of Hazardous Materials* **84**, 57–71.

Eisenberg, G. M. 1943 Colorimetric determination of hydrogen peroxide. *Industrial and Engineering Chemistry* **15** (5), 327–328.

Hassani, A., Çelikdağ, G., Eghbali, P., Sevim, M., Karaca, S. & Metin, Ö. 2017 Reduced graphene oxide ($CoFe_2O_4$-rGO) nanocomposite for the removal of organic dyes from aqueous solution. *Ultrasonics – Sonochemistry*. https://doi.org/10.1016/j.ultsonch.2017.08.026.

Hou, M., Liao, L., Zhang, W., Tang, X., Wan, H. & Yin, G. 2011 Degradation of rhodamine B by Fe (0)-based Fenton process with H_2O_2. *Chemosphere* **83** (9), 1279–1283. https://doi.org/10.1016/j.chemosphere.2011.03.005.

Huang, Y., Huang, Y., Chang, P. & Chen, C. 2008 Comparative study of oxidation of dye-Reactive Black B by different advanced oxidation processes: Fenton, electro-Fenton and photo-Fenton **154**, 655–662. https://doi.org/10.1016/j.jhazmat.2007.10.077.

Islam, M. R. & Mostafa, M. G. 2018 Textile dyeing effluents and environment concerns – a review. *Journal of Environmental Science and Natural Resources* **11**, 131–144.

Karale, R. S., Manu, B. & Shrihari, S. 2014 Fenton and Photo-Fenton oxidation processes for degradation of 3-Aminopyridine from water. *APCBEE Procedia* **9**, 25–29. https://doi.org/10.1016/j.apcbee.2014.01.005.

Karim, A. V., Hassani, A., Eghbali, P. & Nidheesh, P. V. 2022 Nanostructured modified layered double hydroxides (LDHs) -based catalysts: a review on synthesis, characterization, and applications in water remediation by advanced oxidation processes. *Current Opinion in Solid State & Materials Science* **26** (1), 100965. https://doi.org/10.1016/j.cossms.2021.100965.

Khan, E., Wirojanagud, W. & Sermsai, N. 2009 Effects of iron type in Fenton reaction on mineralization and biodegradability enhancement of hazardous organic compounds. *Journal of Hazardous Materials* **161**, 1024–1034. https://doi.org/10.1016/j.jhazmat.2008.04.049.

Kim, T., Park, C., Yang, J. & Kim, S. 2004 Comparison of disperse and reactive dye removals by chemical coagulation and Fenton oxidation. *Journal of Hazardous Materials* **112**, 95–103. https://doi.org/10.1016/j.jhazmat.2004.04.008.

Lapara, T. M., Alleman, J. E. & Pope, P. G. 2000 Miniaturized closed refluux, colorimetric method for the determination of chemical oxygen demand. *Waste Management* **20**, 295–298.

Lojo-López, M., Andrades, J. A., Egea-Corbacho, A., Coello, M. D. & Quiroga, J. M. 2021 Degradation of simazine by photolysis of hydrogen peroxide Fenton and photo-Fenton under darkness, sunlight and UV light. *Journal of Water Process Engineering* **42**, 102115. https://doi.org/10.1016/j.jwpe.2021.102115.

Masomboon, N., Ratanatamskul, C. & Lu, M. C. 2009 Chemical oxidation of 2, 6-Dimethylaniline in the Fenton process. *Environmental Science & Technology* **43** (22), 8629–8634.

Sangami, S. & Manu, B. 2018 Catalytic efficiency of laterite based FeNPs for the mineralization of mixture of herbicides in water. *Environmental Technology*, 1–30. https://doi.org/10.1080/09593330.2018.1449899.

Shokri, A., Bayat, A. & Mahanpoor, K. 2019 Employing Fenton-like process for the remediation of petrochemical wastewater through Box–Behnken design method. *Desalination and Water Treatment* **166**, 135–143. https://doi.org/10.5004/dwt.2019.24634.

Su, S., Guo, W., Leng, Y., Yi, C. & Ma, Z. 2013 Heterogeneous activation of oxone by $CO_xFe_3xO_4$ nanocatalysts for degradation of rhodamine B. *Journal of Hazardous Materials* **244–245**, 736–742. https://doi.org/10.1016/j.jhazmat.2012.11.005.

Tak, B., Tak, B., Kim, Y., Park, Y., Yoon, Y. & Min, G. 2015 Optimization of color and COD removal from livestock wastewater by electrocoagulation process: application of Box–Behnken design (BBD). *Journal of Industrial and Engineering Chemistry* **28**, 307–315.

Tang, Y., Liu, H., Zhou, L., Ren, H., Li, H., Zhang, J., Chen, G. & Qu, C. 2019 Enhanced Fenton-like oxidation of hydroxypropyl guar gum catalyzed by EDTA-metal complexes in a wide pH range. *Water Science and Technology*, 1667–1674. https://doi.org/10.2166/wst.2019.169.

Tekbas, M., Yatmaz, H. C. & Bektas, N. 2008 Heterogeneous photo-Fenton oxidation of reactive Azo dye solutions using iron exchanged zeolite as a catalyst Mesut Tekbas. *Microporous and Mesoporous Materials* **115**, 594–602. https://doi.org/10.1016/j.micromeso.2008.03.001.

Tripathi, P., Chandra, V. & Kumar, A. 2009 Optimization of an azo dye batch adsorption parameters using Box–Behnken design. *Desalination* **249** (3), 1273–1279. https://doi.org/10.1016/j.desal.2009.03.010.

Woods, J. & Mellon, M. 1941 Thiocyanate method for iron: a spectrophotometric study. *Industrial & Engineering Chemistry Analytical Edition* **13** (8), 551–554.

Xue, X., Hanna, K. & Deng, N. 2009 Fenton-like oxidation of Rhodamine B in the presence of two types of iron (II, III) oxide. *Journal of Hazardous Materials* **166**, 407–414. https://doi.org/10.1016/j.jhazmat.2008.11.089.

Zhao, X. U. & Zhu, Y. 2006 Synergetic degradation of Rhodamine B at a porous $ZnWO_4$ film electrode by combined electro-oxidation and photocatalysis. *Environmental Science & Technology* **40** (10), 3367–3372.

Zhou, H., Sun, Q., Wang, X., Wang, L., Chen, J., Zhang, J. & Lu, X. 2014 Removal of 2, 4-dichlorophenol from contaminated soil by a heterogeneous ZVI/EDTA/air Fenton-like system. *Separation and Purification Technology* **132**, 346–353. https://doi.org/10.1016/j.seppur.2014.05.037.

Zhou, W., Zhao, H., Gao, J., Meng, X. & Qin, Y. 2016 Influence of a reagents addition strategy on the Fenton oxidation of Rhodamine B: control of the competitive reaction of OH. *RSC Advances* **6** (3), 108791–108800. https://doi.org/10.1039/c6ra20242j.

First received 3 August 2022; accepted in revised form 10 September 2022. Available online 3 October 2022

doi: 10.2166/wst.2022.364

Electrocoagulation followed by sound agitation for removal of nitrogen and carbon-based pollutants from industrial wastewater

Nibedita Pani[a], Anantha Singh T. S.[b,*], Poornima M. Menon[c], Shrestha Boruah[c], Bhavi Patel[c] and Daya Shankar Kaul[c]

[a] Department of Science, School of Technology, Pandit Deendayal Energy University, Gandhinagar, Gujarat, India
[b] Department of Civil Engineering, National Institute of Technology, Calicut, India
[c] Department of Civil Engineering, School of Technology, Pandit Deendayal Energy University, Gandhinagar, Gujarat, India
*Corresponding author. E-mail: singh87@nitc.ac.in

ASTS, 0000-0002-7735-497X

ABSTRACT

The herculean imprecation of nitrogen-based pollutant like ammoniacal nitrogen (AN) and chemical oxygen demand (COD) on aquatic milieu is now a concern for the dye, pharma and fertiliser industries. Wastewater from these is characterised with high concentration of AN, COD and total dissolved solids (TDS), treatment of which is of utmost importance for a cleaner environment. In the current research work, an attempt was made to apply integrated electro-coagulation (EC) – sonication process for the removal of COD and AN from highly acidic dye intermediate wastewater containing high to very high concentration of COD and AN. Systematic laboratory experiments were conducted for the treatment of dye intermediate wastewater and influences of pH (5–11), applied voltage (0.5–4V) and electrolysis time (30–120 min) were investigated. A Response Surface Methodology (RSM) was used for optimization of major operating parameters for EC. The conditions for minimum fraction remaining (C/C0), was found to be same for both COD and AN, i.e. pH 7, time 90 min and applied voltage 2V. The C/Co value for COD and AN were 0.244 and 0.302, respectively. The C/Co value of COD and AN in combined EC-Sonication process with optimum operating conditions were 0.145 and 0.228 respectively with sonication time 60 min at a frequency of 33 kHz. Thus, EC – sonication process is an efficacious process for their removal from dye industrial wastewater.

Key words: AN, dye intermediate industrial wastewater, electrocoagulation-sonication process, high COD, optimization, response surface methodology

HIGHLIGHTS

- Highly concentrated dye intermediate wastewater was treated with combined advanced oxidation processes.
- Various influencing factors were studied including pH, voltage, and electrolysis time.
- Simultaneous removal of COD and AN was chosen to understand the process efficiency.
- Parametric optimization was conducted to find the optimum operating conditions.

GRAPHICAL ABSTRACT

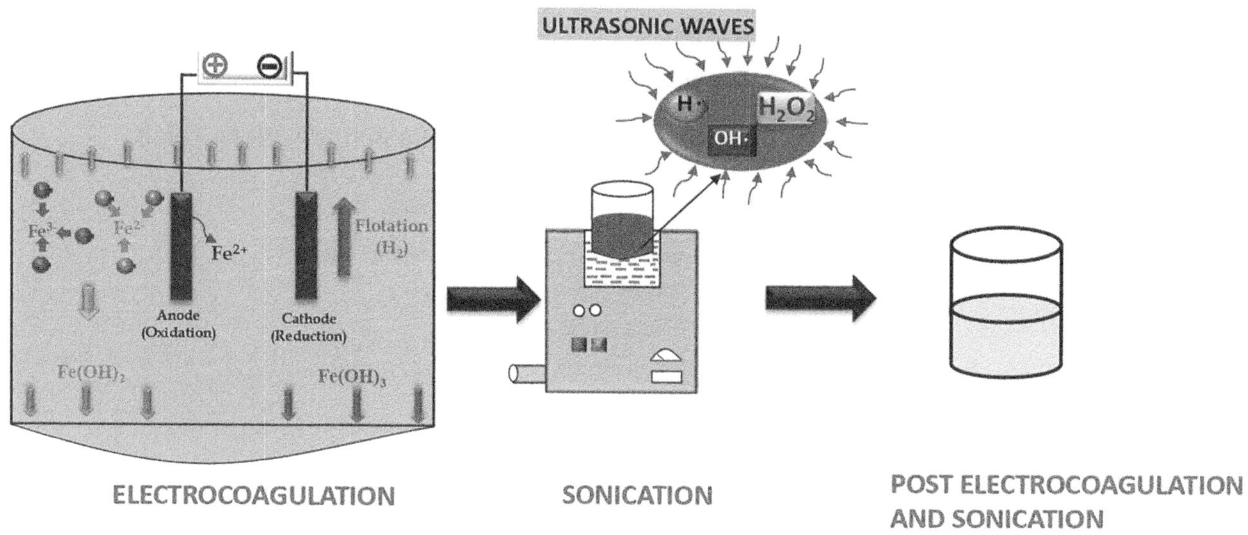

1. INTRODUCTION

The world of textile industries demands varieties of synthetic dyes leading to its numerous intermediate products (Paździor *et al.* 2019). To cater for demand, the dye intermediates are synthesized and manufactured in substantial amount in India. Some regions of Gujarat have grabbed attention at global scale with reference to synthetic pigments and its toxic effluent (Chakraborty & Basu 2019). Dye intermediate manufacturing industries have a wide application in the field of textile, food, leather, rubber, etc. Due to these wide applications, a large amount of wastewater is produced from these industries and are discharged into water bodies with only partial treatment (Hayat *et al.* 2015; Guo *et al.* 2018). The harmful effects of dye intermediate effluents on aquatic environment were discussed in detail by the authors Pani *et al.* (2020).

The effluent includes a variety of hazardous contaminants such as high chemical oxygen demand (COD), ammoniacal nitrogen (AN) (NH_4^+-N), high total organic carbon (TOC), high chloride content, high sulphates, high concentrations of heavy metals such as copper, nickel, chromium, and iron (Yaseen & Scholz 2019; Pani *et al.* 2020).

The alarming concentration of COD and AN in the discharged effluent endangers human health and damages aquatic community by exhausting the DO level in the receiving water bodies. Moreover, toxic algae such as Euglenophyta and Chrysophytaget stimulated by inorganic nitrogen pollution hampers the survival, growth and reproduction of aquatic organisms (Camargo & Alonso 2006; Dong *et al.* 2017; Soliman & Eldyasti 2018).

Several technologies are available for treatment of such industrial wastewaters. The commonly used include physical (Quan *et al.* 2009), chemical (Zhang *et al.* 2009; Huang *et al.* 2017) and biological methods (Phan *et al.* 2017; Miao *et al.* 2018; Zhou *et al.* 2018). Methods such as breakpoint chlorination, precipitation and ion exchange (chemical), ammonia stripping and membrane separation (physical); and nitrification, denitrification (biological) perform poorly in removing AN where high to very high concentration of COD is predominant (Pani *et al.* 2020).

Thus, few researchers have stressed on use of different advanced oxidation processes such as Fenton oxidation process (Xavier *et al.* 2015; Gonzalez-Merchan *et al.* 2016), electrochemical advanced oxidation processes (Candido & Gomes 2011; Zhang *et al.* 2018; Menon *et al.* 2021), electro-Fenton in combination with biological process (Li *et al.* 2018), ultrasonic irradiation (Ozturk & Bal 2015) and photo and electrochemical methods such as electrocoagulation (EC) (Kasmuri & Tarmizi 2018), etc. Electrochemical oxidation processes have seen widespread use over the past two decades and ultrasound technology has gained popularity in recent years as a means of treating wastewater (Hassani *et al.* 2022). The fundamental problem with the EC method is that a passive layer eventually forms on the surface of the electrode. By generating large concentrations of oxidants to remove the contaminants by generating high-pressure sites during the cavitation process, acoustic waves break the sediments generated (Moradi *et al.* 2021). By combining such methods, 99 and 62%, respectively, were determined to be the highest COD and colour removal efficacy for the anode-cathode combination of Fe/Fe (Prajapati 2021).

Further improvement in removal has been achieved by optimizing operating parameters using response surface methodology (RSM) technique. It uses regression analysis to predict the value of dependent variable from combination of independent variables (Huda *et al.* 2017). Various chemical, physical and biological processes for a variety of pollutants have been modelled and optimized using RSM (Asaithambi *et al.* 2016; Mehdipoor & Moosavirad 2020). A few investigators have applied it in Fenton and electro-Fenton process (Akkaya *et al.* 2019; Mohammadi *et al.* 2019a, 2019b).

Fenton oxidation process was effectively applied to remove AN in presence of high COD from dye intermediate wastewater (Pani *et al.* 2020). The removal efficiency of AN and COD were found to be 75.8% and 78.6% respectively at pH 3 and molar dosage of Fe^{2+} and H_2O_2 of 3:3 for 60 min reaction time. But, removal of 75.8% did not reduce their concentration to acceptable values, thus further work on electrochemical processes such as EF and EC were carried out (Menon *et al.* 2021).

EC process is one of the most economical technologies for the treatment of industrial wastewater (Syam-Babu *et al.* 2020). It involves dissolution of sacrificial anodes immersed in wastewater by using current.

The bubbles production in sonication process leads to a series of significant physical and chemical changes in the solution (Soltani *et al.* 2018). Growth and compression happen due to rectified diffusion of micro-bubbles attaining equilibrium size and producing acoustic micro-flows. A tremendous pressure (400 MPa) is generated during bubble collapse which leads to formation of hydroxyl radical as shown in Equations (1)–(5) (Hua & Hoffmann 1997; Villeneuve *et al.* 2009; Wang *et al.* 2019; Menon *et al.* 2021).

$$H_2O \xrightarrow{))))} \bullet OH + H\bullet \tag{1}$$

$$H\bullet + O_2 \rightarrow HO_2\bullet \tag{2}$$

$$2\bullet OH \rightarrow H_2O_2 \tag{3}$$

$$2HO_2\bullet \rightarrow H_2O_2 + O_2 \tag{4}$$

$$H\bullet + O_2 \rightarrow \bullet OH + O\bullet \tag{5}$$

where,)))) represents ultrasonic splitting.

The advantages of EC process include no addition of chemicals, easy maintenance, no harmful by-products, etc. (Priya & Jeyanthi 2019; Lu *et al.* 2022). According to recent studies, combined EC-sonication process is effective in eliminating the disadvantages of EC process like removing the passive film from electrode surface, breaking the flocs in the solution and reducing the particle size thus enhancing the sorption mechanism of EC (Lakshmi & Sivashanmugam 2013; He *et al.* 2016). According to the literatures reported earlier, only few studies are available in which such highly polluted wastewater is treated by EC-sonication process.

Considering the high removal efficiencies in EC and sonication processes as realized by various researchers for several types of wastewaters. The authors of the current study have explored the efficacy of batch EC process and EC followed by sonication process for the removal of AN and COD from dye intermediate manufacturing wastewater. The performance of EC batch process was evaluated by varying major operating parameters like pH, applied voltage and electrolysis time. The EC followed by sonication process was performed to study influence of sonication on the removal efficiency and to compare the performance of EC and EC followed by sonication process. The sonication process was performed on the optimum condition of EC batch process.

To quantify the influence of process parameters, an RSM was executed to predict the optimum values of major operating parameters on treatment efficiency by using MINITAB (version 19) statistical software and ANOVA.

2. MATERIAL AND METHODS

2.1. Materials and analytical procedure

The dye intermediate wastewater was collected in a 20 L container from a dye intermediate manufacturing industry in Ankleshwar, Gujarat, India. The analysis of all the parameters were carried out according to APHA and IS: 3025 methods. All chemicals and solvents used for analysis and experiment were of analytical grade and of Merck brand. The solution pH was analysed using Bench-top pH meter (HACH, sensION+). COD was analysed by closed reflux method and AN analysis was carried out using automatic nitrogen analyser (KELPLUS CLASSIC-DX). Dual DC power supply (Model: ALE100A) with power supply of 0 ± 30 V/5Amp was used for AN. Sonicator (Model: SM50US) operating at a frequency of 33 kHz and capacity of 5 L with UL power 100 W was used for the sonication process. Post treatment sludge analysis was done to study the morphology and shape of a sample, using scanning election microscope (SEM)/energy dispersive X-ray spectroscopy (EDS) analysis.

All parameters at an optimum condition. The physico-chemical characteristics of the wastewater are shown in Table 1.

2.2. Electrochemical reactor setup

The experimental setup of electrocoagulation comprised of a batch reactor with a working volume of 1,000 mL. The experiments were performed using iron electrodes (6*3*0.3 cm) with an inter electrode spacing of 4 cm. The ohmic drop in the reactor results from the ohmic resistance offered by the electrolyte and the spacing between the electrodes, the ohmic resistance is directly proportional to the spacing and inversely proportional to conductivity of the working solution. As the conductivity of the working solution is considerably very high the ohmic resistance offered by the spacing between the electrodes would be insignificant or minimum. Considering aforesaid fact, as well as based on the published articles, authors have considered the spacing between the electrodes as 4 cm in the current study. Electrolysis was conducted in the original conductivity of the dye intermediate wastewater without addition of supporting electrolyte using a digital DC power supply. The solution was mixed during the process with magnetic stirrer at set RPM for homogenisation of the process. The previous studies done by Vanlangendonck *et al.* (2005), Li & Liu (2009) and Mahvi *et al.* (2011) found that the optimum pH range for removal of ammonia is 5.5–10. Neutral pH is suggested for AN removal. In strongly alkaline conditions, the oxidation rate of AN decreases. Thus, the experiments were conducted at different pH ranging from 5 to 11, to study the effect of pH. The effectiveness of the process and the expense of the treatment are both influenced by the voltage, which is one of the key parameters in electrochemical methods of treatment. According to earlier study by Pani *et al.* (2020) and Menon *et al.* (2021), which used the same effluent, the ideal voltage range is from 1 to 4 V. Beyond 4 V, the power consumption will increase drastically, due to which, this study is only limited to maximum of 4 V. By creating alkaline conditions and changing the pH, metal hydroxides are produced, which increases the efficiency of removing contaminants (Shahedi *et al.* 2020). Thus, the experiments were conducted at potentiostat mode (constant voltage) varying the voltage from 0.5 to 4 V. The potentiostat mode consists of the application of a constant potential at the working electrode (WE) throughout the electrolysis time. During potentiostat mode current response value changes quite fast, when compared to the potential response during the galvanostatic mode of functioning.

Effect of electrolysis time was studied by varying the time between 0 and 120 min. At optimum conditions (acquired through experiment) of EC process, the sample was subjected to sonication for 60 min.

The experimental setup and layout of EC and sonication is shown in Figure 1.

Table 1 | Physico-chemical characteristics of dye intermediate wastewater

Parameters	Values in mg/L (except pH)	Standard deviation
pH	0.89	0.0052
Total solids	2,01,790	29.02
Total dissolved solids	1,81,450	28.87
Total suspended solids	20,340	28.86
COD	15,360	184.75
BOD	90	10.39
AN	741.44	5.82
Conductivity	258.15 mS/cm	5.2
Sulphate	2,575	–
Chloride	560	–
TOC	20,040	288.67
Nitrate	1,020	–

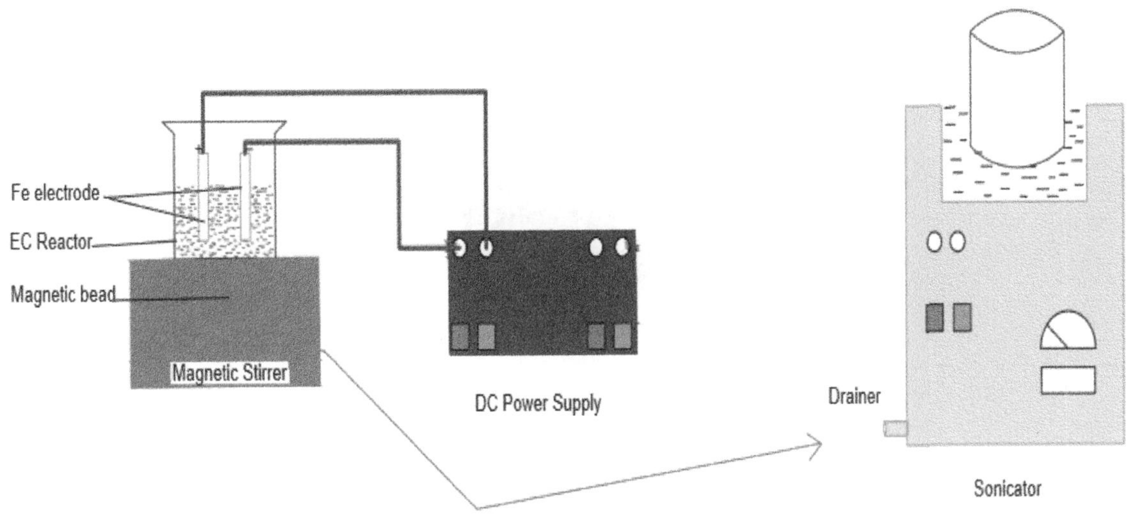

Figure 1 | Experimental setup of electrocoagulation and sonication process.

2.3. Parameter optimization with response surface methodology

RSM applies combination of mathematical and statistical techniques for parameter optimization. The most important use of RSM is in those particular situations when several input parameters potentially influence the performance or quality characteristic of the process. Methodology is to fit a polynomial equation to the experimental data, which describes the behaviour of a data set with the objective of making statistical projections (Bezerra *et al.* 2008). A linear or square polynomial function is employed to describe the system and explore the experimental conditions, finally culminating into parameter optimization (Bezerra *et al.* 2008).

The general equation to which the experimental data was fitted is as follows

$$Y = \beta_0 + \beta_1X_1 + \beta_2X_2 + \beta_3X_3 + \beta_{11}X_{11}^2 + \beta_{22}X_{22}^2 + \beta_{33}X_{33}^2 + \beta_{12}X_1X_2 + \beta_{13}X_1X_3 + \beta_{23}X_2X_3 + \varepsilon \tag{6}$$

where, y is response, β_0, $(\beta_1,\beta_2,\beta_3)$, $(\beta_{11},\beta_{22}\beta_{33})$, and $\beta_{12},\beta_{13},\beta_{23}$ are the regression coefficients for intercept, linear, quadratic and interaction terms; X_1, X_2 and X_3 are the independent variables and ε represents standard error between the observed and modelled response value (Akkaya *et al.* 2019). The quality of fit was evaluated by R^2, and its statistical significance was evaluated by T- and F-test. Minitab software (Minitab) was used for the purpose and results are described in detail in section 5.1.

3. RESULTS AND DISCUSSION

This section is divided into several subsections which explore the effect of pH, voltage and time on treatment in EC process. Then, treatment parameters are optimized which is followed by comparison between EC and EC-sonication process.

3.1. Influence of pH on EC process

The performance of EC process is controlled by pH of the aqueous medium as the amount of Fe^{2+} and Fe^{3+} formation is influenced by pH. Precipitation of pollutants at lower pH is unfavourable due to production of mono and di-hydroxide complex of iron (Singh & Ramesh 2014; Yavuz & Ögütveren 2018). The speciation of various species of iron with respect to pH (Nidheesh & Singh 2017) is provided in Table 1s. According to Nidheesh & Singh 2017, the various forms of ferrous ion and ferric ion can exist in different pH conditions.

The pH of the wastewater was varied between 5 and 11 at an applied voltage of 1V, maintaining the electrode distance of 4 cm and electrode area of 18 cm^2. At various intervals of time, i.e., 30 min, 60 min, 90 min and 120 minutes, the samples were taken for analysis of COD and AN. The most favourable pH value for COD and AN removal was found to be pH 7. At this pH, the fraction remaining (C/C_0) for COD and AN were minimum (i.e., 0.244 and 0.302, respectively) at electrolysis time of 90 min and an applied voltage 1 V (Figure 2(a) and 2(b)). Lower pH (pH < 7) leads to formation of $Fe(OH)^{2+}$ and $Fe(OH)_2^+$ complexes resulting in poor removal efficiency of COD and AN as described elsewhere also (Izadi *et al.* 2018). The

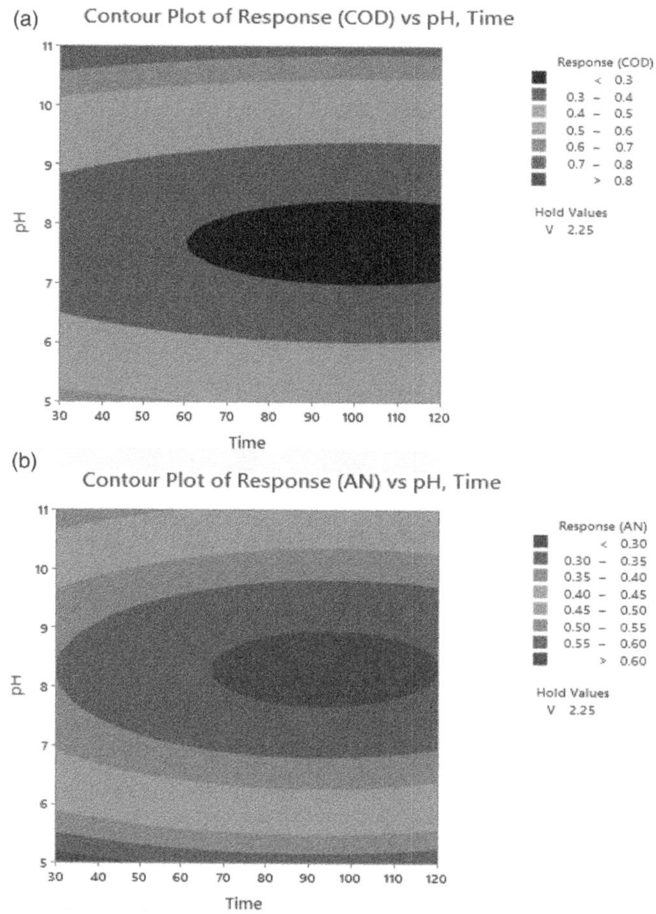

Figure 2 | (a) Fraction remaining for COD at different pH and time interval (experimental conditions: applied voltage-1 V, electrode distance-4 cm, electrode area-18 cm^2). (b) Fraction remaining for AN at different pH and time interval (experimental conditions: applied voltage-1 V, electrode distance-4 cm, electrode area-18 cm^2).

poor coagulants are formed which are highly soluble in solution with no adsorption capacity (Li *et al.* 2018). Cationic monomeric species such as Fe^{+2}, $Fe(OH)_2^+$, $Fe(OH)^{+2}$, $Fe(OH)^{4-}$ are soluble in water in the high and low pH conditions (El-Ashtoukhy *et al.* 2017).

Higher pH (pH > 7) leads to the production of hydroxide ion resulting in more amount of Fe ion generation (Singh & Ramesh 2014). When pH value is increased above 10, due to high hydrogen evolution, the C/C_0 values of COD and AN were increased as shown in Figure 2(a) and 2(b). The oxidation reaction occurs at anode and the water reduction at cathodes which are represented in Equations (1)–(4). When iron electrode is used in EC process, Fe^{2+} is hydrolysed to give $Fe(OH)_2$ and on oxidation gives Fe^{3+} and then hydrolysed to $Fe(OH)_3$. The coagulants such as $Fe(OH)^+$ and $Fe(OH)_2^+$ are formed (Yavuz & Ögütveren 2018) which does not help in adsorption of pollutants.

The species like Fe^{2+}, $Fe(OH)^{2+}$, $Fe(OH)_2$+ and $Fe(OH)_3^-$ exists in low pH range. The $Fe(OH)_2$ and $Fe(OH)_3$ are the coagulants that helps in COD and AN removal. As the main coagulants are present in low concentration, the C/C_0 values for COD and AN have high values leading to less removal efficiency than the neutral and basic pH (Figure 2(a) and 2(b)) as described elsewhere also (Garcia-Segura *et al.* 2017; Mamelkina *et al.* 2019).

At neutral pH, the species like $Fe(OH)_3$ exists as the principal coagulants in the solution. The pollutants are adsorbed effectively by these species and thehighest removal efficiency is achieved as found elsewhere also (Nidheesh & Singh 2017; Syam-Babu *et al.* 2020).

Amorphous flocs produced called sweep flocs have larger surface area which is useful for rapid adsorption of pollutants such as COD and AN and permit colloidal material to precipitate (Boudjema *et al.* 2014; El-Ashtoukhy *et al.* 2017). Fe^{+3} and OH^- ions produced at anode and cathode, respectively, react to form various monomeric species that finally transform into insoluble, stable, and amorphous $Fe(OH)_3$ through complex polymerization kinetics and the adsorption capacity of Fe flocs towards organic molecules increase in neutral pH and, consequently, the adsorption capacity of ferric flocs towards the organic molecule increases. The degradation of the organic contaminants from water occurs by surface complexation and electrostatic attraction that followed by coagulation (Daneshvar *et al.* 2006; El-Ashtoukhy *et al.* 2017).

The effect of major ions such as chloride and sulphate on electrocoagulation process has been analysed. In general, the presence of ions or electrolytes like chloride, sulphates increase the corrosion of metallic species in the anode regime by breaking down the passive anode layer formed by the insulating Ca^{2+} and Mg^{2+} ions present in the wastewater (Anantha-Singh & Ramesh 2014). The presence of high sulphate ions concentration was fairly unfavourable for power utilization and overall process efficacy. Whereas, the sulphate ions can act as an electrolyte that boost the current distribution in the reactor, which has been shown to improve the effectiveness of the process for organic industrial wastewaters treatment (Ghernaout & Ghernaout 2012).

In electrochemical conversion of ammonium to nitrogen gas, chloride plays an integral part. At different pH (5–11), formation of hypochlorous acid (HOCl) and hypochlorite ion (OCl^-) from chloride ions occurs. Regeneration of Cl^- ions takes place through breakpoint chlorination when hypochlorous acid (HOCl) reacts with NH^3 or NH^{4+}. The purpose of the Cl^- ions was to act as a catalyst, hence, after the reaction its concentration remains unchanged (Kabdaşlı *et al.* 2012).

3.2. Influence of applied voltage on EC process

Applied voltage is one of the strong factors affecting the removal of carbon and nitrogen-based pollutants in EC process. The production rate of metal hydroxide flocs increases with applied voltage due to which significant increase in pollutant removal is observed (Khandegar & Saroha 2013).

In this study, the applied voltage was varied from 0.5 to 4 V, i.e., 0.5 V (0.05 A/m^2), 1 V (0.083 A/m^2), 1.5 V (0.012 A/m^2), 2 V (0.02 A/m^2), 2.5 V (0.026 A/m^2), 3 V (0.042 A/m^2), 3.5 V (0.047 A/m^2) and 4 V (0.052 A/m^2) at pH 7, electrode distance of 4 cm. The results are shown in Figure 3(a) and 3(b).

It can be seen that the minimum fraction remaining (C/C_0) was observed at 2 V. Afterwards, the process efficaciousness decreases with increasing applied voltage. Hence, the efficiency proportionate with current density to a certain extend achieving the optimum removal condition and then saturates to a constant value as reported elsewhere also Mollah *et al.* 2001; Deghles & Kurt 2016. The rigorous release of metal ions is initiated by dissolution of anodic material at higher current which is governed by Faraday's law. A series of insoluble metallic hydroxide species evolves through hydrolysis process. Hence, more iron ions are generated which successively react with OH^- to form iron-based hydroxides. N- species are not adsorbed by monomeric as well as polymeric species of iron such as $Fe(OH)^+$, $Fe(OH)^{2+}$, $Fe(OH)_2^+$, $Fe(H_2O)_5OH^{2+}$, $Fe(H_2O)_4\ (OH)_2^+$, $Fe(H_2O)_8(OH)_2^{4+}$ which forms at different pH conditions as shown in Fig. 1s. Additionally, due to elevated current density, the hydrogen gas is produced in cathodic surface which enhances the upward flux resulting in more removal efficiency, similar findings have been reported elsewhere also (Izadi *et al.* 2018; Mohammadi *et al.* 2019a, 2019b). The production rate of coagulant Fe^{2+} or Fe^{3+}, and the size and distribution of bubbles were strongly influenced by current density/ applied voltage. By increasing current density above a certain range, the efficiency reduces as the rate of bubble formation and flocs increases. Also, the liberation of H_2 gas resulted in reduced adsorption on the surface of amorphous iron hydroxides (El-Ashtoukhy *et al.* 2017).

The specific energy consumption for EC process has been calculated using Equation (6), where V is the cell voltage (V), I is the current (A), t is the electrolysis time (min), COD_0 and COD_t are the initial chemical oxygen demands and chemical oxygen demands at time t (mg/L), respectively, similarly the AN was considered.

$$\mathrm{EEC} = \frac{I \times V \times t}{V_R}$$

$$\mathrm{E_{Sono}}\ (\text{for COD}) = \frac{P_{et} \times t}{V \times m}$$

$$\mathrm{E_{Sono}}\ (\text{for AN}) = \frac{P_{et} \times t}{V \times m}$$

where,

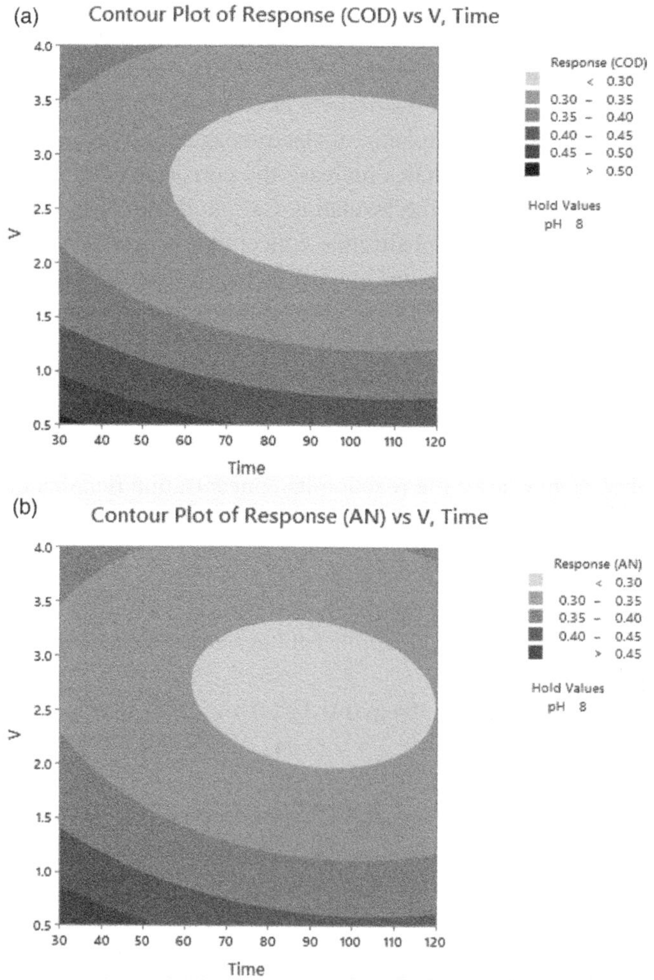

Figure 3 | (a): Fraction remaining (C/C_0) for COD at different applied voltage and time interval (experimental conditions: pH-7, electrode distance-4 cm, electrode area-18 cm^2). (b) Fraction remaining (C/C_0) for AN at different applied voltage and time interval (experimental conditions: pH-7, electrode distance-4 cm, electrode area-18 cm^2).

P_{et} = Rated power in W
t = Treatment time (h)
V = Volume of the wastewater (L)
M = Change in the concentration of the pollutant (gm/L)

The specific energy consumption was found to be 2.01 kWh/m^3 for the COD removal under the EC process. The specific energy of sonication for COD and AN was calculated using Equation (7) and the values were found to be 0.1731 kWhr/kg and 3.8845 kWh/kg, respectively.

3.3. Influence of electrolysis time on EC process

The cost of power consumption is driven by the runtime of the reactor (Sen *et al.* 2019). As the electrolysis time increases, the removal efficiency ascends to a certain extent, giving a significant yield and finally reaches to a constant value (Khandegar & Saroha 2013; Damaraju *et al.* 2017; Manikandan *et al.* 2018). As shown in Figures 2 and 3, this value was observed at 90 min experimentally, where the C/C_0 for COD is 0.244 and AN is 0.302 at the following conditions, i.e., voltage of 2 V, pH value of 7, electrode distance of 4 cm (emerging area = 18 cm^2).

The release of coagulating species happens during dissolution of anodic material. With increase in electrolysis period, the production of the metal ion and the hydroxide flocs also increase. For a longer electrolysis time, there is an increase in generation of flocs and beyond the optimum electrolysis time, the flocs are not available; as a result, there is no further increase in removal efficiency, as found elsewhere (Aoudj *et al.* 2010; Manikandan *et al.* 2018).

An increase occurs in the concentration of ions and their hydroxide flocs occurs when the electrolysis period is increased. A change in colour of the dye solution from dark green to yellow brown indicates the increase in concentration of $Fe(OH)_3$. This formation of flocs due to spontaneous discharge of ferric ions further forms ferric hydroxide (brown flocs) (Ghernaout *et al.* 2008).

3.4. Comparison of EC batch process and EC-sonication process

Sonication process generally works in the frequency 20 kHz–2 MHz. Application of sonication is found to be unproductive alone for removal or degradation of pollutants. Several studies reveal that on integrating sonication with other electrochemical process leads to good results in removal of pollutants (Gągol *et al.* 2018). Sonochemistry works through both the physical and chemical effect. The chemical effect through $OH^{\cdot}$ and physical effect through cavitation increases the treatment efficiency (Nidheesh *et al.* 2021).

The enhancement of removal efficiency in combined electrocoagulation and sonication process is due to generation of •OH, H• and HO_2• radicals by the ultrasound energy which are responsible for the degradation of organic pollutants (Kovatcheva & Parlapanski 1999; Chu *et al.* 2012; Raschitor *et al.* 2014; He *et al.* 2016; Al-Qodah *et al.* 2018; Menon *et al.* 2021.

In this study, at optimum conditions of EC (achieved experimentally as discussed in the previous sections, i.e., pH 7, 2 V and 90 min), the sample was exposed to sonication process for 60 min. It was found that EC-Sonication had better removal efficiency than EC for same operating conditions. The C/C_0 of COD and AN after sonication were found to be 0.145 and 0.228, respectively, as shown in Figure 4. The TOC of the wastewater has been reduced from 20,040 mg/L to 1,536 mg/L after the combined process. This is due to the synergetic effect of sonication. Due to the high rate of mixing and generation of oxidative species, the combined EC and sonication process gives a remarkable increase in removal efficiency. The results observed by the authors in the current study has been compared with the results reported by different researchers and shown in Table 2. The results observed are comparable or better than the results observed by other researchers for different wastewater.

3.5. Sludge characteristics

The sludge generated out of the electrocoagulation-sonication process has been collected and characterized using SEM and the elemental compositions of the samples were observed via EDX spectra. The SEM and EDS analysis of the dried sludge are shown in Figure 5(a) and 5(b). Aggregate shapes with non-uniform distribution of particle sizes varying from 6 to 100 μm of sludge has been observed from the micrograph image. The EDS analysis shows that the sludge contains various metallic

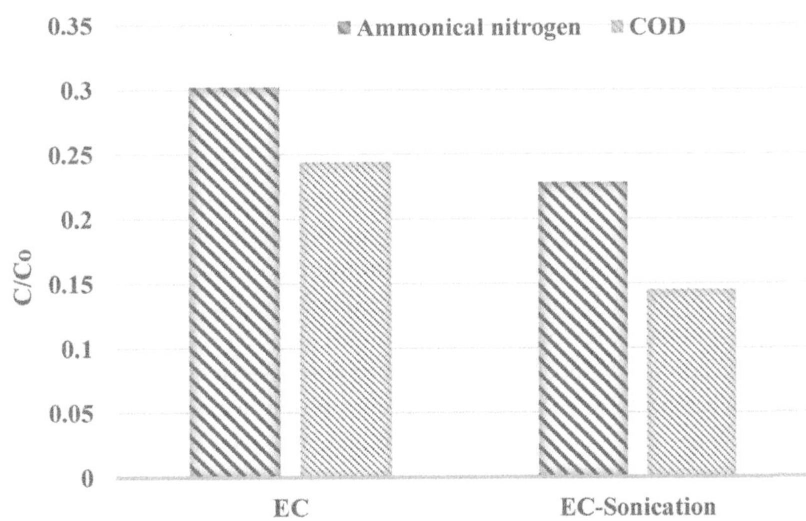

Figure 4 | The fraction remaining (C/C_0) post degradation of COD and AN (with and without sonication).

Table 2 | Comparison of the present study with the treatments of the other industrial wastewater

Sr. No.	Type of electrode and optimum parameters	Type of wastewater	Degradation efficiency	Reference
1	Aluminium electrodes pH = 4.5, current density = 560 A/m^2, electrolysis duration = 65 min	Palm oil mill industrial wastewater	75% [COD] 100% [Colour] 100% [Turbidity]	Bashir *et al.* (2016)
2	Aluminium electrodes pH = 4–7, current density = 400 A/m^2, electrolysis duration = 6 h	Tannery industrial wastewater	81% [COD] 97% [Chromium (III)]	Elabbas *et al.* (2016)
3	Aluminium electrodes pH = 6, current density = 18 A/m^2, electrolysis duration = 36 min	Sewage wastewater	64% [COD] 80% [Turbidity]	Baran *et al.* (2018)
4	Iron electrodes pH = 8.3, current density = 1.5 mA/cm^2, electrolysis duration = 25 min	Paper mill industrial wastewater	>95% [Colour] BOD_5/COD increased from 0.26 to 0.41	Wagle *et al.* (2020)
5	Iron and aluminium electrodes pH = 6, Current density = 0.32 mA/cm^2, NaCl concentration = 1 g/L, Electrolysis duration = 1 min	Biodiesel industrial wastewater	Iron electrode 92% [COD] 92% [TOC] Aluminium electrode 91% [COD] 92% [TOC]	Tanatti *et al.* (2018)
6	Aluminium electrodes pH = 7.8, current density = 1.5 mA/cm^2, electrolysis duration = 20 min	Food industrial wastewater	70% [COD]	Veli *et al.* (2018)
7	Aluminium electrodes pH = 4.84, current density = 100 A/m^2, electrolysis duration = 293 min, electrode distance = 6 cm	Pistachio processing industrial wastewater	44% [COD] 56% [Total phenols]	Ozay *et al.* (2018)
8	Iron electrodes pH = 7, applied voltage = 2 V, electrolysis duration = 90 min	Dye intermediate industrial wastewater	75.6% [COD] 70% [NH_4^+]	Current study

elements with the abundance of C > O > N > Cl > Cu > Fe > Al. Group of metals represent electrode dissolution and floc formed during the process which settled as sludge. EDX shows high metal content in sludge which demonstrate the morphological nature and texture of sludge which shows that the sludge contains both crystalline and amorphous content. As the sludge contains high amounts of C, O, N, Fe, and trace elements, hence it can be used as fertilizer and soil fertility enhancer. This metallic composition of the sludge also enables it to be used in cement manufactory as raw material (Nidheesh & Singh 2017).

3.6. Parametric optimization

RSM analysis was performed on experimental data to derive the second order polynomial equation for the estimation of optimum conditions. The ANOVA results are summarised in Tables 3 and 4 for AN and COD, respectively. The regression equations (i.e., responses) generated for AN and COD are shown by Equations (7) and (8). The quality of fitted model and its statistical significance was evaluated by R^2 and T-Test/F-Test. Values of adjusted R^2 show that the model terms are highly significant. The values of predicted R^2 are the values predicted by the design, which measures the variation in model predicted data (Garg & Prasad 2016). R^2 values of 78.2% for AN and 82.7% for COD, respectively, indicated an acceptable level of accuracy of the fitted model. Model terms were utilized post assessment of *P*-value (with 95% confidence level). *P*-values less than 0.05 implies higher significance whereas greater than 0.1, the lower. *P*-value less than 0.00001 indicate that terms are highly significant. The linear and square terms of pH and voltage terms have higher significance for AN as well as COD, others are not significant. The insignificant terms were removed from the general equation. Response surface methodology was used for estimating optimal values of variable of Equations (7) and (8) at which removal is maximized.

Figure 5 | (a) SEM micrograph of sludge. (b) EDS spectrum of dried sludge.

The response surface for pH, time and voltage is depicted in Figures 6 and 7 for AN and COD. It can be seen that the optimal values of pH, time and voltage was ~8.2, ~90.9 min and ~2.6 V for AN and ~ 7.6, ~100.9 min and ~2.6 V for COD. The C/C_0 was estimated on these optimized values, which were 0.302 (i.e., ~70% removal) and 0.244 (i.e., 76% removal) for AN and COD, respectively.

Regression equation of AN

$$AN\ (response) = 2.462 - 0.4390\text{pH} - 0.00304\text{Time} - 0.1633\text{V} + 0.02659\text{pH}*\text{pH} + 0.02732\text{V}*\text{V} \quad (7)$$

Table 3 | Analysis of variance of AN (DF: degree of freedom; SS: sum of squares; MS: means squares)

Source	DF	Adj SS	Adj MS	F-Value	*P*-Value
Model	8	0.520088	0.065011	22.85	0.000
Linear	3	0.068472	0.022824	8.02	0.000
pH	1	0.029269	0.029269	10.29	0.002
Time	1	0.013255	0.013255	4.66	0.036
V	1	0.034477	0.034477	12.12	0.001
Square	3	0.346074	0.115358	40.55	0.000
pH*pH	1	0.293124	0.293124	103.05	0.000
Time*Time	1	0.010827	0.010827	3.81	0.057
V*V	1	0.031679	0.031679	11.14	0.002
2-Way interaction	2	0.003612	0.001806	0.63	0.534
pH*Time	1	0.000105	0.000105	0.04	0.848
Time*V	1	0.003120	0.003120	1.10	0.300
Error	51	0.145071	0.002845		
Lack-of-fit	47	0.145063	0.003086	1646.11	0.06
Pure error	4	0.000008	0.000002		
Total	59	0.665159			

Table 4 | Analysis of variance of COD (DF: degree of freedom; SS: sum of squares; MS: means squares)

Source	DF	Adj SS	Adj MS	F-Value	P-Value
Model	8	1.45460	0.181825	30.36	0.000
Linear	3	0.19408	0.064692	10.80	0.000
pH	1	0.08868	0.088684	14.81	0.000
Time	1	0.02001	0.020006	3.34	0.043
V	1	0.05879	0.058787	9.82	0.003
Square	3	0.83426	0.278087	46.44	0.000
pH*pH	1	0.75944	0.759440	126.82	0.000
Time*Time	1	0.00698	0.006977	1.17	0.285
V*V	1	0.04695	0.046952	7.84	0.007
2-Way interaction	2	0.00127	0.000634	0.11	0.900
pH*Time	1	0.00014	0.000136	0.02	0.881
Time*V	1	0.00093	0.000934	0.16	0.695
Error	51	0.30540	0.005988		
Lack-of-fit	47	0.30538	0.006497	1,155.10	0.07
Pure error	4	0.00002	0.000006		
Total	59	1.76000			

Regression equation of COD

$$\text{COD (response)} = 3.173 - 0.6552\,\text{pH} - 0.00248\,\text{Time} - 0.1904\,\text{V} + 0.04280\,\text{pH}*\text{pH} + 0.0333\,\text{V}*\text{V} \qquad (8)$$

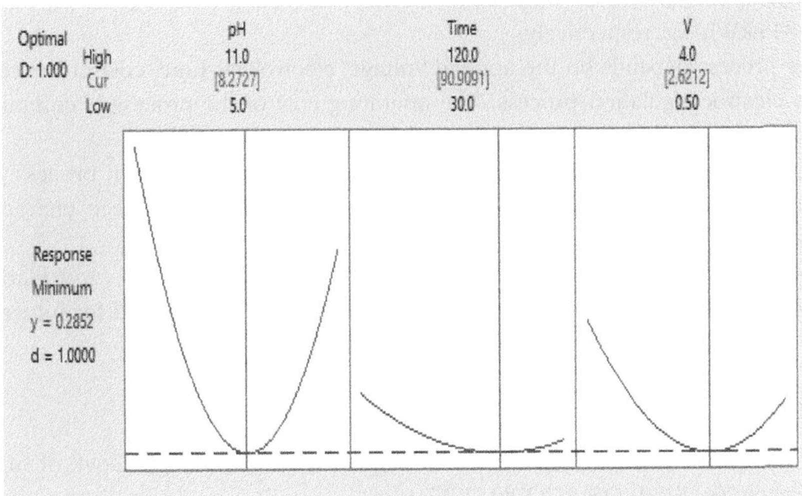

Figure 6 | Response surface plot of optimal condition of removal of AN.

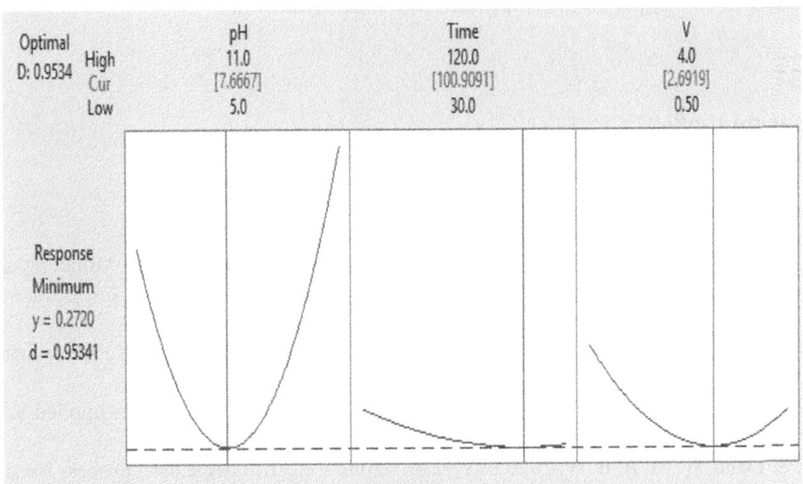

Figure 7 | Response surface plot of optimal condition of removal of COD.

4. CONCLUSION

- In this work, EC and EC-sonication combined was investigated for the removal of ammoniacal nitrogen and COD from dye intermediate wastewater.
- A set of experiment, by varying pH between 5 and 11, applied voltage from 0.5 to 4 V for an electrode area of 18 cm^2 and electrolysis time up to 120 min was performed. The minimum fraction remaining (C/C_0) were 0.244 and 0.302, respectively, for COD and AN, at electrolysis time of 90 min, pH 7 and applied voltage of 2 V.
- ANOVA and response surface methodology were used for estimating optimum conditions for achieving maximum removal efficiency.
- The optimal values of pH, time and voltage was ~8.2, ~90.9 min and ~2.6 V, respectively, for AN and ~7.6, ~100.9 min and ~2.6 V respectively for COD removal.
- In addition, experiments with combination of two processes: electrocoagulation and sonication, for the removal of AN and COD from dye intermediate wastewater was also performed.
- The removal efficiency was inferred using such experimental data only. The C/C_0 for COD and AN after EC-sonication process were 0.145 and 0.228, respectively, with a post EC-sonication time 60 min at a frequency of 33 kHz.

- The specific energy consumption for EC process was 2.01 kWh/m^3 and for sonication process for COD and AN was 0.1731 kWh/kg and 3.8845 kWh/kg, respectively.
- The optimum cost of the process depends on the applied voltage, electrolysis time, cost of the reactor, cost of electrodes, mixing cost, etc. for the electrocoagulation process. The operating cost of the process is computed to be 0.711 USD/m^3 of the wastewater treatment.
- It can be concluded that the application of both the processes, i.e., EC and sonication process, is effective for removing COD and AN from dye intermediate manufacturing wastewater. The sludge has been characterised and the metallic elements with the abundance of C > O > N > Cl > Cu > Fe > Al was observed.
- As the dye wastewater is highly concentrated with some traces of heavy metals, the recovery/treatment of heavy metals can be suggested. Further detailed characterization and disposal of sludge can be extended for future studies.

ACKNOWLEDGEMENT

We would like to convey our gratification to Department of Science and Technology, Govt. of India for their support and financial assistance (Reference No- SR/WOS-A/ET-89/2017).

DATA AVAILABILITY STATEMENT

All relevant data are included in the paper or its Supplementary Information.

CONFLICT OF INTEREST

The authors declare there is no conflict.

REFERENCES

Akkaya, G. K., Erkan, H. S., Sekman, E., Top, S., Karaman, H., Bilgili, M. S. & Engin, G. O. 2019 Modeling and optimizing Fenton and electro-Fenton processes for dairy wastewater treatment using response surface methodology. *Int. J. Environ. Sci. Technol.* **16** (5), 2343–2358.

Al-Qodah, Z., Al-Shannag, M., Bani-Melhem, K., Assirey, E., Yahya, M. A. & Al-Shawabkeh, A. 2018 Free radical-assisted electrocoagulation processes for wastewater treatment. *Environ. Chem. Lett.* **16** (3), 695–714.

Aoudj, S., Khelifa, A., Drouiche, N., Hecini, M. & Hamitouche, H. 2010 Electrocoagulation process applied to wastewater containing dyes from textile industry. *Chem. Eng. Process. Process Intensif.* **49** (11), 1176–1182.

Asaithambi, P., Aziz, A. R. A. & Daud, W. M. A. B. W. 2016 Integrated ozone – electrocoagulation process for the removal of pollutant from industrial effluent: optimization through response surface methodology. *Chem. Eng. Process. Process Intensif.* **105**, 92–102.

Baran, W., Adamek, E., Jajko, M. & Sobczak, A. 2018 Removal of veterinary antibiotics from wastewater by electrocoagulation. *Chemosphere* **194**, 381–389.

Bashir, M. J., Mau Han, T., Jun Wei, L., Choon Aun, N. & Abu Amr, S. S. 2016 Polishing of treated palm oil mill effluent (POME) from ponding system by electrocoagulation process. *Water Sci. Technol.* **73** (11), 2704–2712.

Bezerra, M. A., Santelli, R. E., Oliveira, E. P., Villar, L. S. & Escaleira, L. A. 2008 Response surface methodology (RSM) as a tool for optimization in analytical chemistry. *Talanta* **76** (5), 965–977.

Boudjema, N., Drouiche, N., Abdi, N., Grib, H., Lounici, H., Pauss, A. & Mameri, N. 2014 Treatment of Oued El Harrach river water by electrocoagulation noting the effect of the electric field on microorganisms. *J. Taiwan Inst. Chem. Eng.* **45** (4), 1564–1570.

Camargo, J. A. & Alonso, Á. 2006 Ecological and toxicological effects of inorganic nitrogen pollution in aquatic ecosystems: a global assessment. *Environ. Int.* **32** (6), 831–849.

Candido, L. & Gomes, J. A. C. P. 2011 Evaluation of anode materials for the electro-oxidation of ammonia and ammonium ions. *Mater. Chem. Phys.* **129** (3), 1146–1151.

Chakraborty, J. & Basu, P. 2019 Linking industrial hazards and social inequalities: environmental injustice in Gujarat, India. *Int. J. Environ. Res. Public Health* **16** (1), 42.

Chu, J. Y., Li, Y. R., Li, N. & Huang, W. H. 2012 Treatment of car-washing wastewater by electrocoagulation-ultrasound technique for reuse. *Adv. Mater. Res.* **433–440**. Trans Tech Publi. Ltd., pp. 227–232.

Damaraju, M., Bhattacharyya, D. & Kurilla, K. K. 2017 Removal of recalcitrant carbon from an industrial wastewater using electrocoagulation. *Int. J. Civ. Eng.* **15** (4), 697–703.

Daneshvar, N., Oladegaragoze, A. & Djafarzadeh, N. 2006 Decolorization of basic dye solutions by electrocoagulation: an investigation of the effect of operational parameters. *J. Hazard. Mater.* **129** (1–3), 116–122.

Deghles, A. & Kurt, U. 2016 Treatment of tannery wastewater by a hybrid electrocoagulation/electro dialysis process. *Chem. Eng. Process. Process Intensif.* **104**, 43–50.

Dong, Y., Qu, Y., Li, C., Han, X., Ambuchi, J. J., Liu, J., Yu, Y. & Feng, Y. 2017 Simultaneous algae-polluted water treatment and electricity generation using a biocathode-coupled electrocoagulation cell (bio-ECC). *J. Hazard. Mater.* **340**, 104–112.

Elabbas, S., Ouazzani, N., Mandi, L., Berrekhis, F., Perdicakis, M., Pontvianne, S., Pons, M.-N., Lapicque, F. & Leclerc, J. P. 2016 Treatment of highly concentrated tannery wastewater using electrocoagulation: influence of the quality of aluminium used for the electrode. *J. Hazard. Mater.* **319**, 69–77.

El-Ashtoukhy, E. Z., Amin, N. K., Abd El-Latif, M. M., Bassyouni, D. G. & Hamad, H. A. 2017 New insights into the anodic oxidation and electrocoagulation using a self-gas stirred reactor: a comparative study for synthetic CI Reactive Violet 2 wastewater. *J. Cleaner Prod.* **167**, 432–446.

Gągol, M., Przyjazny, A. & Boczkaj, G. 2018 Wastewater treatment by means of advanced oxidation processes based on cavitation–a review. *Chem. Eng. J.* **338**, 599–627.

Garcia-Segura, S., Eiband, M. M. S., de Melo, J. V. & Martínez-Huitle, C. A. 2017 Electrocoagulation and advanced electrocoagulation processes: a general review about the fundamentals, emerging applications, and its association with other technologies. *J. Electroanal. Chem.* **801**, 267–299.

Garg, K. K. & Prasad, B. 2016 Development of Box Behnken design for treatment of terephthalic acid wastewater by electrocoagulation process: optimization of process and analysis of sludge. *J. Environ. Chem. Eng.* **4** (1), 178–190.

Ghernaout, D. & Ghernaout, B. 2012 Sweep flocculation as a second form of charge neutralisation – a review. *Desalin. Water Treat.* **44** (1–3), 15–28.

Ghernaout, D., Badis, A., Kellil, A. & Ghernaout, B. 2008 Application of electrocoagulation in *Escherichia coli* culture and two surface waters. *Desalination* **219** (1–3), 118–125.

Gonzalez-Merchan, C., Genty, T., Bussière, B., Potvin, R., Paquin, M., Benhammadi, M. & Neculita, C. M. 2016 Influence of contaminant to hydrogen peroxide to catalyzer molar ratio in the advanced oxidation of thiocyanates and ammonia nitrogen using Fenton-based processes. *J. Environ. Chem. Eng.* **4** (4), 4129–4136.

Guo, Y., Xue, Q., Zhang, H., Wang, N., Chang, S., Wang, H., Panh, H. & Chen, H. 2018 Treatment of real benzene dye intermediates wastewater by the Fenton method: characteristics and multi-response optimization. *RSC Adv.* **8** (1), 80–90.

Hassani, A., Malhotra, M., Karim, A. V., Krishnan, S. & Nidheesh, P. V. 2022 Recent progress on ultrasound-assisted electrochemical processes: a review on mechanism, reactor strategies, and applications for wastewater treatment. *Environ. Res.* **205**, 112463.

Hayat, H., Mahmood, Q., Pervez, A., Bhatti, Z. A. & Baig, S. A. 2015 Comparative decolorization of dyes in textile wastewater using biological and chemical treatment. *Sep. Purif. Technol.* **154**, 149–153.

He, C. C., Hu, C. Y. & Lo, S. L. 2016 Evaluation of sono-electrocoagulation for the removal of Reactive Blue 19 passive film removed by ultrasound. *Sep. Purif. Technol.* **165**, 107–113.

Hua, I. & Hoffmann, M. R. 1997 Optimization of ultrasonic irradiation as an advanced oxidation technology. *Environ. Sci. Technol.* **31** (8), 2237–2243.

Huang, H., Liu, J., Zhang, P., Zhang, D. & Gao, F. 2017 Investigation on the simultaneous removal of fluoride, ammonia nitrogen and phosphate from semiconductor wastewater using chemical precipitation. *Chem. Engg. J.* **307**, 696–706.

Huda, N., Raman, A. A. A., Bello, M. M. & Ramesh, S. 2017 Electrocoagulation treatment of raw landfill leachate using iron-based electrodes: effects of process parameters and optimization. *J. Environ. Manage.* **204**, 75–81.

Izadi, A., Hosseini, M., Najafpour Darzi, G., Nabi Bidhendi, G. & Pajoum Shariati, F. 2018 Treatment of paper-recycling wastewater by electrocoagulation using aluminum and iron electrodes. *J. Environ. Health Sci. Eng.* **16** (2), 257–264.

Kabdaşlı, I., Arslan-Alaton, I., Ölmez-Hancı, T. & Tünay, O. J. E. T. R. 2012 Electrocoagulation applications for industrial wastewaters: a critical review. *Environ. Technol. Rev.* **1** (1), 2–45.

Kasmuri, N. & Tarmizi, N. A. A. 2018 The treatment of landfill leachate by electrocoagulation to reduce heavy metals and ammonia-nitrogen. *Int. J. Eng. Technol.* **7**, 109–112.

Khandegar, V. & Saroha, A. K. 2013 Electrocoagulation for the treatment of textile industry effluent–a review. *J. Environ. Manage.* **128**, 949–963.

Kovatcheva, V. K. & Parlapanski, M. D. 1999 Sono-electrocoagulation of iron hydroxides. *Colloids Surf., A* **149** (1–3), 603–608.

Lakshmi, P. M. & Sivashanmugam, P. 2013 Treatment of oil tanning effluent by electrocoagulation: influence of ultrasound and hybrid electrode on COD removal. *Sep. Purif. Technol.* **116**, 378–384.

Li, L. & Liu, Y. 2009 Ammonia removal in electrochemical oxidation: mechanism and pseudo-kinetics. *J. Hazard. Mater.* **161** (2–3), 1010–1016.

Li, X., Chen, S., Angelidaki, I. & Zhang, Y. 2018 Bio-electro-Fenton processes for wastewater treatment: advances and prospects. *Chem. Eng. J.* **354**, 492–506.

Lu, J., Zhuo, Q., Ren, X., Qiu, Y., Li, Y., Chen, Z. & Huang, K. 2022 Treatment of wastewater from adhesive-producing industries by electrocoagulation and electrochemical oxidation. *Process Saf. Environ. Prot.* **157**, 527–536.

Mahvi, A. H., Ebrahimi, S. J. A. D., Mesdaghinia, A., Gharibi, H. & Sowlat, M. H. 2011 Performance evaluation of a continuous bipolar electrocoagulation/electrooxidation – electroflotation (ECEO–EF) reactor designed for simultaneous removal of ammonia and phosphate from wastewater effluent. *J. Hazard. Mater.* **192** (3), 1267–1274.

Mamelkina, M. A., Tuunila, R., Sillänpää, M. & Häkkinen, A. 2019 Systematic study on sulphate removal from mining waters by electrocoagulation. *Sep. Purif. Technol.* **216**, 43–50.

Manikandan, S., Saraswathi, R. & Ansari, A. M. S. 2018 Effect of pH and electrolysis time on removal of reactive black B dye by electrochemical treatment. *Asian J. Eng. Appl. Technol.* **7** (S1), 45–47.

Mehdipoor, M. A. & Moosavirad, S. M. 2020 Effect of Holed Ferrum electrodes (HFE) on the efficiency of the electrocoagulation process for copper recovery and optimization of parameters, using RSM. *Hydrometallurgy* **194**, 105313.

Menon, P., Singh, T. A., Pani, N. & Nidheesh, P. V. 2021 Electro-Fenton assisted sonication for removal of AN and organic matter from dye intermediate industrial wastewater. *Chemosphere* **269**, 128739.

Miao, Y., Peng, Y., Zhang, L., Li, B., Li, X., Wu, L. & Wang, S. 2018 Partial nitrification-anammox (PNA) treating sewage with intermittent aeration mode: effect of influent C/N ratios. *Chem. Eng. J.* **334**, 664–672.

Mohammadi, A., Khadir, A. & Tehrani, R. M. 2019a Optimization of nitrogen removal from an anaerobic digester effluent by electrocoagulation process. *J. Environ. Chem. Eng.* **7** (3), 103195.

Mohammadi, H., Alinejad, A., Khajeh, M., Darvishmotevalli, M., Moradnia, M., Tehrani, A. M., Zare, R. M. & Mengelizadeh, N. 2019b Optimization of the 3D electro-Fenton process in removal of acid orange 10 from aqueous solutions by response surface methodology. *J. Chem. Technol. Biotechnol.* **94** (10), 3158–3171.

Mollah, M. Y. A., Schennach, R., Parga, J. R. & Cocke, D. L. 2001 Electrocoagulation (EC)–science and applications. *J. Hazard. Mater.* **84** (1), 29–41.

Moradi, M., Vasseghian, Y., Arabzade, H. & Khaneghah, A. M. 2021 Various wastewaters treatment by sono-electrocoagulation process: a comprehensive review of operational parameters and future outlook. *Chemosphere* **263**, 128314.

Nidheesh, P. V. & Singh, T. A. 2017 Arsenic removal by electrocoagulation process: recent trends and removal mechanism. *Chemosphere* **181**, 418–432.

Nidheesh, P. V., Scaria, J., Babu, D. S. & Kumar, M. S. 2021 An overview on combined electrocoagulation-degradation processes for the effective treatment of water and wastewater. *Chemosphere* **263**, 127907.

Ozay, Y., Ünşar, E. K., Işık, Z., Yılmaz, F., Dizge, N., Perendeci, N. A. & Yalvac, M. 2018 Optimization of electrocoagulation process and combination of anaerobic digestion for the treatment of pistachio processing wastewater. *J. Cleaner Prod.* **196**, 42–50.

Ozturk, E. & Bal, N. 2015 Evaluation of ammonia–nitrogen removal efficiency from aqueous solutions by ultrasonic irradiation in short sonication periods. *Ultrason. Sonochem.* **26**, 422–427.

Pani, N., Tejani, V., Anantha-Singh, T. S. & Kandya, A. 2020 Simultaneous removal of COD and AN from dye intermediate manufacturing Industrial Wastewater using Fenton oxidation method. *Appl. Water Sci.* **10** (2), 1–7.

Paździor, K., Bilińska, L. & Ledakowicz, S. 2019 A review of the existing and emerging technologies in the combination of AOPs and biological processes in industrial textile wastewater treatment. *Chem. Eng. J.* **376**, 120597.

Phan, T. N., Van Truong, T. T., Ha, N. B., Nguyen, P. D., Bui, X. T., Dang, B. T., Doan, V. T., Park, J., Guo, W. & Ngo, H. H. 2017 High rate nitrogen removal by ANAMMOX internal circulation reactor (IC) for old landfill leachate treatment. *Bioresour. Technol.* **234**, 281–288.

Prajapati, A. K. 2021 Sono-assisted electrocoagulation treatment of rice grain based distillery biodigester effluent: performance and cost analysis. *Process Saf. Environ. Prot.* **150**, 314–322.

Priya, M. & Jeyanthi, J. 2019 Removal of COD, oil and grease from automobile wash water effluent using electrocoagulation technique. *Microchem. J.* **150**, 104070.

Quan, X., Wang, F., Zhao, Q., Zhao, T. & Xiang, J. 2009 Air stripping of ammonia in a water-sparged aerocyclone reactor. *J. Hazard. Mater.* **170** (2–3), 983–988.

Raschitor, A., Fernandez, C. M., Cretescu, I., Rodrigo, M. A. & Cañizares, P. 2014 Sono-electrocoagulation of wastewater polluted with Rhodamine 6G. *Sep. Purif. Technol.* **135**, 110–116.

Sen, S., Prajapati, A. K., Bannatwala, A. & Pal, D. 2019 Electrocoagulation treatment of industrial wastewater including textile dyeing effluent–a review. *Desalin. Water Treat.* **161** (1), 21–34.

Shahedi, A., Darban, A. K., Taghipour, F. & Jamshidi-Zanjani, A. 2020 A review on industrial wastewater treatment via electrocoagulation processes. *Curr. Opin. Electrochem.* **22**, 154–169.

Singh, T. S. A. & Ramesh, S. T. 2014 An experimental study of CI Reactive Blue 25 removal from aqueous solution by electrocoagulation using Aluminum sacrificial electrode: kinetics and influence of parameters on electrocoagulation performance. *Desalin. Water Treat.* **52** (13–15), 2634–2642.

Soliman, M. & Eldyasti, A. 2018 Ammonia-Oxidizing Bacteria (AOB): opportunities and applications – a review. *Rev. Environ. Sci. Biotechnol.* **17** (2), 285–321.

Soltani, R. D. C., Mashayekhi, M., Khataee, A., Ghanadzadeh, M. J. & Sillanpää, M. 2018 Hybrid sonocatalysis/electrolysis process for intensified decomposition of amoxicillin in aqueous solution in the presence of magnesium oxide nanocatalyst. *J. Ind. Eng. Chem.* **64**, 373–382.

Syam-Babu, D., Anantha Singh, T. S., Nidheesh, P. V. & Suresh Kumar, M. 2020 Industrial wastewater treatment by electrocoagulation process. *Sep. Sci. Technol.* **55** (17), 3195–3227.

Tanatti, N. P., Şengil, İ. A. & Özdemir, A. 2018 Optimizing TOC and COD removal for the biodiesel wastewater by electrocoagulation. *Appl. Water Sci.* **8** (2), 1–10.

Vanlangendonck, Y., Corbisier, D. & Van Lierde, A. 2005 Influence of operating conditions on the ammonia electro-oxidation rate in wastewaters from power plants (ELONITA™ technique). *Water Res.* **39** (13), 3028–3034.

Veli, S., Özbay, İ., Özbay, B., Arslan, A. & ÇebĪ, E. 2018 Optimization of process variables for treatment of food industry effluents by electrocoagulation. *Global. Nest J.* **20** (3), 551–557.

Villeneuve, L., Alberti, L., Steghens, J. P., Lancelin, J. M. & Mestas, J. L. 2009 Assay of hydroxyl radicals generated by focused ultrasound. *Ultrason. Sonochem.* **16** (3), 339–344.

Wagle, D., Lin, C. J., Nawaz, T. & Shipley, H. J. 2020 Evaluation and optimization of electrocoagulation for treating Kraft paper mill wastewater. *J. Environ. Chem. Eng.* **8** (1), 103595.

Wang, J., Wang, Z., Vieira, C. L., Wolfson, J. M., Pingtian, G. & Huang, S. 2019 Review on the treatment of organic pollutants in water by ultrasonic technology. *Ultrason. Sonochem.* **55**, 273–278.

Xavier, S., Gandhimathi, R., Nidheesh, P. V. & Ramesh, S. T. 2015 Comparison of homogeneous and heterogeneous Fenton processes for the removal of reactive dye Magenta MB from aqueous solution. *Desalin. Water Treat.* **53** (1), 109–118.

Yaseen, D. A. & Scholz, M. 2019 Textile dye wastewater characteristics and constituents of synthetic effluents: a critical review. *Int. J. Environ. Sci. Technol.* **16** (2), 1193–1226.

Yavuz, Y. & Ögütveren, Ü. B. 2018 Treatment of industrial estate wastewater by the application of electrocoagulation process using iron electrodes. *J. Environ. Manage.* **207**, 151–158.

Zhang, T., Ding, L., Ren, H. & Xiong, X. 2009 Ammonium nitrogen removal from coking wastewater by chemical precipitation recycle technology. *Water Res.* **43** (20), 5209–5215.

Zhang, C., He, D., Ma, J. & Waite, T. D. 2018 Active chlorine mediated ammonia oxidation revisited: reaction mechanism, kinetic modelling and implications. *Water Res.* **145**, 220–230.

Zhou, X., Liu, X., Huang, S., Cui, B., Liu, Z. & Yang, Q. 2018 Total inorganic nitrogen removal during the partial/complete nitrification for treating domestic wastewater: removal pathways and main influencing factors. *Bioresour. Technol.* **256**, 285–294.

First received 3 August 2022; accepted in revised form 28 October 2022. Available online 7 November 2022